ENVIRONMENTAL REMEDIATION TECHNOLOGIES, REGULATIONS AND SAFETY

BIOREMEDIATION

PROCESSES, CHALLENGES AND FUTURE PROSPECTS

ENVIRONMENTAL REMEDIATION TECHNOLOGIES, REGULATIONS AND SAFETY

Additional books in this series can be found on Nova's website under the Series tab.

Additional e-books in this series can be found on Nova's website under the e-book tab.

ENVIRONMENTAL REMEDIATION TECHNOLOGIES, REGULATIONS AND SAFETY

BIOREMEDIATION

PROCESSES, CHALLENGES AND FUTURE PROSPECTS

JESÚS BERNARDINO VELÁZQUEZ-FERNÁNDEZ
AND
SAÉ MUÑIZ-HERNÁNDEZ
EDITORS

nova publishers
New York

For permission to use material from this book please contact us:
Telephone 631-231-7269; Fax 631-231-8175
Web Site: http://www.novapublishers.com

Library of Congress Cataloging-in-Publication Data

Bioremediation : processes, challenges, and future prospects / editors, Velazquez-Fernandez, Jesus Bernardino and Muñiz-Hernandez, Sae (Universidad Autsnoma de Nayarit, Mexico, and others).
pages cm
Includes bibliographical references and index.
ISBN 978-1-62948-513-3 (hardcover)
1. Bioremediation. I. Velazquez-Fernandez, Jesus Bernardino, editor of compilation. II. Muñiz-Hernandez, Sae, editor of compilation.
TD192.5.B55743 2013
628.1'683--dc23
2013041679

Published by Nova Science Publishers, Inc. † New York

CONTENTS

PREFACE

The present book, *Bioremediation: Processes, Challenges and Futures Prospects*, compiles the most important work and the main effort of distinguished scientists around the world about the recent advances on this discipline. Bioremediation are understood as a discipline that uses organisms or their products to reduce or eliminate the adverse effects of pollutants on environment. It is worth to mention that this "novel" discipline or technology arises from an ensemble of several other disciplines, which were traditionally considered as very distant among each other regarding their fields of knowledge. It profits from the expertise of sociologists, molecular and marine biologists, zoologists, botanists, ecologists and microbiologists, also of an analytical character and contribution of chemists, biophysicists, hydrogeologists and biochemists, as well as of prospection and design ability of engineers, physicists and mathematicians. Independently and in a contrast, nature itself takes care of natural attenuation in what has been called natural or intrinsic bioremediation. Evidently, work from a single field would not be enough to understand or to place the most adequate solution for a certain environmental problem; taking what Heisenberg wrote, "what we observe [by research means] is not the nature itself, but the nature exposed to our method of investigation".

Because of multidisciplinary endeavors in bioremediation area at, not only through understanding but taking advantage of nature, our knowledge on biological, chemical or physical features of organisms or biomolecules has been enlarged. This has been used to help nature in coping with chemicals, which we humans have placed in our biosphere. In other words, we can say that bioremediation is a human responsible attempt to ameliorate the damage and the hazard caused by anthropogenic activities. For all of above, its difficult to place in a single book all the knowledge related to bioremediation. We have made our best effort to capture the state-of-art of bioremediation in this book, with the principal goal of present, their processes, challenges, main field applications and future prospects.

Research on bioremediation has shed light on the resilience mechanisms of organisms that allows them to tolerate, adapt or resist to several and specific pollutants. One or more of those mechanisms are employed in processes of bioremediation, even if we have not the precise knowledge about how they occur. Nowadays, it is known that bioremediation could be accomplished by biodegradation, biotransformation, biovolatilization, mobilization, immobilization, confinement, biostimulation or biofiltration. In spite of the lack of complete comprehension of those mechanisms, resilient organisms have been and shall be applied as bioremediators.

The term bioremediator refers to any organism, i.e., bacterium, fungus, alga, plant, rhizosphere (the microecosystem around plant roots) or animal that could achieve bioremediate their environment. Thus, bioremediation strategies could be classified according to the bioremediator used: bacterial bioremediation, mycoremediation, phycoremediation, phytoremediation, rhizoremediation, animal bioremediation, microbioremediation (traditional bioremediation using microorganisms in general), composite bioremediation (using more than one kind of bioremediators) and derivative bioremediation. The latter refers to the use of biomolecules (*e.gr.* enzymes or lipids) derived from organisms to carry out bioremediation. In keeping this in mind, we have arranged the content according to the bioremediator approach, with the aim that it will understandable and ease to search and learn. Indeed, that could be a guide to take decisions or propose bioremediation projects and that it act as a book of support for students or researchers of biotechnological area. Thus, the book is divided in the following sections, 1) bacterial bioremediation, 2) microbioremediation (use of microcosms), 3) phyto- and phycoremediation, 4) composite bioremediation and finally 5) derivative bioremediation. Subsequently and within each section, chapters are disposed by the kind of pollutant: inorganic or organic and then by the complexity of organic mixtures. We hope, dear reader, that you will find this book not only interesting but also useful.

We acknowledge the plausible contribution of authors of each and every chapter. Among them, thanks to *Dr. Mamindy-Pajany, Dr. Constantí, Dr. Rikalovic, Dr. Mondragón, Dr. Agostini, Dr. Bidoia, Dr. Parra-Guevara, Dr. Souza, Dr. Takács, Dr. Costas, Dr. Leitão, Dr. Mazzeo, Dr. Yateem, Dr. Neske and Dr. Rikalovic* with whom we have been kept in touch to succeed in reaching the book completion.

Also, we would like to thank the editorial board from Nova Science Publishers, Inc. for the invitation to be part of this work. We hope that this book will show the relevance, the need, the pragmatism, the feasibility, the efficacy of bioremediation technologies and, worth to mention, that the application of bioremediation technologies is a true act towards a better future world.

Dr. Jesús B. Velázquez-Fernández, Ph. D.
Editor
Researcher Professor
Universidad Autónoma de Nayarit,
Mexico
Tel: 52-311-2118800, ext. 8965
jesusbvf@gmail.com

Dr. Saé Muñiz-Hernández, Ph.D.
Co-editor
Researcher
Instituto Nacional de Cancerologia
sayide@hotmail.com

SECTION 1: BACTERIAL BIOREMEDIATION

In: Bioremediation: Processes, Challenges and Future Prospects ISBN: 978-1-62948-513-3
Editors: Jesús Bernardino Velázquez-Fernández et al.

Chapter 1

BIOREMEDIATION OF HEAVY METALS IN SEDIMENTS

Christine Lors and Yannick Mamindy-Pajany*
Ecole Nationale Supérieure des Mines de Douai, Douai, France
Université Lille Nord de France, Lille, France

ABSTRACT

Sediments in rivers, lakes, and harbors represent the main repository of anthropogenic pollutants. A great variety of pollutants are associated with geochemical components of sediments. Among them, heavy metals are very harmful because of their potential toxicity to human health and the environment. Sediments are one of the most difficult materials to be treated because of their muddy and pasty characteristics and their great geochemical complexity. A promising alternative to traditional treatments, based on chemical extraction or stabilization processes, is technologies of bioremediation. Although microorganisms cannot destroy metals they can alter their chemical properties via several mechanisms (e.g. bioleaching, biosorption, bioreduction, bioprecipitation), some of which can be used to treat metal contamination. In some cases, these processes involve highly specific biochemical pathways that have evolved to protect the microbial cell from toxic heavy metals. This chapter gives a review of metal–bacteria interactions, and describes how they are harnessed to remediate metal-contaminated sediments.

Keywords: Bioremediation, sediments, heavy metals, microorganisms, oxidation/reduction, bioleaching, biovolatilization, biosorption, bioprecipitation

INTRODUCTION

The main anthropogenic heavy metals sources in environment are industrial point sources, including present and former mining activities, foundries and smelters, shipbuilding, chemical industries, metallic industries, and diffuse sources such as combustion by-products.

* Corresponding Author address. Email: yannick.mamindy@mines-douai.fr.

The heavy metals include both metals and metalloids with an atomic density greater than 6 $g \cdot cm^{-3}$ [with the exception of arsenic (As) and selenium (Se)]. This group includes both biologically essential [cobalt (Co), copper (Cu), chromium (Cr), manganese (Mn) and zinc (Zn)] and non-essential [cadmium (Cd), lead (Pb), mercury (Hg)] elements (Luoma & Rainbow, 2008). The essential elements are required in low concentrations for plant, animal and human nutrition whereas the non-essential elements are widely known as "toxic elements" towards living organisms (Tan et al., 2008; Yadav, 2010).

In the aquatic systems, these pollutants may be present as free dissolved species or as organo-metallic complexes with natural organic ligands (Wen, Jiann & Santschi, 2006; Guthrie et al., 2005). Additionally, many metals are associated with particulate matter and become adsorbed or co-precipitated with carbonates, oxyhydroxides, sulfides and clay minerals (Mulligan, Yong & Gibbs, 2001a; Peng et al., 2009). Consequently, sediments accumulate pollutants and may act as sinks for metals in the environment. Within the sediment, pollutants partition between aqueous (pore water and/or water column) and solid phases (sediment, and/or suspended particulate matter or biota) (Luoma & Rainbow, 2008). Sediments from anthropogenic areas, like harbors, can therefore represent a source of contaminants and a threat for aquatic environment and human health (Caplat et al., 2005; Eggleton & Thomas, 2004). In these aquatic areas, periodic dredging activities are necessary to maintain the navigational depth. Also, the large volumes of dredged polluted materials have to be adequately managed. Due to their muddy characteristics and their high geochemical complexity, the most widely used technology for the treatment of these sediments is the separation of the coarse fraction (sands characterized by lower contamination levels), from the fine fraction (silt-clay typically highly contaminated), in order to facilitate subsequent treatments and to minimize the portion that requires dumping (Mulligan, Yong & Gibbs, 2001a; Fettweis et al., 2011). A promising alternative to usual treatments, based on chemical extraction or stabilization, lies on bioremediation technologies (Sakultantimetha et al., 2011; Rocchetti et al., 2012). They can be used to reduce, eliminate, or stabilize hazardous wastes. Over the last decades, it has been widely reported that a large variety of microorganisms (bacteria, fungi, among others) can oxidatively degrade many types of organic pollutants (e.g., PAHs, BTEX) completely into non-toxic products such as CO_2 and water or organic acids and methane or partially into metabolites (Cerneglia, 1992; Lors et al., 2004a; 2010). These degradation processes depend on the physico-chemical environment, microbial communities and nature of the pollutant. This technology includes intrinsic bioremediation, which relies on naturally occurring processes based on the activity of indigenous microorganisms, and on enhanced bioremediation, which improves microbial degradation through the addition of nutrients or adapted microorganisms and the control of pH, moisture and temperature (Namkoong et al., 2002; Perelo, 2010; Lors et al., 2012). Unfortunately, metals cannot be biodegraded. However, microorganisms can interact with these contaminants and transform them from one chemical form to another by changing their oxidation state through oxidation-reduction reactions (Gadd, 2010). In some bioremediation strategies, the solubility of the transformed metal increases, thus increasing the mobility of the contaminant and allowing it released from the environment. In other strategies, the opposite occurs, and the transformed metal may precipitate out of solution, leading to immobilization. Both kinds of transformations present opportunities for bioremediation of metals in the environment - either to immobilize them in place or to accelerate their removal (Gadd, 2004). This review focuses on the use of microorganisms in the remediation of metals in polluted sediments. Following a brief description of sediment properties, the review describes the main biological mechanisms involved in the mobility (mobilization and immobilization) of metals in sediments and associated bioremediation technologies.

SEDIMENTS COMPONENTS AND METALS PARTITIONING

Sediments are composed of various components, as shown in Figure 1. Solid phases in the sediments include organic and inorganic particles (Förstner, 2004; Suresh et al., 2011). Primary minerals include quartz, feldspar, micas, amphiboles, and pyroxenes and are generally found as sand and silt fractions. Secondary minerals are formed by physical, chemical, and/or biological weathering processes. These minerals are layer silicates and are known as phyllosilicates. They include a major fraction of the clay-sized materials in clays. Due to their size and structure, secondary minerals have large specific surface areas and significant surface charges. The major groups of clay minerals include kaolinites, smectites, illites, chlorites, and vermiculites.

Oxides and hydrous oxide minerals include the oxides, hydroxides, and oxyhydroxides of iron, aluminium, manganese, titanium, and silicon. Common crystalline forms of these minerals include anatase, bohemite, gibbsite, hematite, goethite, and quartz. They are different from layer silicate minerals (secondary minerals) because their surfaces consist of broken bonds with hydroxyl (OH^-) groups of disassociated water molecules. The surfaces exhibit pH-dependent charges (Tombacz et al., 2004).

The most common carbonate mineral found in sediments is calcite ($CaCO_3$). Some types of plankton may incorporate calcite into their shell. Some of the other less common forms are magnesite ($MgCO_3$) and dolomite ($CaMg(CO_3)_2$). Gypsum ($CaSO_4 \cdot 2H_2O$) is the most common sulfate mineral found in sediments. Carbonates can retain hazardous materials by the precipitation of heavy metals (Peng et al., 2009).

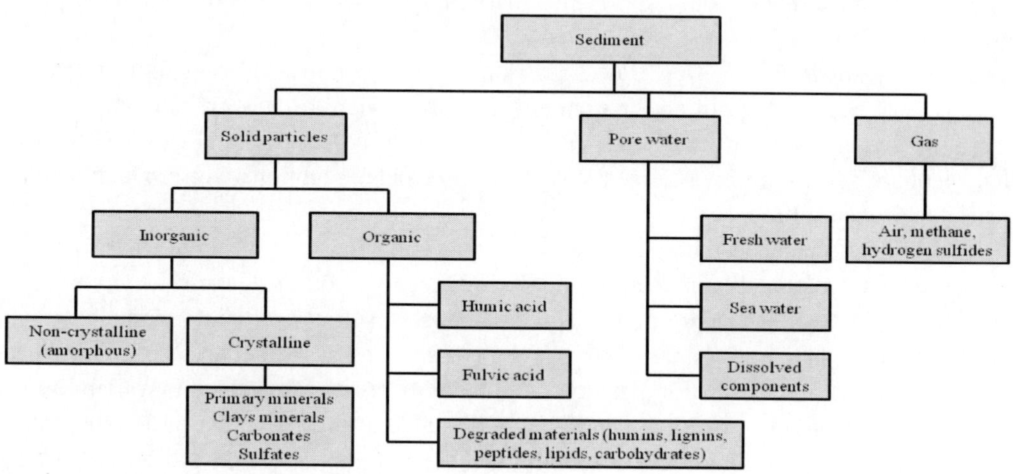

Modified from Mulligan, Fukue & Sato, 2010.

Figure 1. Composition of sediments.

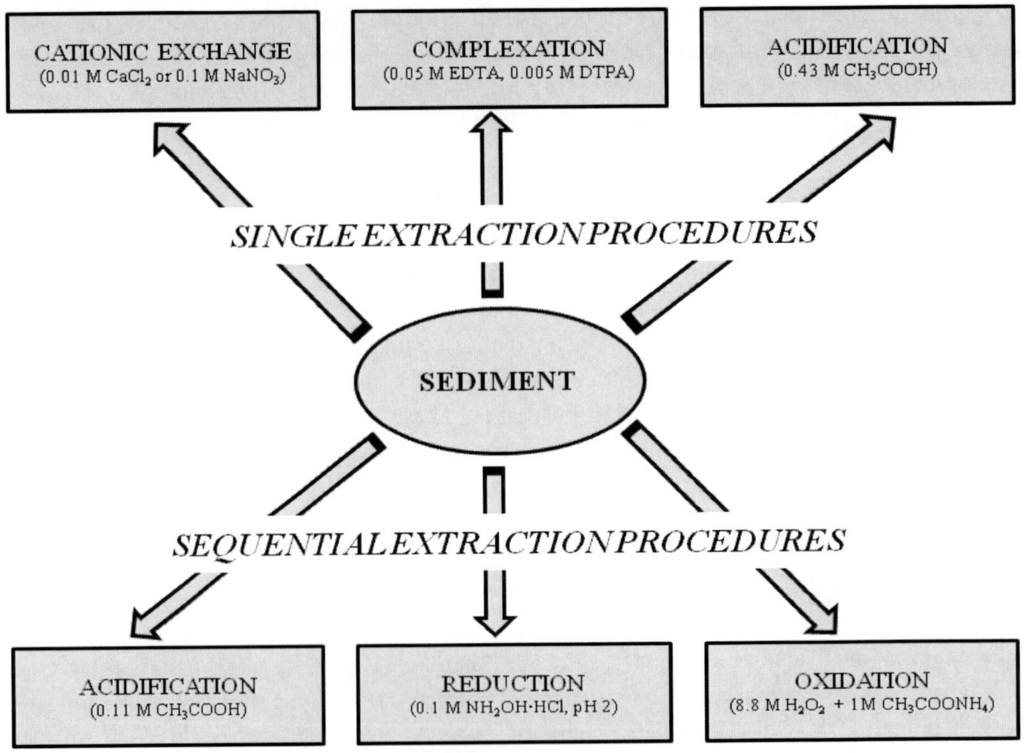

Modified from Sahuquillo, Rigol & Rauret, 2003.

Figure 2. Extraction procedures for assessing heavy metals mobility in sediments.

Organic matter originates from vegetation and animal sources and is generally categorized into humic and nonhumic material. Humic materials are those organics that result from the chemical and biological degradation of nonhumic material. Nonhumic material or compounds, on the other hand, are organics that remain undecomposed or partially degraded (Huber et al., 2011).

Pore water can include freshwater, seawater, or other liquids. The quality of the pore water is very important from an environmental point of view because it can produce toxic effects for living organisms (Eggleton & Thomas, 2004). Pore gas can also be important in characterizing some types of sediments, because anaerobic conditions will produce methane gas and hydrogen sulfide.

In sediments, metals can be bound to the various components in different ways: occluded in amorphous materials; adsorbed on clay surfaces or iron/manganese oxyhydroxides; presented in lattice of secondary minerals like carbonates, sulfates or oxides; complexed with organic matter or lattice of primary minerals such as silicates (Caplat et al., 2005). Since each form has different remobilization potential, and then affects its respective bioavailability and toxicity, the measurement of total metal may not be able to provide exact information about the characteristics of pollution. Consequently, many extracting agents have been proposed in the literature for assessing the metals associated with different sediment phases (Sahuquillo, Rigol & Rauret, 2003). Figure 2 summarizes those most frequently used extracting agents, indicating the extractable fraction with which they are associated. The extractable fractions obtained in each case are usually defined as water soluble, easily exchangeable, bound to carbonates, easily reducible (Fe/Mn oxide fraction), or bound to organic matter and sulfur. The water soluble, carbonate and exchangeable fractions possess the highest mobility, since these metals can be easily released in the aqueous phase. Metals trapped in the Fe/Mn oxide may be leached if redox conditions in the system change. Organic matter always has high affinity towards heavy metals. However, decomposition or oxidation of organic ligands occurs with time, resulting in the exposure of heavy metals. Sulfide fraction is insoluble and thus exhibits lower toxicity. However, chemical states of sulfide are dependent on the system conditions. Metal release happens during the oxidation of sulfide precipitate under oxygen-rich conditions (Caille et al., 2003).

Metal-Microbe Interactions

Bioavailability of heavy metals highly depends on environmental conditions, such as pH, alkalinity, redox potential, and activities of microorganisms (van Hullebusch et al., 2005). Heavy metals cannot be degraded in microbial metabolism/co-metabolism processes and bioremediation of these elements is accomplished by the conversions between inorganic and organic forms or the inorganic valence changes through redox reactions (Gadd, 2010). Table 1 gives an overview on possible reactions and mechanisms resulting in the mobilization of sorbed metals or immobilization of solubilized metals.

Table 1. Main microbial mechanisms involved in mobilization and immobilization of metals

Mechanism	Description of the reaction
Mobilization	
Oxidation-reduction	Metals are microbially oxidized or reduced during oxidation-reduction processes. As result, metal mobility is increased depending on the type of metal and its oxidation state.
Solubilization	Two main mechanisms involved in metal solubilization: The first mechanism consists in increasing metals solubility via proton-induced metal solubilization (acidification process). The second mechanism is the complexation of metals by complexing or chelating agents (organic acids, polymeric substances) produced by microorganisms (chelation process).
Immobilization	
Biosorption	It is a passive process of metal sequestering and concentration by chemical sites (such as carboxyl, sulfonate, phosphate, hydroxyl, amino functional groups) naturally present on the surface of living or dead microbial biomass.
Bioaccumulation	Soluble metals are actively transported through the cell membrane and accumulated within the cells as solid particles or in vacuoles.
Oxidation-reduction	Metals are microbially oxidized or reduced during oxidation-reduction processes. As result, metal mobility is decreased depending on the type of metal and its oxidation state.
Bioprecipitation	Soluble metal species are immobilized (precipitated) by microbially formed complexing agents such as sulfides (metal sulfides with very low solubility products), metal carbonates or metal phosphates.

In summary, solubilization, complexation, and oxidation-reduction reactions are the main mechanisms involved in metals partition within the sediment. Microorganisms may mobilize metals by producing organic or inorganic acids (e.g., citric acid, sulfuric acid), by oxidation-reduction reactions; and by the excretion of complexing agents. Both mechanisms play an important role in metal solubilization and can occur simultaneously in the presence of ligands under acidic conditions (Gadd, 2004).

These general mechanisms are strongly involved in bioremediation techniques used to remediate metals-contaminated sediments (Gadd, 2010). Oxidation-reduction reactions play an important role in all these bioremediation techniques since they may be involved in the mobilization and immobilization of metals. These reactions can occur through direct or indirect reduction mechanisms. The principle of reductive biotransformation (direct reduction) of heavy metals lies on the decrease of mobility and toxicity when metals are reduced to lower redox states (Gadd, 2010). In the direct reduction process, metal-reducing microorganisms use the oxidized form of metals [e.g., Cr(VI), U(VI), Hg(II), Se(VI)) as electron acceptors and transform them into reduced species (i.e., Cr(III), U(IV), Hg(0), Se(IV)].

Reductive biotransformation (Indirect reduction) also facilitates indirect metal immobilization (Prakash et al., 2013). Metal-reducing and sulfate-reducing bacteria are usually involved in this process. Electrons extracted from the oxidation of organic compounds or hydrogen are used to reduce Fe (III), Mn (IV), and SO_4^{2-} to Fe (II), Mn (III), and H_2S, respectively. Heavy metals then interact with these reduced products to form separate or multicomponent insoluble species (van Hullebusch et al., 2005). Indirect enzymatic reduction often occurs in sedimentary and subsurface environments. The most active reduced products are Fe (II) and H_2S. Fe (II) is used as electron donor in the reduction of Cr (VI) to Cr (III) by Fe-reducing bacteria, such as *Geobacter*, *Desulfuromonas*, *Shewanella*, and *Pelobacter* (Coates et al., 1996; Wielinga et al., 2001).

BIOLEACHING

Bioleaching method has been adapted from the mining industry in order to extract metals contained in low grade sulfide ores (Bosecker, 1997). This process transfers the metals in the solution phase while the solid residue is discarded as waste material. Bioleaching is now used in various metal removal operations from groundwaters (Mulligan et al., 2001b), soils (Tichy et al, 1998), sludge (Tyagi et al., 1993; Lombardi & Garcia, 2002) and sediments (Mulligan et al., 2001a). Metal recovery from sulfide minerals is based on the activity of chemolithotrophic bacteria, mainly those belonging to the genera *Acidithiobacillus* and *Leptospirillum*. These bacteria, which are Gram-negative, grow under aerobic conditions and at temperature between 25 and 35 °C. Most *Thiobacilli* species use carbon dioxide from the atmosphere as their carbon source for cellular synthesis. They draw their energy from the oxidation of elemental sulfur, reduced sulfur compounds (like sulfide, thiosulfate). This oxidation reaction leads to the production of ferric ions and sulfuric acid, resulting in the acidification of the sediments and the extraction of heavy metals associated with sediment particles and the solubilization of metal sulfides. *Acidithiobacillus thiooxidans*, *Acidithiobacillus ferrooxidans* and *Leptospirillum ferrooxidans* are the most used in bioleaching because these species grow in highly acidic media. The conditions generated by this process (pH 1.5-3) allow the ions to remain in solution. *A. thiooxidans* which oxidizes rapidly elemental sulfur and other reduced sulfur compounds ($S_2O_3^{2-}$, $S_4O_6^{2-}$) is highly acidophilic (pH 0.5 to 4; optimum pH 2 to 3.5) and decreases the pH in the medium to 1.5 to 1 (Boeseker, 1997). *A. ferrooxidans* and *L. ferrooxidans* also tolerate lower pH values but, unlike other *Thiobacilli, A. ferrooxidans* oxidizes, in addition to reduced sulfur compounds, iron(II) to iron(III) ions whereas *L. ferrooxidans* can oxidize only iron(II) ions (Sand et al., 1992; Bosecker, 1997).

The dissolution of metal sulfide by Fe/S oxiziding bacteria is controlled by two different mechanisms: the thiosulfate and the polysulfide pathways. These pathways depend on the reactivity of metal sulfides with protons (acid solubility). In the case of acid-insoluble metal sulfides (thiosulfate pathway), such as pyrite (FeS_2), molybdenite (MoS_2) and tungsterite (WS_2), iron(III) ions attack metal sulfides by electron extraction and are reduced to the iron(II) ions. This reaction generates metal cations (M^{2+}) and thiosulfate which is oxidized to sulfuric acid. *T. ferrooxidans* and *L. ferrooxidans* catalyze the recycling of iron(III) ions in acidic conditions. Acid-soluble metal sulfides, such as sphaerite (ZnS), galena (PbS), arsenopyrite (FeAsS), chalcopyrite ($CuFeS_2$) and haverite (MnS_2), are solubilized via electron extraction by iron(III) ions and proton attack (polysulfide pathway). Polysulfides, which are the main sulfur intermediates formed, can be oxidized to sulfuric acid by *A. ferrooxidans* and *A. thiooxidans*. The following equations summarize the two mechanisms (Schippers & Sand, 1999):

Thiosulfate pathway (FeS_2, MoS_2 or WS_2):

$$FeS_2 + 6Fe^{3+} + 3H_2O \rightarrow S_2O_3^{2-} + 7Fe^{2+} + 6H^+$$
$$S_2O_3^{2-} + 8Fe^{3+} + 5H_2O \rightarrow 2SO_4^{2-} + 8Fe^{2+} + 10H^+$$

Polysulfide pathway (ZnS, $CuFeS_2$ or PbS):

$$MS + Fe^{3+} + H^+ \rightarrow M^{2+} + 0.5H_2S_n + Fe^{2+}\ (n\geq 2)$$
$$0.5H_2S_n + Fe^{3+} \rightarrow 0.125S_8 + Fe^{2+} + H^+$$
$$0.125S_8 + 1.5O_2 + H_2O \rightarrow SO_4^{2-} + 2H^+$$

Fe/S oxidizing bacteria play an important role in the production and re-cycling of leaching chemical species, mainly as protons and ferric ions (Fonti, Dell'Anno & Beolchini, 2013). The presence of these chemical species contributes to maintain the acidic conditions and the stability of metals in the system.

The leaching process is affected by bacterial adhesion on the mineral surface. Most bacteria are attracted to mineral surfaces by chemotaxis. In the case of *A. ferrooxidans*, iron(III) ions contained into bacterial exopolymers form a positively charged complex with two uronic acid which is able to attach to the negatively charged pyrite (Rohwerder, Gehrke & Sand, 2003).

Bioleaching tests were performed on sediment suspension (Seidel et al., 2006; Akinci & Guven, 2011). Metal solubilization was realized by indigenous Fe/S oxidizing bacteria (Seidel et al., 1998;2006; Lors et al., 2004b), by single pure cultures of *A. thiooxidans* and *A. ferrooxidans* (Lombardi and Garcia, 2002; Kumar & Nagendran, 2007) or by a mixed culture of Fe/S oxidizing bacterial strains (Fonti et al., 2013; Akinci & Guven, 2011). In these conditions, these bacteria were able to solubilize high proportions of metals (> 80%) (Except for lead and arsenic) from contaminated sediments (Karavaiko et al., 1988; Akinci & Guven, 2011). The efficiency of metal solubilization from contaminated sediments in decreasing order is: Zn > Cu > Cr > Pb (Akinci & Guven, 2011). Similar results were obtained by Fonti, Dell'Anno & Beolchini (2013), which confirm the highest solubilization of Zn and the lowest solubilization of Pb.

The amendment of biological sulfur in the leaching suspension yielded considerably better results than technical sulfur (Seidel et al., 2006). The equilibrium in acidification, sulfur oxidation and metal solubilization is reached after 10-14 days of leaching, depending on the amount of sulfur added. These results are attributable to the hydrophilicity of the biologically produced sulfur particles and their high specific area resulting in an increased bioavailability for sulfur-oxidizing bacteria. So, the application of biological sulfur allows the efficiency of suspension leaching experiments to be improved by shortening the leaching time. Conversely, in the column experiments (solid bed leaching carried out in a laboratory percolator system), the metal removal is almost the same for both types of sulfur. Therefore, under these conditions, the rate of sulfur oxidation and metal solubilization are mainly affected by transport phenomena rather than microbial induced conversion processes attributed to the different physico-chemical properties of sulfur sources. Generally, more time is necessary to achieve a similar degree of metal removal with solid-bed bioleaching.

Percolation field tests conducted by Seidel et al. (1998) on contaminated aquatic sediments showed that the properties and the redox state of sediment have a significant effect on leaching efficiency. So, it is necessary to adapt the process operation to specific sediment properties. Under practical conditions of percolation, 62% of metal tested (Zn, Cd, Ni, Co and Mn) were removed after 120 days for oxic contaminated sediments, leading to permissible levels of metals in soil. Conversely, in freshly dredged sediments (anoxic conditions), only 9% of the metals were released. So, anoxic sediments are less suitable for the percolation bioleaching treatment and must undergo before a pretreatment to improve mass transfer and to stimulate the activity of Fe/S oxidizing bacteria.

The application of optimized process parameters is important for an effective bioleaching. The optimum temperature is linked to that of indigenous sulfur-oxidizing bacteria between 30 and 35 °C. The addition of 2% S^0 is considered as an optimum dosage (Löser et al., 2001). Higher elemental sulfur content leads to a more intensive solubilization of heavy metals but also to a mobilization of undesirable compounds, such as Ca, Al and Fe, which interfere with the bioleaching treatment and increase costs.

The maintenance of acid conditions is crucial in the bioleaching process to allow an optimum activity of Fe/S oxidizing bacteria and to keep metals stable in the solution phase. So, geochemical characteristics of sediments can affect the performance of bioleaching. In particularly, calcite and organic matter contents contribute to neutralize the acidity produced during the bio-oxidation of sulfur minerals contained in the sediment and reduce the increase of the pH. In fact, Lors et al. (2004b) showed that the acid production was almost totally buffered by the dissolution of the calcite present in the sediment. So, if the operating conditions lead to a high buffering capacity, it is necessary to add chemical species (such as H^+ and Fe^{3+}) to acidify the system (Fonti, Dell'Anno & Beolchini, 2013).

Overall, all these factors must be considered to establish effective and cost-feasible bioremediation strategies applied *ex situ* and *in situ*.

Biovolatilization

With respect to mercury, remediation strategies applied on soils, waters and sediments are based on the transformation of highly toxic mercury compounds, such as ionic mercury (Hg^{2+}) and organomercurials (mono and dimethylmercury), to elemental mercury, which is less toxic and highly volatile.

In environments contaminated with heavy metals, bacteria develop specific mechanisms of resistance allowing the mercury detoxification (mercury resistance operon) (Barkay et al., 2003). This mechanism is based on intracellular reduction of Hg^{2+} to non-toxic Hg^0 by the mercuric reductase enzyme and diffusional loss of Hg^0 from the cell. The *merA* gene coding for mercuric reductase is responsible for the reduction of inorganic Hg whereas the *merB* gene coding for cytosolic mercuric lyase enzymes cleaves the C-Hg bond of many organomercurials. Bacterial strains, involved in this mechanism, can be isolated from the environment (Sabry et al., 1997; Iohara et al., 2001) or are genetically modified ones (Bruncke et al., 1993; Wagner-Döbler, 2003). A bacterial community Hg-resistant isolated from polluted sediments is composed by five strains belonging to different genera such as *Bacillus*, *Pseudomonas*, *Psychrobacter*, *Halomonas*, *Luteimonas* and *Micrococcus* (Pepi et al., 2011).

As elementary mercury is highly volatile, some specific treatments of the gas phase are necessary to immobilize it. A process can rely on a three-phase fluidized bed reactor (Deckwer et al., 2004). It is based on the reduction ionic mercury (10 $mg \cdot L^{-1}$) contained in effluent streams by a genetically engineered *Pseudomonas putida* strain. The reactor is aerated with air and the mercury is biotransformed on the alginate immobilized biocatalyst. The produced Hg^0 by volatization is removed into the gas phase and is recovered through fast oxidative absorption. The same technology can be applied on soils and sediments, which are suspended in a liquid phase (in generally water) and mixed with a mercury-reducing bacterium of genus *Bacillus* (Bestetti, Gandolfi & Franzetti, 2009). Mercury in its elementary form was carried out by a gas flow, which was transferred it to a trap containing a material to immobilize mercury, such as activated carbon filter.

Other strategies consist in avoiding Hg^0 volatilization by trapping it in dense porous material in which a biofilm of mercury-resistant bacteria grow. Packed bed reactors are based on an active biofilter, which is formed by mercury-reducing bacteria immobilized on porous carrier material. This biofilter transform enzymatically the Hg(II) or organomercurials to Hg^0 which is not volatilized but captured in the packed bed. This process based on mer operon-carrying bacteria was developed for the retention of mercury from waste streams 20 years ago (Bruncke et al., 1993). Wagner-Döbler (2003) demonstrated that this process applied on an industrial wastewater allowed capturing quantitatively the Hg^0 produced in the carrier material of the packed bed with a high efficiency (90% removal of initial mercury).

Bioimmobilization

Divalent Metals

Microbial biomass can immobilize metals by biosorption to cell walls or by extracellular substances. Intracellular accumulation can provoke metal precipitation in viable cells (Gadd, 2000; 2010). These immobilization mechanisms are induced by bacteria to reduce heavy metals concentration (Lloyd & Lovley, 2001). In some bacteria, metal tolerance is the outcome of their metabolism or is an intrinsic property related to their cell wall structure or the presence of extra-cellular polymeric substances. Other bacteria have developed specific resistance mechanisms to protect themselves against the toxic effects of heavy metals. These mechanisms include active transport, mediated by efflux pumps, intra- and extracellular sequestration, enzymatic transformation to less toxic chemical species by redox reactions, methylation, and reduction in the sensitivity of cellular targets to metals (Gadd, 2000).

Precipitation of metals as metal sulfides or phosphates is also an alternative way of increasing the resistance of microorganisms to metals. In sediments, sulfate reduction is mediated by various genera of bacteria (e.g., *Desulfovibrio sp.*, *Geobacter sulfurreducens*, among others) that can utilize SO_4^{2-} as a terminal electron acceptor and are collectively referred to as sulfate-reducing bacteria. These microorganisms are heterotrophic bacteria requiring strictly anaerobic conditions. Sulfate-reducing bacteria couples the oxidation of organic compounds or molecular H_2 with the reduction of sulfate as an external electron acceptor under anaerobic conditions, a process known as dissimilatory sulfate reduction (Park et al., 2011). The reduced S species (H_2S and HS^-) produced by this reaction react rapidly with Fe and many metals (Zn, Cd, Cu, Pb, Ni, Co and Hg) to produce low-solubility metal sulfide compounds, resulting in a lower sulfide concentration and a reduced metal toxicity. These metal sulfide (MeS) precipitates exhibit extremely low solubility and are relatively stable in environments under low redox conditions (Hao, 2000).

Chromium

Both aerobic and anaerobic microorganisms are able to reduce Cr(VI) to Cr(III). In aerobic conditions, it is possible to observe the bio-reduction of Cr(VI) that can be obtained directly as a result of microbial metabolism (Losi & Frankenberger, 1994). In the presence of oxygen, microbial reduction of Cr(VI) is commonly catalyzed by soluble enzymes. Cr(VI)-resistant microorganisms represent an important opportunity to have safe, economical and environmentally friendly methods to reduce Cr(VI) to Cr(III) (Raspor et al., 2000). The reduction of Cr(VI) to Cr(III) is used in *ex situ* and *in situ* treatments of Cr(VI)-contaminated solid matrices such as soils and sediments (Lovley & Coates, 1997). Taking into account that the low solubility of Cr(III) facilitates its precipitation and removal, the biotransformation of Cr(VI) to Cr(III) has been considered as an alternative process to treat Cr(VI)-contaminated wastes (Ganguli and Tripathi, 2002; Kiran et al., 2007; Xu et al., 2011). A wide range of bacteria reduces the highly soluble chromate ion to Cr(III), which under appropriate conditions precipitates as $Cr(OH)_3$. A number of Cr(VI)-reducing bacterial strains have been isolated from chromate-contaminated waters and sediments, including *Escherichia coli*,

Pseudomonas putida, *Desulfovibrio sp.*, *Bacillus sp.*, *Shewanella sp.*, *Arthrobacter sp.*, *Microbacterium sp.*, and *Cellulomonas sp.* (Focardi et al., 2013). Published works on the Hanford 100H area adjacent to the Columbia River in Washington (USA) demonstrated the feasibility of Cr(VI) bioimmobilization (Faybishenko et al., 2008; Chakraborty, Wu & Hazen, 2012). In this area, Hydrogen Release Compound (HRC) was injected to mediate bioimmobilization of Cr(VI) by stimulating indigenous microbial flora. Monitoring studies determining biogeochemical transformations associated with Cr(VI) bioremediation after HRC injection revealed that Cr(VI) levels decreased from 100 $\mu g.L^{-1}$ to below background levels within a year.

Arsenic

Microorganisms can use arsenic compounds as electron donors or electron acceptors and possess arsenic detoxification mechanisms, with pumps at the membrane level able to remove As(III) from cells, eventually after As(V) reduction (Ahmann et al., 1994; Newman et al., 1997a). The first report of an arsenic-respiring bacterial strain concerned the strain MIT-13, a microorganism named *Geospirillum arsenophilus* (Ahmann et al., 1994; Lovley and Coates, 1997). Some works have shown that anaerobic bacteria were able to reduce concomitantly As(V) to As(III) and sulfate to elemental sulfur to produce arsenite sulfide precipitates (Newman et al., 1997b; Lee et al., 2007). Recently, Focardi et al. (2010) described the microbiologically mediated reduction of As(V) to As(III) and the formation of As-S precipitates in polluted sediments. Arsenic and sulfur detected in the precipitates from the isolated bacterial strain were mainly present on the surface of the bacterial cells. Also, similar behavior has previously been reported for other arsenic and sulfur-precipitating bacterial cells (Ledbetter et al., 2007; Lee et al., 2007). Therefore, the concomitant reductions of sulfate and arsenate seem to be an effective mechanism for the precipitation of arsenic and a useful bioremediation process in reducing arsenic mobility and toxicity in sediments. In other chapter in the present book, bacterial arsenic-resistant mechanisms are further reviewed.

Conclusion

Due to their geochemical complexity, sediments have to be treated by combining various approaches: physical separation, chemical extraction, chemical or thermal stabilization. Immobilization and mobilization in metal-contaminated sediments can be achieved through the following bioremediation techniques: bioleaching, biovolatilization and bioimmobilization. Redox conditions play an important role in metals bioremediation techniques, since they can allow microorganisms to mobilize or immobilize metals, metalloids, and organo-metallic compounds by reduction and oxidation processes. Other parameters, such as pH, temperature, organic carbon, iron, and sulfate contents, sediment properties, can also influence metals mobility mechanisms. In addition, the way a particular metal is mobilized depends on the metal chemical and physical characteristics. Precipitation of metal species by reduction is generally limited to anaerobic processes and is ineffective for single-oxidation-state metals whereas precipitation by other means is not limited to reducible metals and has the advantage of producing chemically stable forms of metals. Industries often use a combination of more than one bioremediation approach to properly treat metal-contaminated sediments. The combination of the various approaches can be cost-effective, in particular on dredged materials (Fang et al., 2011).

REFERENCES

Ahmann, D., Roberts, A. L., Krumholz, L. R., Morel, F. M., 1994. Microbe grows by reducing arsenic. *Nature* 371, 750.

Akinci, G., & Guven, D. E., 2011. Bioleaching of heavy metals contaminated sediment by pure and mixed cultures of *Acidithiobacillus* spp. *Desalination* 268, 221-226.

Barkay, T., Miller, S. M., Summers, A. O., 2003. Bacterial mercury resistance: from atoms to ecosystems. *FEMS Microbiol. Rev*. 27, 355-384.

Bestetti, G., Gandolfi, I., Franzetti, A., 2009. Microbiological mercury removal from contaminated materials. US Patent WO 2009/125341 A2.

Bosecker, K., 1997. Bioleaching: metal solubilization by microorganisms. *FEMS Microbiol. Rev.* 20, 591-604.

Brunke, M., Deckwer, W. D., Frischmuth, A., Horn, J. M., Lünsdorf, H., Rhode, M., Röhricht, M., Timmis, K.N., Weppen, P., 1993. Microbial retention of mercury from waste streams in a laboratory column containing merA gene bacteria. *FEMS Microbiol. Rev.* 11, 145-152.

Caille, N., Tiffreau, C., Leyval, C., Morel, J. L., 2003. Solubility of metals in an anoxic sediment during prolonged aeration. *Sci. Tot. Environ.* 301, 239-250.

Caplat, C., Texier, H., Barillier, D., Lelievre, C., 2005. Heavy metals mobility in harbour contaminated sediments: The case of Port-en-Bessin. *Mar. Pollut. Bull.* 50, 504-511.

Cerniglia, C. E., 1992. Biodegradation of polycyclic aromatic hydrocarbons. *Biodegradatio*n 3, 77-126.

Chakraborty, R., Wu, C. H., Hazen, T. C., 2012. Systems biology approach to bioremediation. *Curr. Opin. Biotechnol.* 23, 1-8.

Coates, J. D., Phillips, E. J. P., Lonergan, D. J., Jenter, H., Lovley, D. R., 1996. Isolation of *Geobacter* species from diverse sedimentary environments. *Appl. Environ. Microbiol.* 62, 1531-1536.

Deckwer, W. D., Becker, F. U, Ledakowicz, S., Wagner-Döbler, I., 2004. Microbial removal of ionic mercury in a three-phase fluidized bed reactor. *Environ. Sci. Technol*. 38, 1858-1865.

Eggleton, J., & Thomas, K. V., 2004. A review of factors affecting the release and bioavailability of contaminants during sediment disturbance events. *Environ. Int.* 30, 973-980.

Fang, D., Zhang, R., Zhou, L., Li, J., 2011. A combination of bioleaching and bioprecipitation for deep removal of contaminating metals from dredged sediment. *J. Hazard. Mater.* 192, 226-233.

Faybishenko, B., Hazen, T. C., Long, P. E., Brodie, E. L., Conrad, M. E., Hubbard, S. S., Christensen, J. N., Joyner, D., Borglin, S. E., Chakraborty, R., et al., 2008. *In situ* long-term reductive bioimmobilization of Cr(VI) in groundwater using hydrogen release compound. *Environ. Sci. Technol.* 42, 8478-8485.

Fettweis, M., Baeye, M., Francken, F., Lauwaert, B., Van den Eynde, D., Van Lancker, V., Martens, C., Michielsen, T., 2011. Monitoring the effects of disposal of fine sediments from maintenance dredging on suspended particulate matter concentration in the Belgian nearshore area (southern North Sea). *Mar. Pollut. Bull.* 62, 258-269.

Focardi, S., Pepi, M., Ruta, M., Marvasi, M., Bernardini, E., Gasperini, S., Focardi, S. E., 2010. Arsenic precipitation by an anaerobic arsenic-respiring bacterial strain isolated from the polluted sediments of Orbetello Lagoon, Italy. *Lett. Appl. Microbiol.* 51, 578-585.

Focardi, S., Pepi, M., Focardi, S. E., 2013. Microbial reduction of hexavalent chromium as a mechanism of detoxification and possible bioremediation applications. InTech publisher, 370 p.

Fonti, V., Dell'Anno, A., Beolchini, F., 2013. Influence of biogeochemical interactions on metal bioleaching performance in contaminated marine sediment. *Water Res.* in press.

Förstner, U., 2004. Traceability of sediment analysis. *TrAC-Trend Anal. Chem.* 23, 217-236.

Gadd, G. M., 2000. Bioremedial potential of microbial mechanisms of metal mobilization and immobilization. *Curr. Opin. Biotechnol.* 11, 271-279.

Gadd, G. M., 2004. Microbial influence on metal mobility and application for bioremediation *Geoderma*, 122, 109-119.

Gadd, G. M., 2010. Metals, minerals and microbes: geomicrobiology and bioremediation. *Microbiology* 156, 609-643.

Ganguli, A., & Tripathi, A. K., 2002. Bioremediation of toxic chromium from electroplating effluent by chromate-reducing *Pseudomonas aeruginosa* A2Chr in two bioreactors. *Appl. Microbiol. Biotechnol.* 58, 416-420.

Guthrie, J. W., Hassan, N. M., Salam, M. S. A., Fasfous, I.I., Murimboh, C. A., Murimboh, J., Chakrabarti, C. L., Grégoire, D. C., 2005. Complexation of Ni, Cu, Zn, and Cd by DOC in some metal-impacted freshwater lakes: a comparison of approaches using electrochemical determination of free-metal-ion and labile complexes and a computer speciation model, WHAM V and VI. *Anal. Chim. Acta*, 528, 205-218.

Hao, O. L., 2000. Metal effects on sulfur cycles bacteria and metal removal by sulfate reducing bacteria." In Environmental Technologies to Treat Sulfur Pollution: Principles and Engineering, IWA Publishing, London, pp 393-414.

Huber, S. A., Balz, A., Abert, M., Pronk, W., 2011. Characterisation of aquatic humic and non-humic matter with size-exclusion chromatography – organic carbon detection – organic nitrogen detection (LC-OCD-OND). *Water Res.* 45, 879-885.

Iohara, K., Liyama, R., Nakamura, K., Silver, S., Sakai, M., Takeshita, M., Furukawa, K., 2001. The *mer* operon of a mercury-resistant *Pseudoalteromonas haloplanktis* strain isolated from Minamata bay, Japan. *Appl. Microbiol. Biotechnol.* 56, 736-741.

Karavaiko, G. I., Rossi, G., Agate, A. D., Groudev, S. N., and Avakyan, Z. Z., 1988. Biogeotechnology of metals. Manula UNEP USSR Commission for UNEF, Centre for International Projects, Moscow.

Kiran, B., Kaushik, A., Kaushik, C. P., 2007. Biosorption of Cr (VI) by native isolate of *Lyngbya putealis* (HH-15) in the presence of salts. *J. Hazard. Mater.* 141, 662-667.

Kumar, R. N., Nagendran, R., 2007. Influence of initial pH on bioleaching of heavy metals from contaminated soil emploring indigenous *Acidothibacillus thiooxidans*. *Chemosphere* 66, 1775-1781.

Ledbetter, R. N., Connon, S. A., Neal, A. L., Dohnalkova, A., Magnuson, T. S. 2007. Biogenic mineral production by a novel arsenic-metabolizing thermophilic bacterium from the Alvord Basin, Oregon. *Appl. Environ. Microbiol.* 73, 5928-5936.

Lee, J. H., Kim, M. G., Yoo, B., Myung, N. V., Maeng, J., Lee, T., Dohnalkova, A.C., Fredrickson, J. K.et al., 2007. Biogenic formation of photoactive arsenic-sulfide nanotubes by Shewanella sp. strain HN-41. *Proc. Natl. Acad. Sci.* 104, 410-415.

Lloyd, J. R., & Lovley, D. R., 2001. Microbial detoxification of metals and radionuclides. *Curr. Opin. Biotechnol.* 12, 248-253.

Lombardi, A. T., & Garcia, O., 2002. Biological leaching of Mn, Al, Zn, Cu and Ti in an aerobic sludge effectuated by *Acidithiobacillus ferrooxidans* and its effect on partitioning. *Water Res.* 36, 3193-3202.

Lors, C., Barbé, P., Mossmann, J. R., 2004a. Phenotypic responses of the soil bacterial community to polycyclic aromatic hydrocarbon contamination in soils. *Polycyclic Aromat. Compd.* 24, 21-36.

Lors, C., Tiffreau, C., Laboudigue, A., 2004b. Effects of bacterial activities on the release of heavy metals from contaminated dredged sediments. *Chemosphere* 56, 619-630.

Lors, C., Ryngaert, A., Périé, F., Diels, L., Damidot, D., 2010. Evolution of the bacterial community during bioremediation of PAHs in a coal tar contaminated soil. *Chemosphere* 81, 1263-1271.

Lors, C., Damidot, D., Ponge, J. F., Périé, F., 2012. Comparison of a bioremediation process of PAHs in a PAH-contaminated soil at field and laboratory scales. *Environ. Pollut.* 165, 11-17.

Löser, C., Seidel, H., Hoffmann, P., Zehnsdorf, A., 2001. Remediation of heavy metal-contaminated sediments by solid-bed bioleaching. *Environ. Geol.* 40, 643-650.

Losi, M. E., Frankenberger, W. T., 1994. Chromium-resistant microorganisms isolated from evaporation ponds of a metal processing plant. *Water Air Soil Pollut.* 74, 405-413.

Lovley, D. R., & Coates, J. D., 1997. Bioremediation of metal contamination. *Curr. Opin. Biotechnol.* 8, 285-289.

Luoma, S. N., & Rainbow, P. S., 2008. Metal Contamination in Aquatic Environments: science and lateral management. Cambridge University Press, Cambridge, p 600.

Mulligan, C. N., Yong, R. N., Gibbs, B. F., 2001a. An evaluation of technologies for the heavy metal remediation of dredged sediments. *J. Hazard. Mat.* 85, 145-163.

Mulligan, C. N., Yong, R. N., Gibbs, B. F., 2001b. Remediation technologies for metal-contaminated soils and grounwater: an evaluation. *Eng. Geol.* 60, 193-207.

Mulligan, C. N., Fukue, M., Sato, Y., 2010. Sediments contamination and sustainable remediation. CRC Press, USA, p 305.

Namkoong, W., Hwang, E., Park, J., Choi, J., 2002. Bioremediation of diesel-contaminated soil with composting. *Environ. Pollut.* 119, 23-31.

Newman, D. K., Kennedy, E. K., Coates, J. D., Ahmann, D., Ellis, D. J., Lovley, D. R. and Morel, F. M. M., 1997a. Dissimilatory arsenate and sulphate reduction in *Desulfotomaculum auripigmentum sp. Arch. Microbiol.* 168, 380-388.

Newman, D. K., Beveridge, T., Morel, F., 1997b. Precipitation of arsenic trisulfide by Desulfotomaculum auripigmentum. *Appl. Environ. Microbiol.* 63, 2022-2028.

Park, J. H., Lamb, D., Paneerselvam, P., Choppala, G., Bolan, N., Chung, J. W., 2011. Review: role of organic amendments on enhanced bioremediation of heavy metal(loid)contaminated soils. *J. Hazard. Mater.* 185, 549-574.

Peng, J., Song, Y., Yuan, P., Cui, X., Qiu, G., 2009. The remediation of heavy metals contaminated sediment. *J. Hazard. Mater.* 161, 633-640.

Pepi, M., Gaggi, C., Bernardini, E., Focardi, S., Lobianco, A., Ruta, M., Nicolardi, V., Volterrani, M., Gasperini, S., Trinchera, G., Renzi, P., Gabellini, M., Focardi, S.E., 2011. Mercury-resistant bacterial strains *Pseudomonas* and *Psychrobacter* spp. Isolated from sediments of Orbetello Logoon (Italy) and their possible use in bioremediation process. *Int. Biodeterior. Biodegrad.* 65, 85-91.

Perelo, L. W., 2010. Review: *In situ* and bioremediation of organic pollutants in aquatic sediments. *J. Hazard. Mater.* 177, 81-89.

Prakash, D., Gabani, P., Chandel, A. K., Ronen, Z., Singh, O.V., 2013. Bioremediation: a genuine technology to remediate radionuclides from the environment. *Microbial Biotechnol.* 6, 349-360.

Raspor, P., Batic, M., Jamnik, P., Josic, D., Milacic, R., Pas, M., Recek, M., Rezic-Dereani, V., Skrt, M., 2000. The influence of chromium compounds on yeast physiology. *Acta Microbiol. Immunol. Hung.* 47, 143-173.

Rocchetti, L., Beolchini, F., Hallberg, K. B., Johnson, D. B., Dell'Anno, A., 2012. Effects of prokaryotic diversity changes on hydrocarbon degradation rates and metal partitioning during bioremediation of contaminated anoxic marine sediments. *Mar. Pollut. Bull.* 64, 1688-1698.

Rohwerder, T., Gehrke, T., Sand, W., 2003. Bioleaching review part A: Progress in bioleaching: fundamentals and mechanisms of bacterial metal sulfide oxidation. *Appl. Microbiol. Biotechnol.* 63, 239-248.

Sabry, S. A., Ghozian, H. A., Abou-Zeid, D. M., 1997. Metal tolerance and antibiotic resistance patters of a bacterial population isolated from sea water. *J. Appl. Microbiol.* 82, 245-252.

Sahuquillo, A., Rigol, A., Rauret, G., 2003. Overview of the use of leaching/extraction tests for risk assessment of trace metals in contaminated soils and sediments. *Trac-Trend Anal. Chem.* 22, 152-159.

Sakultantimetha, A., Keenan, H. E., Beattie, T. K., Bangkedphol, S., Cavoura, O., 2011. Bioremediation of tributyltin contaminated sediment: Degradation enhancement and improvement of bioavailability to promote treatment processes. *Chemosphere*, 83, 680-686.

Sand, W., Rohde, K., Sobotke, B., Zenneck, C., 1992. Evaluation of *Leptospirillum ferrooxidans* for leaching. *Appl. Environ. Microbiol.* 58, 85-92.

Schippers, A., & Sand, W., 1999. Bacterial leaching of metal sulfides proceeds by two indirect mechanisms via thiosulfate or via polysulfides and sulfur. *Appl. Environ. Microbiol.* 65, 319-321.

Seidel, H., Ondruschka, J., Morgenstern, P., Stottmeister, U., 1998. Bioleaching of heavy metals from contaminated aquatic sediments using indigenous sulfur-oxidizing bacteria: a feasibility study. *Wat. Sci. Tech.* 37, 387-394.

Seidel, H., Wennrich, R., Hoffmann, P., Löser, C., 2006. Effect of different types of elemental sulfur on bioleaching of heavy metals from contaminated sediments. *Chemosphere* 62, 1444-1453.

Suresh, G., Ramasamy, V., Meenakshisundaram, V., Venkatachalapathy, R., Ponnusamy, V., 2011. A relationship between the natural radioactivity and mineralogical composition of the Ponnaiyar river sediments, India. *J. Environ. Radioact.* 102, 370-377.

Tan, F., Wang, M., Wang, W., Lu, Y., 2008. Comparative evaluation of the cytotoxicity sensitivity of six fish cell lines to four heavy metals in vitro. *Toxicol. In Vitro* 22, 164-70.

Tichy, R., Rulkens, W. H., Grotenhuid, J. T. C., Nydl, V., Cuypers, C., Fajtl, J., 1998. Bioleaching of metals from soils or sediments. *Water Sci. Technol.* 37, 119-127.

Tombácz, E., Libor, Z., Illés, E., Majzik, A., Klumpp, E., 2004. The role of reactive surface sites and complexation by humic acids in the interaction of clay mineral and iron oxide particles. *Org. Geochem.* 35, 257-267.

Tyagi R. D., Blais J. F., Auclair J. C., Meunier N., 1993. Bacterial leaching of toxic metals from municipal sludge: influence of sludg characteristics. *Water Env. Res.* 65, 196-204.

Van Hullebusch, E. D., Peerbolte, A., Zandvoort, M. H., Lens, P. N., 2005. Sorption of cobalt and nickel on anaerobic granular sludges: isotherms and sequential extraction. *Chemosphere* 58, 493-505.

Wagner-Döbler, I., 2003. Pilot plant for bioremediation of mercury-containing industrial wastewater. *Appl. Microbiol. Biotechnol.* 62, 124-133.

Wen, L. S., Jiann, K. T., Santschi, P. H., 2006. Physicochemical speciation of bioactive trace metals (Cd, Cu, Fe, Ni) in the oligotrophic South China Sea. *Mar. Chem.* 101, 104-129.

Wielinga, B., Mizuba, M. M., Hansel, C. M., Fendorf, S., 2001. Iron promoted reduction of chromate by dissimilatory iron-reducing bacteria. *Environ. Sci. Technol.* 35, 522-527.

Xu, L., Luo, M., Li, W., Wei, X., Xie, K., Liu, L., Jiang, C., Liu, H., 2011. Reduction of hexavalent chromium by Pannonibacter phragmitetus LSSE-09 stimulated with external electron donors under alkaline conditions. *J. Hazard. Mater.* 185, 1169-1176.

Yadav, S. K., 2010. Heavy metals toxicity in plants: An overview on the role of glutathione and phytochelatins in heavy metal stress tolerance of plants. *S. Afr. J. Bot.* 76, 167-179.

In: Bioremediation: Processes, Challenges and Future Prospects ISBN: 978-1-62948-513-3
Editors: Jesús Bernardino Velázquez-Fernández et al.

Chapter 2

Microbial Arsenite Biotransformation and its Potential Applications in Bioremediation

Armando Quintero-Castañeda[1], Guadalupe Pacheco-González[1], Jesús Bernardino Velázquez-Fernández[2], Leticia Mónica Sánchez-Herrera[2] and Verónica Alejandra Mondragón-Jaimes[3]*

[1]Unidad Académica de Agricultura,
Universidad Autónoma de Nayarit. Tepic Nayarit, México
[2]Secretaría de Investigación y Posgrado,
Universidad Autónoma de Nayarit. Tepic Nayarit, México
[3]Unidad Académica de Ciencias Químico Biológicas y Farmacéuticas,
Universidad Autónoma de Nayarit, Tepic, Nayarit, México

Abstract

The presence of different chemical species of arsenic in almost every environment has been crucial to find microorganisms that, in spite of being phylogenetically distant, have the ability to oxidize arsenite (As III) to arsenate (As V). They would contribute to both, the distribution and speciation or arsenic throughout ecosystems. Arsenite is oxidized by microorganisms possessing *aox* operon (now *aio*) in their genome. This operon contains genes encoding the arsenite oxidase enzyme itself, regulatory and accessory proteins. In this chapter, we review knowledge on arsenite-oxidizing bacteria: since they were first observed, the discovery and characterization of the arsenite oxidase enzyme, its relation to energy production as well as its potential applications in bioremediation.

---

* Corresponding Author address. Email: veronicam@nayar.uan.mx.

Keywords: Arsenite oxidizing, arsenite oxidase, *aoxB* genes, bioremediation of arsenic, *aio* operon

INTRODUCTION

Arsenic (As) is a trace element, toxic for most forms of life. It is widely spread though unevenly, in the atmosphere, water, soils, sediments, and living beings (Cullen and Reimer, 1989). Natural sources of arsenic include volcanic and hydrothermal activity, erosion of sedimentary and some igneous rocks. Human action such as melting of minerals, mining of precious metals, burning of fossil fuels, and the use of agrochemicals are also important (Oremland and Stolz, 2003).

This metalloid is environmentally ubiquitous as As^{+5} (arsenate), As^{+3} (arsenite), As^{0} (arsenic) and As^{-3} (arsine) (Oremland and Stolz, 2005). The anions arsenite [H_3AsO_3] and arsenate [$H_2AsO_4^-$, y $HAsO_4^{-2}$] are the most commonly form of As found in waters (Oremland and Stolz, 2003; Brown and Ross, 2002). The As content in earth's crust is between 1.5 y 2 mg/Kg, ranking arsenic 20 in the list of the most abundant elements (Committee on Medical and Biological Effects of Environmental Pollutants, 1977). However, arsenic pollution is considered to be mainly due to human activity (Smedley and Kinniburgh, 2002). All the diverse forms of arsenic contaminate water for human use, thus they have caused notorious arsenicosis cases worldwide (Nordstrom, 2002; Ravenscroft et al., 2009).

Interaction with Microorganisms

Because of their metabolism, bacteria participate directly in the bio-geochemical cycle of arsenic (D´Imperio et al., 2007), playing the main role on its speciation and mobilization throughout the environment (Oremland and Stolz, 2005). Thus, bacteria are able to oxidize, reduce and methylate it, diminishing or enhancing its toxicity (Bhattacharjee and Rosen, 2007; Mukhopadhyay et al., 2002; Oremland and Stolz, 2003).

Arsenate and arsenite are both toxic to microorganisms; arsenite affects the function of several enzymes and arsenate acts as an analog of phosphate both in its process of transport into the cell, as well as in its utilization (Tamaki and Frankenberger, 1992). However, some organisms have developed mechanisms that allow them to survive in the presence of this element, among them are the increase in the specificity for the entrance of phosphate (Cervantes et al., 1994), the oxidation of arsenite through the enzyme arsenite oxidase (Anderson et al., 1992; Muller et al., 2003; Oremland and Stolz, 2005; Inskeep et al., 2007; Santini et al., 2002).

The Study of the Arsenite Oxidase Enzyme

The first observation that bacteria from cattle immersion baths resisted up to 1% arsenic trioxide and that that they oxidized arsenite to arsenate occurred in South Africa in 1918 (Green, 1918). Later in 1949, Turner isolated five different bacteria with similar capacity.

Moreover, They succeeded in characterizing the arsenite dehydrogenase and its associated cytochromes (Turner, 1949). In 1976, Phillips and Taylor isolated and described 13 strains of *Alcaligenes faecalis* able to oxidize arsenite. Among them, one strain showing the highest activity was deeply studied (Phillips and Taylor, 1976).

However, it was until 1992 that the arsenite oxidase was purified for the first time from *A. faecalis*. It was demonstrated that it is stimulated by the presence of As (III) and that it is able to oxidize arsenate (As V) (Anderson et al., 1992). In 2001, the 3D structure of arsenite oxidase from the same species was found to consist of two subunits: a large 825-amino acid protein named AoxB (or AroA) and a small protein of 134 residues called AoxA (or AroB) (Ellis et al., 2001).

In *Agrobacterium tumefaciens* 5A, the structural genes of arsenite oxidase, aoxA y aoxB, encode a small subunit Rieske, and a large subunit molibdopterine, respectively. The product of the aoxB (AoxB) gene is the catalytic subunit of arsenite oxidase, it has molibdene as cofactor with a 3Fe-4S cluster (Stolz et al., 2006). Both genes are found under the control of the regulatory genes aoxS and aoxR which are upstream (Kashyap et al., 2006). This also was described in *Rhizobium* sp. NT-26 (Sardiwal et al., 2010). The aoxS gene encodes for a histidine kinase (HK) and aoxR which regulate a system with two components. Besides those genes, downstream are located aoxC and aoxD which encode for cytochrome c and for an enzyme involved in the synthesis molibdopterine. Those genes are co-transcribed forming part of an operon responding to quorum sensing signals (Kashyap et al., 2006). These same genes have been described in *Herminiimonas arsenicoxydans* and *Ochrobactrum tritici* (Branco et al., 2009). In *H. arsenicoxydans*, proteins RpoN and DnaJ participate in the control of the oxidation of arsenite. Finally, within this operon, the aoxX gene encodes a protein that can bind oxi-anions (Cai et al., 2009).

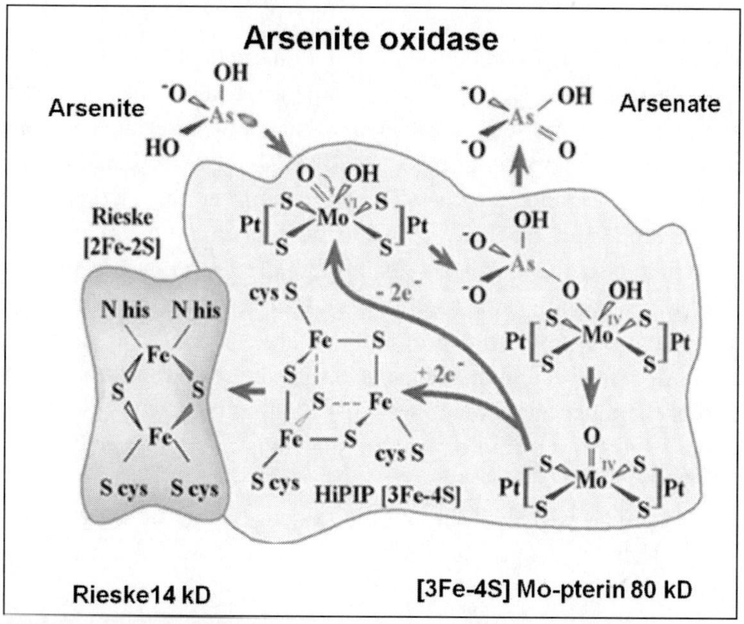

Modified after Silver and Phung, 2005.

Figure 1. Models for heterodimer arsenite oxidase. Arsenite is internalized through a cone shaped opening found on the surface of the enzyme, making direct contact with the molybdenum Mo VI.

Table 1. Previous and new names of genes involved in arsenite oxidation

Protein	Gene name		Reference
	Gene	Previous Nomenclature	
Small subunit of arsenite oxidase	*aioB*	*aoxA*	Muller *et al.*, 2003.
		aroB	Santini and Vanden Hoven 2004.
		asoB	Silver and Phung 2005.
Large subunit of arsenite oxidase	*aioA*	*aoxB*	Muller *et al.*, 2003.
		aroA	Santini and Vanden Hoven 2004.
		asoA	Silver and Phung 2005.
Sensor histidine kinase	*aioS*	*aoxS*	Kashyap *et al.*, 2006.
		aroS	Sardiwal *et al.*, 2010.
Transcriptional regulator	*aioR*	*aoxR*	Kashyap *et al.*, 2006.
		aroR	Sardiwal *et al.*, 2010.
Conjugation proteins to oxy-anion	*aioX*	*aoxX*	Cai *et al.*, 2009

* Modified after Lett et *at.*, 2012.

Taken from Lett et al., 2012.

Arsenite oxidase is as periplasmic protein found on the outside of the internal membrane, coupled functionally to the respiratory chain (Ahmann et al., 1994). AoxA possess a TAT signal (Twin-Arginine Transporter) at the end of the amino terminal; it sorts the folded heterodimeric enzyme into the periplasmic space through the cytoplasm (Muller et al., 2003). Arsenite is internalized through a cone-shaped opening of the enzyme, then it contacts directly with the molybdenum ion (Mo VI) inside the enzyme. Immediately, two electrons are transferred and as final step, the arsenate exits through the same cone. In the process, Mo (VI) is reduced to Mo (IV). Thereafter, the electrons are transferred to the 3Fe-FS group found 1.5 nm from the Mo within the Mo-pterine group of the large subunit, and later onto the [2Fe-2S] cluster found in the small subunit. Subsequently, the electrons are transferred to the respiratory chain of the inner membrane. It is possible that this occurs firstly to an azurine or a cytochrome and eventually, to the oxygen as the final electron acceptor (Anderson et al., 1992, Ellis et al., 2001).

The enzyme described is part of an aerobic system (using an oxygen molecule as final electron acceptor). There is also an anaerobic arsenite oxidase coded by *arxA,* which was discovered in 2010 (Zargar et al., 2010).

To have consensus on the designation of genes involved in the oxidation of arsenite, a nomenclature (Table 1) was proposed in the year 2012. In such proposal, genes specifically involved in oxidation were designated as *aio.*

Horizontal Gene Transfer and Phylogenetic Diversity

In 1984, a short study on the role played by the horizontal gene transfer in the acquisition of resistance to arsenic was published. In strains of *Corynebacterium flaccumfaciens,* Hendrick et al. (1984) found a conjugative plasmid conferring the capacity to survive in the presence of arsenite salts, arsenate and antimony. This phenomenon explains why certain bacteria have a resistance to metals and antibiotics in environments never exposed to these contaminants (Morozzi et al., 1986). Also, it demonstrates that their presence could be conditioned by a concentration gradient of the metalloid in polluted sites (Quéméneur et al., 2010).

The contamination generated by metals contributes to the permanence and distribution of both types of genes (Goñi-Urriza et al., 2000) through the interchange of plasmids among unrelated bacteria in natural environments (Kruse and Sorum, 1994). The contaminants, such as arsenic serve as selection agents in the proliferation of the resistance. Several mechanisms exist that permit this process of co-selection, co-resistance (resistance conferred by different elements) and crossed resistance (same genetic determinant responsible for both resistances to antibiotics and to metals). For this reason, this type of contamination represents a process of persistent and generalized selective pressure that potentially contributes to maintain and expand the factors involved in antibiotic resistance (Baker-Austin et al., 2006; Mondragón et al., 2011).

The same phenomenon of horizontal transfer has placed a crucial role in the distribution of the *aoxB* in prokaryotic communities. By analyzing the published sequences of strains from arsenic-polluted soils, sediments and sites of geothermal activity, it was demonstrated that the distribution of bacterial *aoxB* genes is wider than previously thought (Heinrich-Salmeron et al., 2011).

The ecological and phylogenetic diversity of the sequences of *aoxB* genes obtained from bacteria found in sites of geothermal activity showed that such genes are widely distributed in the bacterial domain and extended in soil-water systems containing arsenic (Inskeep et al., 2007). These genes have been identified in 25 isolates of bacteria as well as Archaea, the majority of which belong to Alpha, Beta or Gamma proteobacteria (Heinrich-Salmeron et al., 2011). The phylogenetic analyses of nucleotide sequences from microorganisms which are evolutionary distant indicate that even though there is a marked diversity (Stolz et al., 2006), there is a common origin found before the divergence between Archaea and Bacteria (Lebrun et al. 2003). Some well-known prokaryotes oxidize arsenite to arsenate in aerobic conditions: *Herminiimonas arsenicoxydans*, *Thiomonas* spp. and *Rhizobium* spp., strain NT26 and in anaerobic conditions *Alkallilimnicola ehrlichii* was found as part of the process of detoxification (Heinrich-Salmeron et al., 2011). Hamamura et al. studied the diversity and distribution of Mo-pterines of arsenite oxidases along with *aoxB* type genes were studied in water bodies, in 2010. This represents an important step in comprehension of evolution of the *aoxB* genes diversity of bacterial arsenite oxidases (Hamamura et al., 2010).

Relationship to Generation of Energy

Certain strains of *Agrobacterium* and *Rhizobium* have been reported to exhibit chemolithotrophic growth in minimal media in the presence of arsenite as donator of electrons, oxygen as acceptor and bicarbonate-carbon dioxide as carbon source. Furthermore, it has been demonstrated that the oxidation of arsenite is catalyzed by the periplasmic arsenite oxidase at an optimum pH of 5.5 (Santini et al., 2000).

Also, in the year 2002, Oremland et al., reported the presence of a chemoautotrophic bacterium from the Mono Lake, California. It is capable of surviving by utilizing arsenite as donator of electrons and nitrate as acceptor. When arsenic is absent, this bacterium obtains electrons from hydrogen (H_2) and sulphur (S^{2+}). The strain also managed heterotrophic growth in acetate and nitrate under both aerobic and anaerobic conditions. However, in spite of this behavior, the presence of the *aox* operon was not studied.

An interesting observation is the reversible dynamic behavior in the process of oxidation-reduction of arsenic. *Cladophora* sp., strains collected in residual suburban waters without detectable As, were capable of reducing As (V) to As (III) under anoxic conditions and also oxidized As (III) to As (V) under aerobic incubation (Kulp et al., 2004).

Some bacteria possess the *aoxB* gene and have the ability to use the perchlorate anion (ClO_3^-) as acceptor of electrons during the process of oxidation of arsenite (Wenjie, 2010). Similarly, other bacteria use selenate (SeO_4^{2-}) in anaerobic conditions with the same efficacy as other microorganisms that do so with nitrate in the medium (Fisher and Hollibaugh, 2008).

Regarding the use of oxygen to reduce or oxidize arsenic, it seems not to be necessary. Hoeft et al., 2010 found that process similar to *Cladophora* sp. (see above), could take place in the presence of light by the anoxigeic photosynthesis. Contrarily, same isolates reduced arsenate in darkness by oxidation at a comparable magnitude.

It should also be noted that the reduction of arsenate to arsenite occurred in the same way in the presence of light. This demonstrates a co-occurrence of both processes as examples of coupled arsenotrophy. Surprisingly, no PCR amplification products were detected for arsenite oxidase genes, even though aerobic bacteria biofilm oxidized arsenite. Considering that this happened in the presence of light, it should be said that it is photo-heterotrophy and not chemoheterotrophy, which is occurring. Evidently, there must be other genetic determinants that are responsible for this behavior, which also suggests think that diverse forms of life explored the use of arsenic, perpetuating it among many bacterial species.

Habitats of Arsenite Oxidating Strains

The bacterial strains that have the capacity to oxidize have come from diverse sources. They have been isolated from soils, residual suburban water, soil-water systems (Kulp et al., 2004, Rhine et al., 2006, Inskeep et al., 2007), biofilms with geothermal activity (Hoeft et al., 2010), soil from mines and their sewers (Osborne et al., 2010; Santini et al., 2000; Santini et al., 2002; Casiot, 2003; Bruneel, 2003), also from lakes and rivers (Fisher, 2008; Oremland et al., 2002, Quéméneur et al., 2010). Although these species have been identified in environments with moderate to high temperatures (Donahoe-Christiansen et al., 2004; Rhine et al., 2005; D´Imperio et al., 2007; Inskeep et al., 2007; Quéméneur et al., 2008; Quéméneur et al., 2010; Hoeft et al., 2010), it does not imply that such microorganisms are limited to these conditions. Indeed, arsenite-oxidant microorganisms have even been isolated and characterized from sub-Arctic subterranean sites where the temperature fluctuates between 4 y 10 °C (Osborne et al., 2010).

There is controversy regarding the physical chemical conditions needed for arsenite oxidation activity to occur. It has been described that optimum pH range falls between 7-10 and temperature of 25-35°C (Ike et al., 2008). Donahoe-Christiansen et al., 2004 had reported a strain isolated from Yellowstone National Park with an optimal temperature of 55 a 60°C at a pH of 3.0. The presence of H_2S inhibits the oxidation of arsenite by *Hydrogenobaculum* sp. If H_2S concentration is above ~5.0 μM, the process is blocked. Because of this, the reaction occurs only when H_2S concentration is low or zero, and is not influenced by temperature or oxygen concentration (Donahoe-Christiansen et al., 2004, D´Imperio et al., 2007).

Potential Bioremediation Applications

The process of bioremediation depends on the microbial enzymatic activity to degrade the contaminants and convert them to innocuous substances. It should also be considered in the context of the environmental conditions. The growth of the microorganisms and the enzymatic activities are determined by these parameters.

Presently bioprocesses for the treatment of waters contaminated by arsenic are being developed, but few have reached the industrial stage. The US Environmental Protection Agency (EPA) has reviewed the best available technologies. The candidate methods for water treatment included filtration, coagulation, ionic interchange and inverse osmosis. In the first stage, most of them achieved oxidation of As (III) to As (V) through the application of strong oxidants such as potassium permanganate, hydrogen peroxide or ozone. However, the costs of these technologies made it impossible to apply to large volumes. The biological and microbiological methods turned out to be economically viable, besides being innocuous to the environment. Because many bacteria play an important role in the geochemical cycle of arsenic, it is logical to consider them to treat water contaminated with arsenic. The knowledge of physiology and metabolism of these bacteria have been used to develop diverse innovative systems according to diverse environmental scenarios (Lièvremont et al., 2009).

The application of biological processes to the oxidation and extraction of iron turned out to be also efficient in the elimination of arsenic. In this case, it was determined that As (III) partially oxidizes to As (V) in a 10 month treatment process (Zouboulis and Katsoyiannis, 2005).

It was also observed that biofilm systems are particularly adequate for the treatment of corrosive compounds due to its high microbial biomass and the capacity to immobilize compounds. In this case, bioremediation is enhanced with the increase of the concentration of contaminants for degradation as a result of bacterial chemotaxis. The strategies to enhance the efficiency of bioremediation include genetic engineering to improve the strains, the chemotactic capacity of the mixed populations and the optimization of physical chemical conditions (Singh et al., 2006). A good example of a mixed culture with arsenite oxidation activity was reported by Ike et al. (2008) who worked with strains of *Haemophillus, Micrococcus* and *Bacillus* obtained from a soil sample free of contamination. In these experiments, they used activated alumina and determined an improved arsenic extraction by the bacterial oxidation from As (III) to As (V).

After the oxidation of As (III) to As (V), the arsenic is easily removed because As (V) has more affinity to absorbent material than As (III) (Ike et al., 2008). Mokashi and Paknikar in 2002 took advantage of this property to remove the As (V) by using three methods: neutral iron, activated carbon and ferric chloride. In all of them, up to 50 mmol/L of As (III) was oxidized by *Microbacterium lacticum* strains isolated from municipal wastewater. To accomplish it, bacteria were immobilized on brick chunks and then bottled in a glass column. With this, fast complete oxidation of As (III) of subterranean waters at a neutral pH and ambient temperature were achieved by using methanol as substrate (Mokashi and Paknikar, 2002). Additional materials have been used with a double purpose: adsorption of the As (V) generated after the oxidation process and immobilization of the microorganisms. In 2003, Lièvremont et al., studied the oxidative properties of ULPAs1 when cultivated in batch reactors in the presence of two solid phases: cabasite and kutnahorite. In parallel, the retention efficiency of these solid phases toward arsenic ions and particularly arsenate was studied. Pure quartz sand was the material used as reference. Kutnahorite absorbed As (V) efficiently, while cabasite allowed only the oxidation of As (III), but not the absorption of As (V). Pure quartz sand did not absorb at all. The arsenite oxidative properties of ULPAs1 were conserved when they were cultured in quartz or cabacite (Lièvremont et al., 2003). In another study using CASO1, an autotrophic arsenite oxidant derived from a mine, its efficiency was evaluated also in reactors. The arsenite oxidation carried out by this population turned out to be particularly interesting for de-contamination of arsenic residues from subterranean waters, as they exhibited high As (III) oxidative activity with no need of organic nutrients (Battaglia-Brunet et al., 2002).

In other studies, the anoxic oxidation of arsenite coupled to the chemolithotrophic denitrification demonstrated to be feasible in continuous bioreactors, when maintained during long continuous (3 year) periods with a arsenic conversion yield higher than 92% with high volumes (Sun et al., 2010). Another species identified that *Pseudomonas stutzeri* strain GIST-BDan2 possessing the *aoxB* and *azor* genes, could play an important role in the oxidation of arsenite to arsenate. This strain completely oxidized As (III) 1mM to As (V) in 25 to 30 hours (Chang et al., 2010).

Other methods have been used to diminish the concentration of arsenic in polluted waters, such as plant species with the capacity to accumulate this metalloid (Sung et al., 2009). Few botanical species are naturally tolerant to As, among them, *Pteris vittata* and many other members of the Pteridaceae hyperaccumulate arsenic (Ma et al., 2001; Visoottiviseth et al., 2002; Ellis et al., 2006; Pickering et al., 2006; Zhao et al., 2009, Finnegan and Chen, 2012). Using microorganisms in bioremediation represents an advantage because it is an easier system to manage with short replication cycles and can be improved by genetic engineering.

Arsenic Oxidant Bacteria As Biosensors

Besides being used in bioremediation of polluted areas, microbe sensors can be used as tools to monitor heavy metals or other contaminants. Such a system is a simple cost-effective and sensitive measure for evaluating presence or concentration of pollutants. Such systems indicate the presence of a target molecule by expressing a reporter protein, such as green fluorescence, luciferase or galactosidase. Several arsenite biosensors have been constructed to contain a reporter gene joined to the operator-promotor of the ars operon (Cai and DuBow, 1997; Ramanathan et al., 1997; Scott et al., 1997; Stocker et al., 2003; Tauriainen et al., 1997; Trang et al., 2005; included in: Yoshida et al., 2008), such as the carotenoid biosynthesis gene, *crtA*, wich can be used as a novel reporter in whole-cell biosensors made from *Rhodovulum sulfidophilum* (Yoshida et al., 2008). It is notable that to this day, there are no reports that involve the use of *aoxB* gene or its products, either for bioremediation or as biosensor.

The *aoxB* gene has important characteristics of molecular markers: (i) It has been found in all aerobic arsenite oxidating bacteria studied until now, (ii) the conserved regions of this gene have permitted the design of valuable recognition sequences, (iii) the studied region is sufficiently large to obtain genetic variation permitting a discrimination of phylogenetic groups. Based on these observations, the use of *aoxB* gene should indeed be an ideal specific functional marker of arsenite oxidants in environmental diversity studies. Future work should be centered in evaluating the relationships among speciation, arsenic abundance, and *aoxB* diversity (Quéméneur et al., 2008).

In general, these microorganisms can be used as an interesting model to explore fundamental questions of microbial ecology, which allows understanding the relationship between functional attributes and their environment (D´Imperio et al., 2007).

Conclusion

Bacteria capable of metabolizing toxic elements represent an attractive tool for restoring polluted sites. The bacterial process of the arsenite oxidation has substantial importance and determines arsenic environmental movement and speciation through the environment. It involves the arsenite oxidase enzyme, which has been deeply studied in structure and function. It is closely related to certain mechanisms generating energy. Its activation depends on bacteria and the environment where they are found. Indeed, AoxB protein posses a high potential to be applied in bioremediation processes, either by using the complete organisms in such case, all their physiological characteristics should be considered, or by using the purified enzyme in a reactor, or as biosensors of contamination. The biological methods available to treat arsenic polluted soil or water are very promising. However, studies so far have been mostly applied to laboratory scale and should be tested on the field and/or in an industrial context to evaluate their actual feasability.

References

Abdrashitova, S. A., Abdullina, G. G., Ilyaletdinov, A. N., 1986. Role of arsenates in lipid peroxidation in *Pseudomonas putida* cells oxidizing arsenite. *Mikrobiologia.* 55,212-216.

Abdrashitova, S. A., Abdullina, G. G., Mynbaeva, B. N., Ilyaletdinov, A. N., 1990. Oxidation of iron and manganese by arsenic-oxidizing bacteria. *Mikrobiologia.* 59,85-89.

Ahmann, D., Roberts, A. L., Krumholz, L. R., Morel, F. M., 1994. Microbe grows by reducing arsenic. *Nature.* 371, 750.

Anderson, G. L., Williams, J., Hille, R., 1992. The purification and characterization of arsenite oxidase from *Alcaligenes faecalis*, a molybdenum-containing hydroxylase. *J. Biol. Chem.* 267, 23674–23682.

Baker-Austin, C., Wright, M. S., Stepanauskas, R., McArthur, J. V., 2006. Co-selection of antibiotic and metal resistance. *TRENDS in Microbiology*.14, 176-182.

Battaglia-Brunet, F., Dictor, M.C., Garrido, F., Crouzet, C., Morin, D., Dekeyser, K., Clarens, M., Baranger, P., 2002. An arsenic (III)-oxidizing bacterial population: selection, characterization, and performance in reactors. *J. Appl Microbiol.* 93, 656-67.

Bhattacharjee, H., Rosen, B. P., 2007. Arsenic metabolism in prokaryotic and eukaryotic microbes. In: Dietrich H. Nies, Simon Silver. *Molecular Microbiology of Heavy Metals: Microbiology Monographs. Springer-Verlag Berlin Heidelberg.* 6, 371-406.

Branco, R., Francisco, R., Chung, A. P., Vasconcelos, P., 2009. Identification of an *aox* system that requires cytochrome *c* in the highly arsenic-resistant bacterium *Ochrobactrum tritici* SCII24. *Appl. Environ. Microbiol.* 75, 5141-5147.

Brown, K., Ross, G., American Council on Science and Health., 2002. Arsenic, drinking water, and health: a position paper of the American Council on Science and Health. *Regul. Toxicol. Pharmacol.* 36, 162–174.

Bruneel, O., 2003. Mediation of arsenic oxidation by *Thiomonas* sp. In acid-mine drainage (Carnoule`s, France). *J. Appl. Microbiol.* 95, 492–499.

Cai, L., Rensing, C., Li, X., Wang, G., 2009. Novel gene clusters involved in arsenite oxidation and resistance in two arsenite oxidizers: *Achromobacter* sp. SY8 and *Pseudomonas* sp. TS44. *Appl. Microbiol. Biotechnol.* 83, 715-725.

Casiot, C., 2003. Bacterial immobilization and oxidation of arsenic in acid mine drainage (Carnoule`s Creek, France). *Water Res.* 37, 2929–2936.

Cervantes, C., Ji, G., Ramirez, J., Silver, S., 1994. Resistance to arsenic compounds in microorganisms. *FEMS Microbiol Rev.* 15, 355-367.

Chang, J. S., Yoon, I. H., Lee, J. H., Kim, K. R., An, J., Kim, K. W., 2010. Arsenic detoxification potential of *aox* genes in arsenite-oxidizing bacteria isolated from natural and constructed wetlands in the Republic of Korea. *Environ. Geochem. Health.* 32, 95-105.

Committee on Medical and Biological Effects of Environmental Pollutants, National Research Council (CMBEEP). 1977. Arsenic: Medical and biologic effects of environmental pollutants. *The National Academies Press, National Academy of Sciences.* 16-79.

Cullen, W. R., Reimer, K. J., 1989. Arsenic speciation in the environment. *Chem. Rev.* 89, 713–764.

D´Imperio, S., Lehr, C. R., Breary, M., McDermott, T. R., 2007. Autecology of an arsenite chemolithotroph: sulfide constraints on function and distribution in a geothermal spring. *Appl. Environ. Microbiol.* 73, 7067-7074.

Donahoe-Christiansen, J., D'Imperio, S., Jackson, C. R., Inskeep, W. P., McDermott, T. R., 2004. Arsenite-Oxidizing *Hydrogenobaculum* strain isolated from an acid-sulfate-chloride geothermal spring in Yellowstone National Park. *Appl. Environ. Microbiol.* 70, 1865-1868.

Ehrlich, H. L., 1997. Microbes and metals. *Appl. Microbiol. Biotechnol.* 48, 687-692.

Ellis, D. R., Gumaelius, L., Indriolo, E., Pickering, I. J., Banks, J. A., Salt, D.E., 2006. A novel arsenate reductase from the arsenic hyperaccumulating fern *Pteris vittata*. *Plant Physiol.* 141, 1544–1554.

Ellis, P. J., Conrads, T., Hille, R., Kuhn, P., 2001. Crystal structure of the 100 kDa arsenite oxidase from *Alcaligenes faecalis* in two crystal forms at 1.64 A and 2.03 A. *Structure.* 9, 125–132.

Finnegan, P. M., Chen, W., 2012. Arsenic toxicity: the effects on plant metabolism. *Front. Physiol.* 3, 182-200.

Fisher, J. C., Hollibaugh, J. T., 2008. Selenate-dependent anaerobic arsenite oxidation by a bacterium from Mono Lake, California. *Appl. Environ. Microbiol.* 74, 2588-2594.

Goñi-Urriza, M., Capdepuy, M., Arpin, C., Raymond, N., Caumette, P., Quentin, C., 2000. Impact of an urban effluent on antibiotic resistance of riverine *Enterobacteriaceae* and *Aeromonas* spp. *Appl. Environ. Microbiol.* 66, 125–132.

Green, H. H., 1918. Isolation and description of a bacterium causing oxidation of arsenite to arsenate in cattle-dipping baths. *Rep. Dir. Vet. Res. S. Afr.* 6, 593-599.

Hamamura, N., Macur, R. E., Liu, Y., Inskeep, W. P., Reysenbach, A. L., 2010. Distribution of Aerobic Arsenite Oxidase Genes within the Aquificales. *Interdisciplinary Studies on Environmental Chemistry -Biological Responses to Contaminants. TERRAPUB, Tokyo, Japan*, 47-55.

Heinrich-Salmeron, A., Cordi, A., Brochier-Armanet, C., Halter, D., Pagnout, C., Abbaszadeh-Fard, E., Montaut, D., Seby, F., Bertin, P. N., Bauda, P., Arsène –Ploetze, F., 2011. Unsuspected diversity of arsenite-oxidizing bacteria as revealed by widespread distribution of the *aoxB* Gene in Prokaryotes. *Appl. Environ. Micrbiol.* 77, 4685-4692.

Hendrick, C. A., Haskins, W. P., Vidaver, A. K., 1984. Conjugative plasmid in *Corynebacterium flaccumfaciens* subsp. *Oortii* that confers resistance to arsenite, arsenate, and antimony (III). *Appl. Environ. Microbiol.* 48, 56-60.

Hoeft, S. E., Kulp, T. R., Han, S., Lanoil, B., Oremland, R. S., 2010. Coupled arsenotrophy in a hot spring photosynthetic biofilm at Mono Lake, California. *Appl. Environ. Microbiol.* 76, 4633-4639.

Ike, M., Miyazaki, T., Yamamoto, N., Sei K., Soda, S., 2008. Removal of arsenic from groundwater by arsenite-oxidizing bacteria. *Water Science & Technology—WST.* 58, 1095–1100.

Inskeep, W., Macur, R., Hamamura, N., Warelow, T., Ward, S., Santini, J., 2007. Detection, diversity and expression of aerobic bacterial arsenite oxidase genes. *Environ. Microbiol.* 9, 934–943.

Kashyap, D. R., Botero, L. M., Franck, W. L., Hassett, D. J., McDermott, T. R., 2006. Complex regulation of arsenite oxidation in *Agrobacterium tumefaciens*. *J. Bacteriol.*188, 1081–1088.

Kruse, H., Sorum, H., 1994. Transfer of multiple drug resistance plasmids between bacteria of diverse origins in natural microenvironments. *Appl. Environ. Microbiol.* 60, 4015–4021.

Kulp, T. R., Hoeft, S. E., Oremland, R. S., 2004. Redox transformations of arsenic oxyanions in periphyton comunities. *Appl. Environ. Microbiol.* 70, 6428-6434.

Lebrun, E., Brugna, M., Baymann, F., Muller, D., Lie`vremont, D., Lett, M. C., Nitschke, W., 2003. arsenite oxidase, an ancient bioenergetic enzyme. *Mol. Biol. Evol.* 20, 686–693.

Lett, M. C., Muller, D., Lièvremont, D., Silver, S., Santini, J., 2012. Unified nomenclature for genes involved in prokaryotic aerobic arsenite oxidation. *J. Bacteriol.* 194, 207–208.

Lièvremont, D., Bertin, P. N., Lett, M. C., 2009. Arsenic in contaminated waters: biogeochemical cycle, microbial metabolism and biotreatment processes. *Biochimie.* 91, 1229–1237.

Lièvremont, D., N'negue, M. A., Behra, P., Lett, M. C., 2003. Biological oxidation of arsenite: batch reactor experiments in presence of kutnahorite and chabazite. *Chemosphere*. 51, 419-28.

Ma, L. Q., Komar, K. M., Tu, C., Zhang, W. H., Cai, Y., Kennelley, E. D., 2001. A fern that hyperaccumulates arsenic. *Nature*. 409, 579.

Mokashi, S. A., Paknikar, K. M., 2002. Arsenic (III) oxidizing *Microbacterium lacticum* and its use in the treatment of arsenic contaminated groundwater. *Lett. Appl. Microbiol*. 34, 258-62.

Mondragón, V. A., Llamas, D. F., González, G. E., Márquez-González, A. R., Padilla-Noriega R., Durán-Avelar, Mde. J., Franco B., 2011. Identification of *Enterococcus faecalis* bacteria resistant to heavy metals and antibiotics in surface waters of the Mololoa River in Tepic, Nayarit, Mexico. *Environ. Monit. Assess*. 183, 329-40.

Morozzi, G., Cenci, G., Scardazza, F., Pitzurra, M., 1986. Cadmium uptake by growing cells of gram-positive and gram-negative bacteria. *Microbios*. 48, 27–35.

Mukhopadhyay, R., Rosen, B. P., Phung, L., Silver, S., 2002. Microbial arsenic: from geocycles to genes and enzymes. *FEMS Microbiol. Rev*. 26, 311–325.

Muller, D., Lièvremont, D., Simeonova, D. D., Hubert, J. C., Lett, M. C., 2003. Arsenite oxidase *aox* genes from a metal-resistant β-Proteobacterium *J. Bacteriol*. 185, 135-141.

Nordstrom, D. K., 2002. Worldwide occurrences of arsenic in ground water. *Science*. 296, 2143-2145.

Oremland, R. S., Hoeft, S. E., Santini, J. M., Bano, N., Hollibaugh, R. A., Hollibaugh, J. T., 2002. Anaerobic oxidation of arsenite in Mono Lake water and by facultative, arsenite-oxidizing chemoautotroph, strain MLHE-1. *Appl. Environ. Microbiol*. 68, 4795–4802.

Oremland, R. S., Stolz, J. F., 2003. The ecology of arsenic. *Science*. 300, 939–944.

Oremland, R. S., Stolz, J. F., 2005. Arsenic, microbes and contaminated aquifers. *Trends Microbiol*. 13, 45–49.

Osborne, T. H., Jamieson, H. E., Hudson, K. A., Nordstrom, D. K., Walker, S. R., Ward, S. ., Santini, J. M., 2010. Microbial oxidation of arsenite in a subarctic environment: diversity of arsenite oxidase genes and identification of a psychrotolerant arsenite oxidizer. *BMC Microbiology*. 10, 205.

Phillips, S. E., Taylor, M. L., 1976. Oxidation of arsenite to arsenate by *Alcaligenes faecalis*. *Appl. Environ. Microbiol*. 32, 392-399.

Pickering, I. J., Gumaelius, L., Harris, H. H., Prince, R. C., Hirsch, G., Banks, J. A., Salt, D. E., George, G. N., 2006. Localizing the biochemical transformations of arsenate in a hyperaccumulating fern. *Environ. Sci. Technol*. 40, 5010–5014.

Quéméneur, M., Cébron, A., Billard, P., Battaglia-Brunet, F., Garrido, F., Leyval, C., Joulian, C., 2010. Population structure and abundance of arsenite-oxidising bacteria along an arsenic pollution gradient in waters of the Upper Isle River Basin, France. *Appl. Environ. Microbiol*. 76, 4566-4570.

Quéméneur, M., Heinrich-Salmeron, A., Muller, D., Lièvremont, D., Jauzein, M., Bertin, P. N., Garrido, F., Joulian, C., 2008. Diversity surveys and evolutionary relationships of *aoxB* genes in aerobic arsenite-oxidising bacteria. *Appl. Environ. Microbiol*. 74, 4567-4573.

Ravenscroft, P., Brammer, H., Richards, K., 2009. Arsenic pollution: a global synthesis. *UK: Wiley-Blackwell*.

Rhine, E., Garcia-Dominguez, E., Phelps, C. D., Young, L. Y., 2005. Environmental microbes can speciate and cycle arsenic. *Environ. Sci. Technol.* 39, 9569-9573.

Rhine, E., Phelps, C., Young, L., 2006. Anaerobic arsenite oxidation by novel denitrifying isolates. *Environ. Microbiol.* 8, 899–908.

Santini, J. M., Sly, L. I., Shnagl, R. D., Macy, J. M., 2000. A new chemolithoautotrophic arsenite-oxidizing bacterium isolated from a gold mine: phylogenetic, physiological, and preliminary biochemical studies. *Appl. Environ. Microbiol.* 66, 92-97.

Santini, J. M., Sly, L. I., Wen, A., Comrie, D., De Wulf-Durand, P., Macy, J. M., 2002. New arsenite-oxidizing bacteria isolated from Australian gold mining environments: phylogenetic relationships. *Geomicrobiol.* 19, 67–76.

Santini, J. M., Vanden-Hoven, R. N., 2004. Molybdenum-containing arsenite oxidase of the chemolithoautotrophic arsenite oxidizer NT-26. *J. Bacteriol.* 186, 1614 –1619.

Sardiwal, S., Santini, J. M., Osborne, T. H., Djordjevic, S., 2010. Characterization of a two-component signal transduction system that controls arsenite oxidation in the chemolithoautotroph NT-26. *FEMS Microbiol. Lett.* 313, 20-28.

Silver, S., Phung L. T., 2005. A bacterial view of the periodic table: genes and proteins for toxic inorganic ions. *J. Ind. Microbiol. Biotechnol.* 32, 587–605.

Silver, S., Phung L. T., 2005. Genes and enzymes involved in bacterial oxidation and reduction of inorganic arsenic. *Appl. Environ. Microbiol.* 71, 599–608.

Singh, R., Paul, D., Jain, R. K., 2006. Biofilms: implications in bioremediation. *Trends Microbiol.* 14, 389-97.

Smedley, P. L., Kinniburgh, D. G., 2002. A review of the source, behaviour and distribution of arsenic in natural waters. *Appl. Geochem.* 17, 517-568.

Stolz, J. F., Basu, P., Santini, J. M., Oremland, R. S., 2006. Arsenic and selenium in microbial metabolism. *Annu. Rev. Microbiol.* 60, 107-130.

Sun, W., Sierra-Alvarez, R., Hsu, I., Rowlette, P; Field, J. A., 2010. Anoxic oxidation of arsenite linked to chemolithotrophic denitrification in continuous bioreactors. *Biotechnol. Bioeng.* 105, 909-17.

Sung, D. Y., Kim, T. H., Komives, E. A., Mendoza, D. G., Schroeder, J. I., 2009. ARS5 is a component of the 26S proteasome complex, and negatively regulates thiol biosynthesis and arsenic tolerance in *Arabidopsis*. *Plant J.* 59, 802-13.

Tamaki, S., Frankenberger, W., 1992. Environmental biochemistry of arsenic. *Rev. Environ. Cont. Toxicol.* 124, 79-110.

Turner, A. W., 1949. Bacterial oxidation of arsenite. *Nature (London).* 164, 76-77.

Visoottiviseth, P., Francesconi, K., Sridokchan, W., 2002. The potential of Thai indigenous plant species for the phytoremediation of arsenic contaminated land. *Environ. Pollut.* 118, 453–461.

Wenjie, S., Sierra-Alvarez, R., Milner, L., Field, J., 2010. Anaerobic oxidation of arsenite linked to chlorate reduction. *Appl. Environ. Microbiol.* 76, 6804-6811.

Yoshida, K., Inoue, K., Takahashi, Y., Ueda, S., Isoda, K., Yagi, K.,Maeda, I., 2008. Novel carotenoid-based biosensor for simple visual detection of arsenite: characterization and preliminary evaluation for environmental application. *Appl. Environ. Microbiol.* 74, 6730-6738.

Zargar, K., Hoeft, S., Oremland, R., Saltikov, C. W., 2010. Identification of a novel arsenite oxidase gene, *arxA*, in the haloalkaliphilic, arsenite-oxidizing bacterium *Alkalilimnicola ehrlichii* strain MLHE-1. *J. Bacteriol.* 192, 3755-3762.

Zhao, F. J., Ma, J. F., Meharg, A. A., McGrath, S. P., 2009. Arsenic uptake and metabolism in plants. *New Phytol*. 181, 777–794.

Zouboulis, A. I., Katsoyiannis, I. A., 2005. Recent advances in the bioremediation of arsenic-contaminated groundwater. *Environ. Int.* 31, 213-9.

In: Bioremediation: Processes, Challenges and Future Prospects ISBN: 978-1-62948-513-3
Editors: Jesús Bernardino Velázquez-Fernández et al.

Chapter 3

BIODEGRADATION OF METHYL *TERT*-BUTYL ETHER: ANALYSIS OF NEWLY IDENTIFIED SOIL MICROORGANISMS AND PROTEIN EXPRESSION

Magda Constantí*
Departament d'Enginyeria Química,
Universitat Rovira i Virgili, Tarragona, Spain

ABSTRACT

Methyl *tert*-butyl ether (MTBE) is a fuel ether that has been widely used as a gasoline oxygenate, and which has become a soil and water contaminant. Several methods are being developed to remove it. Of these, bioremediation – the addition of microbial cultures that degrade the compound – seems to be particularly promising. This study describes how twelve microbial strains were isolated from gasoline-contaminated soils and selected for their capacity to grow in the presence of MTBE. The strains were screened for their ability to consume MTBE aerobically in a simple mineral solution and identified by 16S/ITS rDNA gene sequencing. MTBE degradation was detected by solid phase microextraction and gas chromatography. High levels of MTBE biodegradation were obtained using resting cells of the bacteria *Achromobacter xylosoxidans* MCM1/1 (78 %), *Enterobacter cloacae* MCM2/1 (50 %), and *Ochrobactrum anthropi* MCM5/1 (52 %) and the fungus *Exophiala dermatitidis* MCM3/4 (14 %). Our phylogenetic analysis clearly showed that bacterial MTBE biodegraders belong to the clade of Proteobacteria. To optimize the biodegradation process, *A. xylosoxidans* MCM1/1 cells were immobilized in Na-alginate beads and removal rates were similar. For further insight, we also analyzed the effects of MTBE on the biology of *A. xylosoxidans* MCM1/1. Its proteomic profile after incubation with MTBE was compared with that of unchallenged bacteria. The 2D proteomic analysis showed that the following four proteins were induced by MTBE: amino acid-binding periplasmic protein, 50S ribosomal protein L10, endoribonuclease L, and ATP synthase. This shed light on the survival/adaptation mechanisms of bacteria that could lead to biodegradation. Also, this knowledge allows us to improve bioremediation process.

---

* Email: magda.constanti@urv.cat.

Keywords: MTBE biodegradation, *Achromobacter xylosoxidans*, *Enterobacter cloacae*, *Ochrobactrum anthropi*, *Exophiala dermatitidis*, Na-alginate immobilization, proteome

INTRODUCTION

Fuel oxygenates have been added to gasoline since the 1970s mainly as octane enhancers to increase combustion efficiency and to reduce such toxic air emissions as lead compounds or carbon monoxide. Methyl *tert*-butyl ether (MTBE) has been the most commonly used because it is high in octane, low in cost and easy to blend with gasoline (Barceló, 2007). However, groundwaters have become contaminated with MTBE after reformulated gasoline has been spilt during handling or when it has leaked from storage and fuel tanks. The problems with MTBE are that: (1) it is highly soluble in water and poorly absorbed by soil particles; therefore it can spread rapidly through the soil ahead of the gasoline plume, and (2) it can be detected by taste and smell at very low concentrations (15 ppb) (Lyew et al., 2007). Conventional treatment methods such as air stripping and carbon adsorption do not remove MTBE as effectively as other volatile organic compound (VOC) contaminants (Baek, Yang & Yang, 2003). In recent years, bioremediation has emerged as an alternative technique for removing MTBE because it is cost-effective and environmentally friendly. It can completely mineralize MTBE to harmless products such as carbon dioxide, water and biomass. Most studies on MTBE biodegradation have focused on the degradability of MTBE, despite the fact that early studies suggested that it is resistant to microbial attack (Sequillace et al., 1997).

Various research groups have shown that several microorganisms can degrade MTBE. Initial studies showed that MTBE could be degraded under both aerobic and anaerobic conditions (Salanitro et al., 1994; Yeh & Novak, 1994). Recently, one bacterium, *Mycobacterium austroafricanum* IFP 2012, previously isolated from activated sludge at the Institut Français du Pétrole, was shown to be able to use MTBE in a fixed-bed reactor (Maciel et al., 2008). *Aquincola tertiaricarbonis* (Muller, Rohwerder & Harms, 2008) and several bacteria isolated by Purswani et al. (2008) used MTBE as their sole energy and carbon source. Chen, Chen & Zhong (2009) enhanced the ability of *Methylibium petroleiphilum* PM1 to degrade MTBE by adding yeast extract. Finally, Zhong et al. (2011) showed that a closed algal-bacterial symbiotic system degraded MTBE.

Our research group isolated several cultures capable of degrading MTBE from a gasoline-contaminated soil; degradation activity was as high as 78 % in just five days when resting cells were used (Barberà et al., 2011). These cultures were characterized using the 16S/ITS rDNA protocol and their phylogenetic analysis showed that three of the cultures belong to the Proteobacteria phylum. Four microorganisms showed a good capacity to biodegrade MTBE: *Achromobacter xylosoxidans* MCM1/1, *Enterobacter cloacae* MCM2/1, *Ochrobactrum anthropi* MCM5/1 and the fungi *Exophiala dermatitidis* MCM3/4. In order to optimize their biodegradation capability, *A. xylosodixans* MCM1/1 was chosen for further studies. Cells of this bacterium were entrapped in sodium alginate beads to increase biodegradation activity and facilitate its reuse. Immobilization can eliminate the costly processes of cell recovery and creates cells that are more stable than free cells for various applications (Quan, Fan & Ohta, 2003). On the other hand, a proteomic analysis was carried out on *A. xylosoxidans* MCM1/1 while biodegrading MTBE (Eixarch & Constantí, 2010).

The main enzymes involved in the process may be isolated for use as single biodegradation units. Thus, analysing the mechanism of MTBE use may help improve strains and increase their efficiency as biocatalysts.

MATERIALS AND METHODS

Materials

MTBE-degrading microorganisms were isolated from five samples of soil contaminated with gasoline taken from an industrial area in Valencia, Spain.

The following minimal medium (MM) was used for bacteria cultivation: KH_2PO_4 1 g/L, $MgCl_2{\cdot}6H_2O$ 1 g/L, NH_4Cl 2 g/L, $CaCl_2$ 0.001 g/L, $FeSO_4{\cdot}7H_2O$ 0.001 g/L (Constantí, Giralt & Bordons, 1996). It was supplemented with yeast extract (0.5 – 0.1 g/L), MTBE (25-100 mg/L) or agar (20 g/L) when needed.

MTBE (99.8 % purity) was purchased from Sigma-Aldrich. All other chemicals were of the highest purity available.

Selection and Isolation of Microorganisms

Each soil sample (5 g) was incubated in 250 mL of MM at 30 ºC and shaken at 200 rpm. The medium was supplemented with 0.5 g/L yeast extract (MMYE) and 100 mg/L MTBE. After several incubation transfers in the MMYE, the samples were inoculated into MM containing MTBE (100 mg/L) and yeast extract in gradually decreasing concentrations (0.5 g/L, 0.25 g/L, 0.1 g/L). Several colonies were selected by plating the samples onto MM agar with 100 mg/L of MTBE and MTBE saturated air.

Identification of the Microorganisms

Biochemical identification. Firstly, morphological and biochemical tests were used to identify the isolated microorganisms. The morphology of the bacteria was determined by Gram staining and optical microscopy (Nikon, Eclipse E600-FN Microscope provided by IZASA-REGO). Initial biochemical tests for oxidase, catalase and fermentation of carbohydrates (growth on Triple Sugar Iron Agar from Difco) were conducted and then miniaturized biochemical tests (from Biomerieux) were carried out: API20®E for carbohydrate fermenting bacteria and API20®NE for carbohydrate non-fermenting bacteria. APIWEB software was used for the identification, and was considered acceptable when the probability was 80 % or greater. On the other hand, direct microscopic observations of micelles were carried out on isolated fungi with lactophenol and lactophenol blue to view the fungal wall better.

Molecular identification of the microorganisms. Bacterial genomic DNA was extracted using the GenElute Bacterial Genomic DNA Kit (Sigma). Partial 16S rDNA sequences (1000 bp) of bacterial strains were amplified by PCR using the universal primers 616V (5´- AGA

GTT TGA TYM TGG CTC AG -3´) and 699R (5´- RGG GTT GCG CTC GTT –3´). PCR reaction mixtures contained 3 mM $MgCl_2$, 200 μM (each) dNTP, 0.4 μM of each primer, 100 ng of genomic DNA, 1 unit of GoTaq (Promega) and 1x GoTaq Flexibuffer (Promega), in a total volume of 50 μL. The optimized PCR conditions were as follows: initial denaturation at 95 °C for 2 minutes; 25 cycles of denaturation at 95 °C for 30 seconds, annealing at 55 °C for 25 seconds; extension at 72 °C for 1 minute; and a final extension at 72 °C for 5 minutes. The PCR products were sent for sequencing to Macrogen. Isolates 1/1, 2/1, 2/2, 4/2 and 5/1 were also corroborated by their 16S rRNA gene sequencing by Colección Española de Cultivos Tipo (CECT, http://www.cect.org, Spain).

The fungal strain was identified by internal transcribed spacer (ITS) of the nuclear rDNA gene sequence analysis. Total fungal DNA (tDNA) was extracted using the Fast DNA kit protocol (Bio101, Vista, USA). PCR amplification of the ITS nuclear rDNA gene was performed using the primers ITS4 and ITS5 (White et al. 1990). The PCR reaction mixture (25 μL) contained 20-60 ng of template DNA, 50 mM KCl, 2.5 mM $MgCl_2$, 100 μM (each) dNTP, 1 μM of each primer, 1.5 units of AmpliTaq DNA polymerase (Roche) and 10 mM Tris-HCl (pH 8.3). The PCR conditions were as follows: initial denaturation at 94 °C for 5 minutes followed by 35 cycles of denaturation at 95 °C for 30 seconds, annealing at 50 °C for 1 minute, extension at 72 °C for 1 minute and a final extension step at 72 °C for 7 minutes. The PCR products were purified by the GFXTM PCR DNA purification kit (Pharmacia Biotech, Spain). The fungal sequence was resolved using an Abi Prism 3730 automated sequencer (Applied Biosystems, USA).

Phylogenetic Analysis of DNA Sequences

The rDNA sequences in the phylogenetic tree were first aligned by BioEdit and then manually inspected and edited using the software Geneious (http://www.geneious.com/) (Barberà et al., 2011). Only those positions that were unambiguously aligned were manually included in the final analysis, resulting in a total of 1035 nucleotide positions. Maximum likelihood phylogenetic trees were estimated by Raxml using a GTR model of evolution and with a gamma distribution (8 categories) (GTR+Γ) (Stamatakis, 2006). Statistical support was obtained from 1000 bootstrap replicates using the phyML program (Guindon & Gascuel, 2003) following a GTR+Γ model of evolution with four rate categories.

Biodegradation Experiments with Resting Cells

As described in Barberà et al. (2011) each strain was grown in MMYE supplemented with MTBE (80 mg/L) until the mid-exponential phase growth. Then, cells were collected by centrifugation, washed twice with 0.9 % NaCl, and resuspended in 1 mL of 0.9 % NaCl. A total of 5 % (v/v) resting cells was inoculated in 10 mL 0.9 % NaCL in glass vials (20 mL) plus 80 mg/L or 100 mg/L of substrate to proceed with the investigation of MTBE biodegradation. The vials were sealed with aluminum crimp seals with PTFE/silicone septa, incubated at 30 °C and shaken at 200 rpm for 5 days. The control samples were prepared using the same procedure but without adding cell suspension. Solid phase microextraction

(SPME) was applied for 30 min at 55 °C using polydimethyl siloxane/divinylbenzene fibre (PDMS/DVB, Sigma-Aldrich) to detect the levels of MTBE biodegradation. Then, MTBE concentrations were determined by gas chromatography using a HP6890 capillary gas chromatograph (Hewlett Packard ®) with FID detector and injector temperatures of 250 °C, an initial column temperature of 50 °C and helium as carrier gas. The column temperature increased to 85 °C at a rate of 7 °C/min.

Statistical analysis. The Student's t test was applied for two independent variables to analyze the removal of MTBE during biodegradation.

Biodegradation Experiments with Entrapped Cells

Immobilization of cells. A. xylosoxidans MCM1/1 cells were grown in MM supplemented with 0.5 g/L yeast extract and 100 mg/L of MTBE (hereafter named as MMYE-MTBE medium) until the mid-exponential phase. The cells were spun for 15 min, 4075 *g*, 4 °C, then harvested by centrifugation, washed and suspended, and finally mixed with 2 % (w/v) sodium alginate. The slurry containing the cells (1 mL, containing 0.02 g microorganism cells) was dropped by micropipette into a 2 % $CaCl_2$ solution under magnetic stirring. The beads were stirred in this solution for 2 h. They were then collected by filtration, washed three times with distilled water and stored in a 2 % (w/v) $CaCl_2$ solution at 4 °C.

Biodegradation experiments. In immobilized microorganism studies, the salts containing K^+, Mg^{2+} and PO_4^{3-} ions dissolved calcium-alginate beads. Therefore, the composition of the biodegradation medium was optimized to improve the stability of the beads (Dursun & Aksu, 2002). The reaction vials contained a biodegradation medium (KH_2PO_4, 0.035 g/L; $(NH_4)_2SO_4$, 0.3 g/L; $MgSO_4{\cdot}7H_2O$, 0.05 g/L), 5% (v/v) of immobilized cells, MTBE at the desired concentration (10 mL of the final volume), as indicated in the results section, and a magnetic stirrer. The vials were sealed with aluminum crimp seals with PTFE/silicone septa, incubated at 30 °C, and shaken (200 rpm) for 0 h, 1 day, 3 days and 5 days. Control experiments in a microorganism-free medium (including MTBE and other constituents) were also carried out in order to evaluate how much MTBE could be removed through volatilization. Polydimethyl siloxane/divinylbenzene (PDMS/DVB) fiber was used to carry out solid phase microextraction (30 min at 55 °C) and analyze the level of MTBE degradation. MTBE concentrations were determined as indicated in resting cell experiments.

Oxygen Uptake Conditions

A. xylosoxidans MCM1/1 was grown in MMYE-MTBE until the mid exponential phase. The cells were spun twice for 15 min, 4075 *g*, at 4 °C, washed with sterilized 0.9 % NaCl, and then re-suspended in 0.9 % NaCl. Oxygen uptake was measured using a Clark oxygen electrode (YSI 5300A). Experimental data were obtained for bacteria alone, as a control, and bacteria in the presence of MTBE.

Cytochrome P450 Activity

Individual colonies of *A. xylosoxidans* MCM1/1 were taken from a Petri plate, inoculated into 10 mL MMYE and incubated at 30 °C, by shaking at 250 rpm. After 24 h, 1 mL of inoculum was added to 50 mL MMYE containing 100 mg/L of MTBE and incubated at 30 °C by shaking at 250 rpm. Bacteria were grown until they reached exponential phase and then pellets were obtained by centrifugation for 15 min at 2795 *g*. The pellets were resuspended in 0.5 mL phosphate buffer 0.1 M, pH 8.0 and 0.625 g of silica beads was added. The sample was shaken vigorously with the help of an MBB-8 Mini Bead Beater (Biospec) for 30 s and then kept on ice for other 30 s. This process was repeated at least five times. After that, the sample was centrifuged for 10 min at 18894 *g* at 4 °C. The supernatant was recovered and frozen at -20 °C until needed for cytochrome P450 activity studies.

Different volumes of cleared lysate (*A. xylosoxidans* MCM1/1) and phosphate buffer (final volume of lysate plus buffer being 70 μL) were transferred to a 96-well microtiter plate and 10 μL of styrene (100 mM in methanol) was added using the method by Alcalde, Farinas & Arnold (2004). The reaction was initiated by adding 10 μL of NADPH (20 mM in phosphate buffer 0.1 M, pH 8.0). The final concentrations in the reaction mixture were 10 mM Styrene, 2 mM NADPH, and 10 % methanol. The mixture was shaken and incubated at room temperature for 5 min. Next, 25 μL of NBP (γ-(4-nitrobenzyl)pyridine) solution (0.4 g in 10 mL ethanol) was added, and the plate was shaken and sealed. The plate was heated in an oven at 80 °C for 30 min and then chilled on ice for 10 minutes. The film was removed, and 150 μL of dimethylformamyde and 25 μL of 1 M K_2CO_3 were added in rapid succession. The plate was shaken again, and the absorbance at 600 nm was immediately measured with a plate reader. A styrene oxide calibration curve (from 0 to 500 μM) was performed using the same conditions as in the reaction protocol, with styrene oxide added to the solution. The final reaction contained 2 mM NADPH, 10 % methanol, 10 % cleared lysate, and a known amount of styrene oxide. After shaking the plate, 25 μL of NBP solution was added, and the remaining steps of the assay were performed as described above. Each point was done in triplicate.

Proteomic Analysis

The protocol was performed in Eixarch & Constanti (2010) as follows.

Cell extract preparation. An isolated colony of *A. xylosoxidans* MCM1/1 was inoculated in MM with 0.5 g/L yeast extract. Cells were harvested after 8 hours of incubation, at the exponential phase of growth (A_{440} was 0.3). They were centrifuged at 2795 *g*, at 4 °C for 15 min and washed once with MM without yeast extract. Cells were then resuspended in 10 mL of MM (Control) or 10 mL MM with 100 mg/L MTBE and incubated for 36 h at 30 °C, 250 rpm. After incubation, cells were harvested by centrifugation at 18894 *g*, 4 °C for 5 min, the pellet was resuspended in 160 μL of cold ddH_2O with 1 mM PMSF and 0.2 g of silica beads was added. Cells were disrupted with a MiniBeadbeater-8 cell disrupter (Biospec). Then, 20 μL of buffer I (0.5 M Tris-HCl, pH 8.8; 50 % β-mercaptoethanol; 3 % SDS) was added, and the sample was incubated for 5 min at 100 °C and placed on ice. Finally, 20 μL of buffer II (0.5 M Tris-HCl, pH 8.8; 25 mM $MgCl_2$; 1 mg/mL DNase I; 0.25 mg/mL RNase A) was added and the sample was centrifuged at 18894 *g*, at 4 ºC for 10 min. The supernatant was recovered and the total protein in the sample was quantified in accordance with Bradford method (1976).

Two dimensional gel electrophoresis analysis. First dimension IEF was performed on 18 cm strips (Immobiline drystrip pH 3-10 NL, GE Healthcare) in an Ettan IPGphor IEF unit (GE Healthcare). A total of 45 µg of protein was mixed with rehydration buffer (7 M urea, 2 M thiourea, 1 % NP-40, 1 % Bio-Lyte 3/10 ampholytes and traces of bromophenol blue) to a final volume of 500 µL, and incubated at 37 ºC for 30 min, followed by centrifugation at 18894 *g* for 10 min. The strips were rehydrated with the sample at room temperature for 16-20 h. IEF was carried out according to the following program: gradient to 500 V for 5 h; constant 500 V for 5 h; a gradient to 3500 V for 9.5 h; constant 3500 V for 5 h. After IEF, the strips were subjected to a two-step equilibration. The strips were incubated in equilibration buffer I (2 % SDS, 50 mM Tris-HCl, pH 8.8; 6 M urea; 30 % glycerol; 1 % (w/v) DTT; traces of bromophenol blue) for 15 min at room temperature and then in buffer II (2 % SDS, 50 mM Tris-HCl, pH 8.8; 6 M urea; 30 % glycerol; 2.5 % (w/v) iodoacetamide; traces of bromophenol blue) for another 15 min. The second dimension was performed on a 1.0 mm-thick 12.5 % SDS-PAGE in an Ettan™ DALT*six* Electrophoresis Unit (GE Healthcare) at 12 ºC, applying 1 W/gel, 600 mA, 400 V for 15.5 h. Samples were run a minimum of three times.

Silver staining. Gels were fixed for a minimum time of 2 h with 50 % ethanol and 10 % acetic acid. After two washes with Milli-Q water (20 min each), gels were incubated with 5.2 mg/L DTT for 30 min. For staining, 2.05 g/L of silver nitrate was used, and gels were incubated in this solution for 30 min. After washing with 34.7 g/L Na_2CO_3, 0.05 % formaldehyde, gels were left to develop in a fresh solution of the same composition. Development was stopped with 3 % acetic acid.

Analysis of gel images. The stained gels were scanned, and the images were analyzed with the PDQuest Basic 2D-Analysis software (BioRad). The protein spot intensities were normalized against the total spot intensities of the gel. A spot was analyzed only if it was present in at least two of the three gels of each condition. The protein quantity for each protein spot was obtained and an average of the replicates was calculated. The relative change in protein abundance for MTBE *vs.* control was calculated by dividing the average spot quantity of each protein spot in the MTBE gels by the average quantity of the same spot in the control gels. A spot from the MTBE-treated gel group was considered to be significantly overexpressed or underexpressed when there was at least a 1.5-fold difference with the control group. A one-tailed Student's test was performed to determine if the relative change in the protein spots was statistically significant. A change was considered to be statistically significant if $P < 0.05$.

Tryptic in-gel digestion. Protein spots were excised manually and then digested automatically using a Proteineer DP protein digestion station (Bruker-Daltonics, Bremen, Germany). The digestion protocol used was the one developed by Schevchenko et al. (2006).

Protein identification by MS/MS. An aliquot of the above digestion solution was mixed with an aliquot of α-cyano-4-hydroxycinnamic acid in 33 % aqueous acetonitrile and 0.25 % trifluoroacetic acid. This mixture was deposited onto a 600 µm AnchorChip prestructured MALDI probe (Bruker-Daltonics) and allowed to dry at room temperature. MALDI-MS(/MS) data were obtained in an automated analysis loop using an Ultraflex time-of-flight mass spectrometer (Bruker-Daltonics) equipped with a LIFT-MS/MS device (Suckau et al., 2003).

Spectra were acquired in the positive-ion mode at a laser frequency of 50 Hz, and 100 to 1000 individual spectra were averaged. For fragment ion analysis in the tandem time-of-flight (TOF/TOF) mode, precursors were accelerated to 8 kV and selected in a timed ion gate. Fragment ions generated by laser-induced decomposition of the precursor were further accelerated by 19 kV in the LIFT cell and their masses were analyzed after they had passed the ion reflector. An automated analysis of mass data was performed using flexAnalysis software (Bruker-Daltonics). MALDI-TOF mass spectra were internally calibrated using two trypsin autolysis ions with m/z = 842.510 and m/z = 2211.105; for MALDI-MS/MS, calibrations were performed with fragment ion spectra obtained for the proton adducts of a peptide mixture covering the 800-3200 m/z region.

Database analysis. MALDI-MS and MS/MS data were combined through the BioTools program (Bruker-Daltonics) to search a nonredundant protein database (NCBInr; or SwissProt) using the Mascot software (Matrix Science) (Perkins et al., 1999). MALDI-MS/MS spectra and database search results were manually inspected in detail using the above programs.

RESULTS AND DISCUSSION

Selection and Identification of Isolated Microbial Strains

The microorganisms were selected for their growth in liquid MM plus MTBE and decreasing amounts of yeast extract. When microorganisms were grown in liquid MM supplemented with 100 mg/L MTBE but no yeast extract they showed extremely low growth (data not shown). It should be taken into account that the MTBE structure contains an ether bond that is difficult to cleave, so under these conditions MTBE alone may not be a sole energy and carbon source (Muñoz-Castellanos et al., 2006). Then solid MM with 20 g/L agar, supplemented with 100 mg/L MTBE was used to plate samples in an MTBE-saturated atmosphere. In these conditions, twelve different colonies grew after 15 days of incubation at 30 °C.

The colonies were microscopically analyzed and eleven of the isolates were shown to be bacteria and one a fungus (isolate 3/4). The genus and species of the bacterial isolates were identified by API tests and 16S rDNA sequencing, as can be seen in Table 1 (Barberà et al., 2011). On the other hand, whether API20®NE or API20®E tests were used depended on the carbohydrate-fermenting capacity of the isolates. These biochemical tests clearly identified only four isolates and all others were identified by 16S rRNA gene sequencing. The resulting 16S rRNA gene sequences were compared in a BLAST search with those in the database of the Ribosomal Database Project (RDP-II) website (Maidak, Cole & Parker, 1999). The results in Table 1 show the best hit obtained for each isolate and the scores. According to the molecular identification, most of the bacterial isolates belong to the phylum Proteobacteria with representatives of the three classes (*alpha, beta and gamma*). The isolates belonging to the same genus were considered to be different strains since comparison of the 16S rDNA sequences showed some variation (data not shown). When both methodologies were compared, we observed some discrepancy in the identification. Bosshard et al. (2006) also obtained ambiguous results when these two techniques were applied to environmental samples. One year later, Awong-Taylor et al., concluded that 16S rDNA sequencing is more accurate at identifying Gram-negative non-fermenters than API 20NE. Our results confirmed these previous studies and corroborated that 16S rRNA gene sequencing is the most accurate method for screening the microorganism composition of environmental soil samples.

The fungus (isolate 3/4) was identified as the black yeast *Exophiala dermatitidis*, a human fungal pathogen present in plant debris and soil (Kennes & Veiga 2004).

Biodegradation of MTBE by Resting Cells

Resting cells of all the isolates were used so that their ability to degrade MTBE in a simple mineral solution (only 0.9 % NaCl and MTBE) could be analyzed. Resting cells have been shown to consume pollutants at higher rates than growing cells (Chen et al., 2008). Under our experimental conditions, four of the twelve isolates were capable of biodegrading MTBE (Figure 1) (Barberà et al., 2011). The removal yield (or removal efficiency, i.e., % MTBE removal) was defined as the ratio of the concentration of the MTBE removed to the initial concentration of MTBE.

Table 1. Molecular and Biochemical identification of soil strains

16S/ ITS rDNA sequence identification				API identification		
Isolate	**Identified strain**[a] (Genus specie and strain)	**Sequence**[b]	**Score**[c]	**Identified strain**[d] (Genus and specie)	**Method**	**(%)**
1/1	*Achomobacter xylosoxidans subps denitrificans DSM 30026*	AJ278451.1	0.995	Unidentified	API20NE	
2/1	*Enterobacter cloacae* ATCC 13047T	AJ251469	0.980	Unidentified		
2/2	*Achromobacter sp.* LMG 6003	AY170847	0.986	Unidentified	—	
3/1	*Ochrobactrum sp.* S1	AJ002812	0.694	*Ochrobactrum anthropi*	API20NE	99.9%
3/2	*Ochrobactrum sp.* S1	AJ002812	1.000	*Pseudomonas putida*	API20NE	84.7%
3/3	*Achromobacter xylosoxidans strain B8L*	DQ466568.1	0.980	*Alcaligenes denitrificans*	API20NE	82.2%
3/4	*Exophiala dermatitidis* strain UWFP985[d]	AY213651.1	1.000			
3/5	*Pantoea agglomerans* A65	AF130934	1.000	Unidentified	API20E	
3/6	*Pantoea agglomerans* A65	AF130934	1.000	Unidentified	API20E	
4/1	*Bacillus thuringiensis* WS 2617	Z84585	1.000	*Ochrobactrum anthropi*	API20NE	99.9%
4/2	*Burkholderia graminis* C4D1M	U96939	0.993	Unidentified	API20NE	
5/1	*Ochrobactrum anthropi GH1568*	AJ276036	1.000	Unidentified	API20NE	

[a] As obtained from the Sequence Match feature of the RDP-II Web site and corroborated by Genebank Blast.

[b] Nearest hit identified.

[c] S_ab score, similarity score.

[d] Fungi sequence was compared using the Basic Local Alignment Search Tool (BLAST) network service in Genebank.

From Barberà et al., 2011.

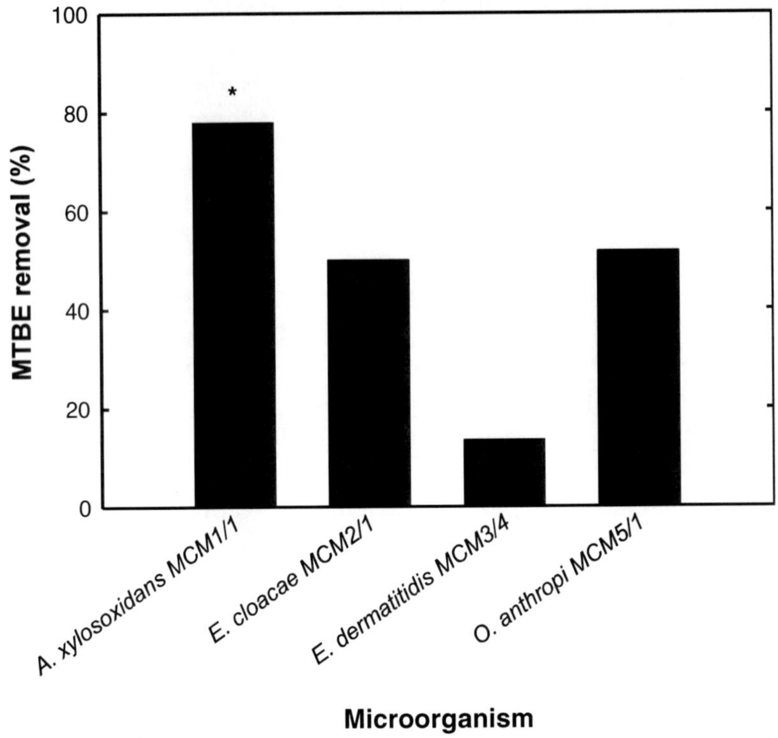

From Barberà et al., 2011.

Figure 1. MTBE removal by resting cells of *A. xylosoxidans* MCM1/1, *E. cloacae* MCM2/1, *E. dermatitidis* MCM3/4 and *O. anthropi* MCM5/1. Asterisk indicates that the removal is statistically significant at a significance level of 0.05. Values are the means from duplicate samples.

Three of the biodegradative isolates were bacteria, *A. xylosoxidans* MCM1/1 (isolate 1/1), *E. cloacae* MCM2/1 (isolate 2/1) and *O. anthropi* MCM5/1 (isolate 5/1) and one was a filamentous fungi *E. dermatitidis* MCM3/4 (isolate 3/4). The biodegradation occurred aerobically in a simple mineral medium, in the absence of any other compound except MTBE. Actually, MTBE removal was quite high after just five days of incubation for *A. xylosoxidans* MCM1/1, which showed a statistically significant difference (Student's t test) with a P value of 0.0027, at a significance level of 0.05. *E. cloacae* MCM2/1 (P value = 0.0793) and *O. anthropi* MCM5/1 (P value = 0.0845) showed a significant decrease at a significance level of 0.1. Figure 1 shows the biodegradation of MTBE by resting cells of these microorganisms. Previous studies with *A. xylosoxidans*, *E. cloacae* and *O. anthropi* (Andreoni et al., 2004; Iwashita et al., 2004; Lacayo-Romero et al., 2005; Müller et al., 1998; Olaniran et al., 2006) have already demonstrated that they can degrade a wide range of other compounds such as phenanathrene, dichloroethylene, toxaphene and chlorophenols. No biodegradation activity was observed in the other isolates. However, we cannot completely discount that these strains play a role in MTBE biodegradation.

E. dermatitidis MCM3/4 degraded MTBE with a P value of 0.176. Some species of *Exophiala* are considered promising fungal biocatalysts (Kennes & Veiga, 2004) and they have been extensively studied for their ability to biodegrade petrol compounds. In our experiments, *E. dermatitidis* MCM3/4 showed a lower biodegradation capability (14%) than bacteria although it demonstrated the best growth on solid MM supplemented with MTBE (data not shown). Although fungi have considerable potential as biocatalysts in pollution remediation (Kennes & Veiga, 2004), their use in bioremediation has not been as extensively studied as bacteria. Only two filamentous fungi species have been reported as MTBE biodegraders: *Graphium sp.* (Hardison et al. 1997) and *Fusarium solani* (Magaña-Reyes, Morales & Revah, 2005).

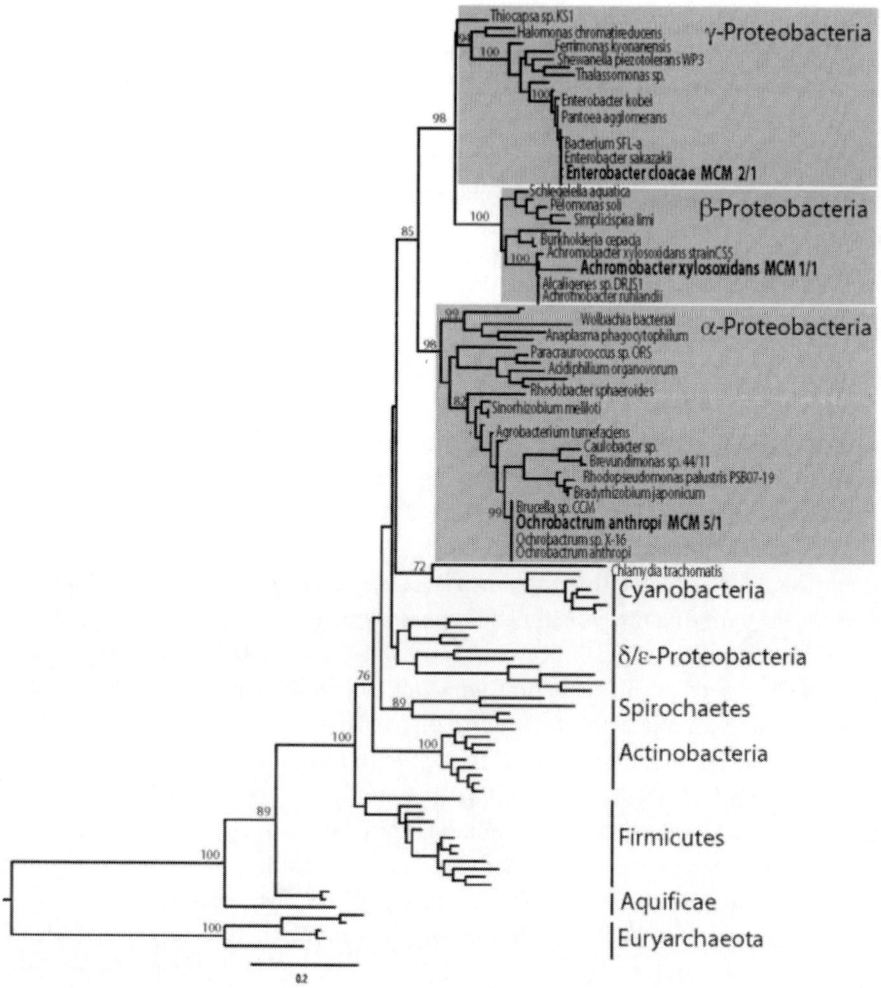

From Barberà et al., 2011.

Figure 2. Maximum likelihood phylogenetic tree of the 16S rRNA gene. Taxon sampling includes the three proteobacteria strains able to biodegrade MTBE (in bold) and several representatives from the major eubacterial and archaebacterial lineages. Statistical support was obtained by 1000 ML bootstrap replicates in phyML. Bootstrap values >70% are indicated at branch nodes. Major taxonomic assignment, rather than species names, is shown for some clades for the sake of clarity.

We also performed phylogenetic analysis to characterize the bacterial isolates that biodegraded MTBE. Closely related species of each of the biodegrading bacteria were included in the alignment (Figure 2) (Barberà et al., 2011). Our maximum likelihood (ML) phylogenetic analysis based on 16S rRNA gene sequences confirmed that the three strains belong to classes α, *β*, and *γ*-Proteobacteria (Figure 2). The overall topology of the tree is in accordance with previously published molecular phylogenies (Woese, 1987). The three proteobacterial clades have very good statistical support; boostrap values (BV) of 94 % for the gamma-proteobacteria (except *Thiocapsa*), 100% for the beta-proteobacteria, and 99 % for the alpha-proteobacteria clade. *E. cloacae* MCM2/1 clearly grouped within the *Enterobacter* + *Pantoea* clade (BV= 100 %). Similarly, the *A. xylosoxidans* MCM1/1 clearly grouped within the *Alcaligenes* + *Achromobacter* clade (BV= 100 %), whereas the *O. anthropi* MCM5/1 clearly grouped within the *Ochrobactrum* + *Brucella* clade (BV = 99 %).

Although MTBE bacterial biodegraders are a phylogenetically diverse group (François et al., 2002; Hanson, Ackerman & Scow, 1999; Lee & Cho 2009; Liu et al., 2009; Mo et al., 1997; Smith & Hyman 2004), and strains with the capacity to grow on and degrade MTBE belong to different bacterial orders, the pure cultures that actively degraded MTBE also belong like ours, to the Proteobacteria group (Hanson et al., 1999; Müller, Rohwerder & Harms, 2008; Smith & Hyman, 2004; Streger et al., 2002; Zaitsev, Volita & Häggblom, 2007).

Biodegradation of MTBE by Immobilized *A. Xylosoxidans* MCM1/1

For this experiment and others we chose *A. xylosoxidans* MCM1/1, the strain that showed the highest biodegradation capability with resting cells. The ability of Na-alginate immobilized *A. xylosoxidans* MCM1/1 to biodegrade MTBE in liquid media was tested with different initial MTBE concentrations. The medium did not contain any other chemical as a carbon source. For this part of the study, initial MTBE concentration was changed from 0 to 100 mg/L for each of the experiments, which were repeated twice.

The course of biodegradation capability was followed for 25, 50 75, 100 mg/L initial MTBE concentrations for 5 days of incubation with immobilized *A. xylosoxidans* MCM1/1 cells. Figure 3 shows an example with 100 mg/L of MTBE. At the end of five days, *A. xylosoxidans* MCM1/1 had degraded 68.11 % of the initial MTBE, although we also observed that a rapid biodegradation had occurred, since 44.10 % of the MTBE disappeared in only 1 day of incubation. Our analysis, using gas chromatography and mass spectrometry, did not detect metabolic intermediates of the MTBE pathway. Further studies are currently ongoing in our laboratory to identify the biodegradation products. It was observed that at the end of five days free *A. xylosoxidans* MCM1/1 cells removed a slightly higher percentage of MTBE than did Na-alginate immobilized *A. xylosoxidans* MCM1/1 (68.11 %). Pimentel-González et al. (2008) also observed that a microbial consortium entrapped in a water-in-oil in-water emulsion had a lower MTBE biodegradation rate than free cells. However, in our case, immobilized bacteria removed a higher percentage of MTBE than did free bacteria at three days (51.43 %). This means that immobilized bacteria needed less time to consume this compound, probably because the Ca alginate beads protected the bacteria against the chemical. A previous study (Barberà et al., 2011) showed that MTBE decreased the growth of *A. xylosoxidans* MCM1/1 by 5.5 %. To date, only a few studies have immobilized other pure bacteria for MTBE biodegradation. Lyew et al. (2007), for example, chose perlite to immobilize *Mycobacterium africanum* IFP2012, whereas Chen et al. (2011) and Pannier et al. (2010) selected Ca-alginate to immobilize *Methylibium petroleiphilum* PM1 and *Aquincola tertiaricarbonis* L108, respectively.

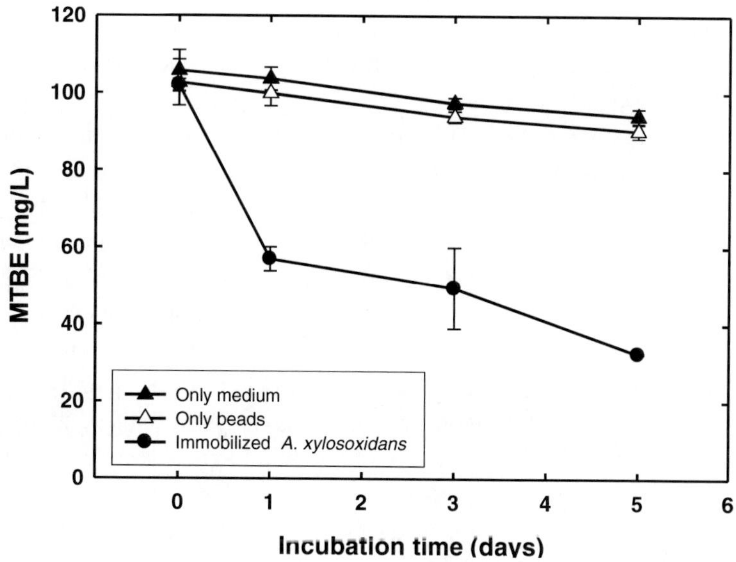

Figure 3. Evolution of MTBE in a solution containing 100 mg/L of initial MTBE under different conditions: only minimal medium, in the presence of alginate beads, in the presence of immobilized *A. xylosoxidans* MCM1/1. Error bars indicate standard deviation.

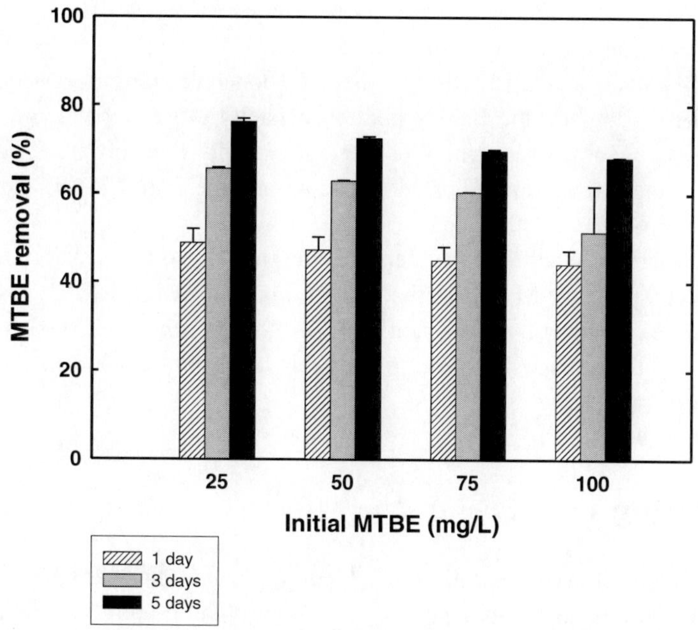

Figure 4. Removal yields of MTBE obtained on changing initial MTBE concentrations for immobilized *A. xylosoxidans* MCM1/1, after different incubation times. Error bars indicate standard deviation.

Figure 3 also shows the stability of MTBE at 30 °C, because, under this control condition and without bacteria, the concentration of MTBE is almost the same throughout the incubation period. At five days of incubation we obtained 94.03±1.99 mg/L MTBE from initial 100 mg/L. In the same way, MTBE concentration was 23.74±0.31, 46.53±2.04 and

68.02±0.99 mg/L from initial 25, 50, and 75 mg/L, respectively. This showed that MTBE decreased 6 % on average. At shorter incubation times, the percentage was equal or even lower (data not shown).

Figure 3 shows another control experiment in which Na alginate beads are incubated in the presence of MTBE, without bacteria, to analyze MTBE adsorption to the beads. In the same figure, we observe 90.33±1.99 mg/L MTBE at five days. For the other concentrations we obtained 21.42±0.44, 44.27±2.25 and 64.55±1.91 mg/L after the incubation of initial 25, 50, and 75 mg/L MTBE, respectively. Therefore, an average of 12.3 % MTBE was adsorbed to the beads. As above, at shorter incubation times the percentages were the same or much lower. These values showed a lower percentage of MTBE decrease than when immobilized *A. xylosoxidans* MCM1/1 was used to biodegrade MTBE. Consequently, *A. xylosoxidans* MCM1/1 could be used as an immobilized biocatalyzer with this kind of bead to boost biodegradation. In the future, this immobilization would facilitate the reuse of biocatalysts for removing MTBE, thus improving the process.

Figure 4 shows the removal yields of MTBE obtained using immobilized *A. xylosoxidans* MCM1/1 in alginate beads for different initial MTBE concentrations. The initial MTBE concentration affected the ability of immobilized *A. xylosoxidans* MCM1/1 to biodegrade MTBE. The biodegradation of MTBE increased over time, whereas the opposite occurred as the initial MTBE concentration is increased up to 100 mg/L. Degradation of MTBE depends on its initial concentration and the incubation time. These parameters directly affect the biodegradation capacity of immobilized *A. xylosoxidans* MCM1/1. The experimental results indicated that immobilized *A. xylosoxidans* MCM1/1 also has considerable potential for removing MTBE over a wide range of MTBE concentrations. Consequently, for this immobilized microorganism, the biodegradation capacity is greater when concentrations of MTBE are lower. MTBE removal can be very rapid and efficient if the operating characteristics of the process are adjusted (for example, the initial MTBE concentration and the biomass quantity).

In practical terms, when an accident occurs and gasoline leaks into the soil, the immediate concentrations of MTBE in the soil are high; however, MTBE is not adsorbed on the soil over time. Even so, the capacity of *A. xylosoxidans* MCM1/1 to degrade high concentrations of MTBE in a short time makes this bacterium a good potential biodegrader for bioremediation processes.

Oxygenase Activity in *A. Xylosoxidans* MCM1/1

Oxygen uptake of A. xylosoxidans MCM1/1 in the presence of MTBE. The extraction of MTBE by SPME followed by its analysis on gas chromatography is a laborious and time consuming technique. To test a simpler methodology we also chose *A. xylosoxidans* MCM1/1. We performed oxygen uptake experiments with an oxygen electrode. As Figure 5 indicates, *A. xylosoxidans* MCM1/1 shows a 22 % use of oxygen in the presence of MTBE after 30 minutes of incubation. This result indicates bacterial activity against this xenobiotic compound, which suggests that oxygenases are taking part in the process. Oxygen uptake experiments reveal the bacterial capacity to oxidize this compound, although they do not show the removal of MTBE. The experiments were performed in duplicate.

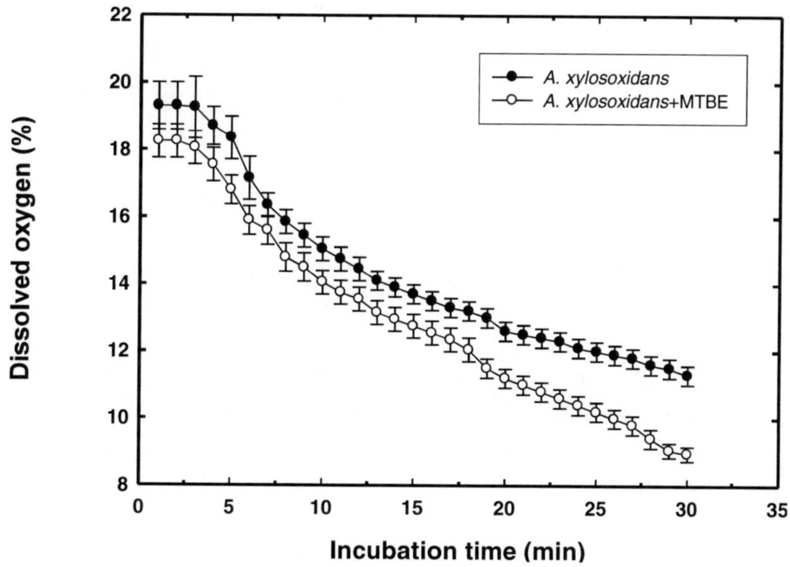

Figure 5. Oxygen uptake of *A. xylosoxidans* MCM1/1 in the presence of MTBE. Error bars indicate standard deviation.

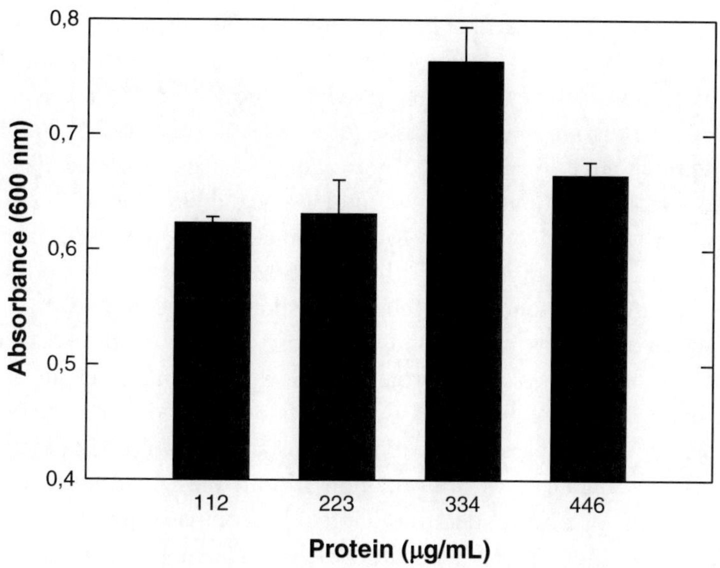

Figure 6. Cytochrome P450 activity in *A. xylosoxidans* MCM1/1. Error bars indicate standard deviation.

Cytochrome P450 activity. Cytochrome P450 exhibits high activity towards a variety of substrates (Alcalde, Farinas & Arnold, 2004; Urlacher & Schmid, 2006). To check for the presence of cytochrome P450 in our bacterial strain, different concentrations of lysate were tested for the enzyme activity using the colorimetric method, as explained above.

From Figure 6, it can be concluded that as the volume of the bacterial lysate increases (from 112 μg/mL to 334 μg/mL of protein), there is an increase in the colorimetric value. And as the concentration of the protein increases further, the absorbance starts decreasing. This clearly demonstrates the active presence of cytochrome P450 in the bacteria and that 334 μg/mL of the bacterial protein is the ideal amount to best detect cytochrome P450 activity.

Protein Induction by MTBE in *A. Xylosoxidans* MCM1/1

Two factors contribute to bacteria making effective use of MTBE (Rittmann, 2004): a high oxygen concentration and the presence of oxygenase activity. Our assay conditions meet these requirements: cells were grown in the presence of oxygen, and oxygenase activity was detected in the bacterial strain by measuring the bacterial oxygen uptake in the presence of MTBE (Figure 5) and the cytochrome P450 activity versus styrene following the method described by Alcalde, Farinas & Arnold (2004) (Figure 6). Thus, as these factors were present, MTBE is degrading although it may also have a toxic effect on bacteria as it did in *Pseudomonas veronii* T1/1 (Bartos et al., 2008).

We then moved on to study the adaptations that might be taking place at the molecular level during MTBE biodegradation. We used resting cells to compare protein extracts from control cells (without MTBE) and bacteria in the presence of MTBE (Eixarch & Constantí, 2010). These cells were often protected against extracellular insults by nonspecific resistance mechanisms (Krayl et al., 2003). Thus, only absolutely necessary protein expression changes were expected to take place (that is, changes related to the presence of MTBE). Total cell extracts were resolved by 2D gel electrophoresis and gel images were acquired and analyzed. As expected, very few changes were observed: of the 225 protein spots considered, 18 showed increased expression and 25 decreased expression. Figure 7 (Eixarch & Constantí, 2010) shows representative images of the gels and the selected spots. Considering that our aim was to find a protein that might be able to be used as a biocatalyst (that is, a protein that might be artificially produced and added to contaminated soils) we used only overexpressed proteins for protein identification. MALDI and MS/MS spectra were performed (Figure 8 shows example spectra of protein 1) and four of the spots were identified.

The four selected proteins were found to enclose peptides from 50S ribosomal protein L10, amino acid-binding periplasmic protein, ATP synthase delta chain and a conserved hypothetical protein with an 85 % homology with endoribonuclease L. The results of protein identification are shown in Table 2 (Eixarch & Constanti, 2010). It should be noted that the available databases contain little information about the *Achromobacter xylosoxidans* proteome, and protein spots were identified largely by comparison with other strains. In our study, all matches were compared with the *Bordetella* genus, which is closely related to *Achromobacter* in the phylogenetic tree (von Wintzingerode, Böcker & Schlötelburg et al., 2002).

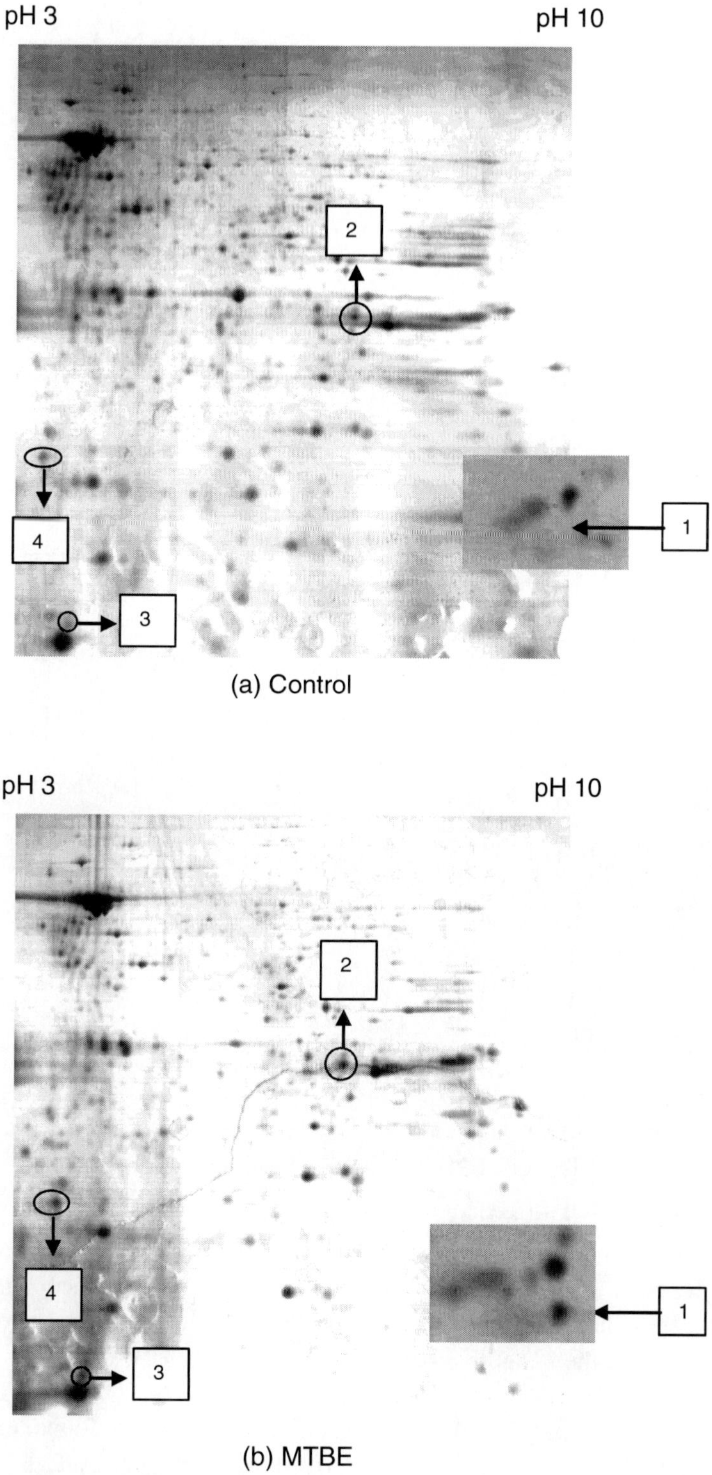

From Eixarch & Constantí, 2010.

Figure 7. Protein patterns of *A. xylosoxidans* MCM1/1 total extracts resolved in 2-DE gels of pH 3-10. Circled proteins indicate overexpressed proteins in cells grown in the presence of MTBE with respect to control cultures.

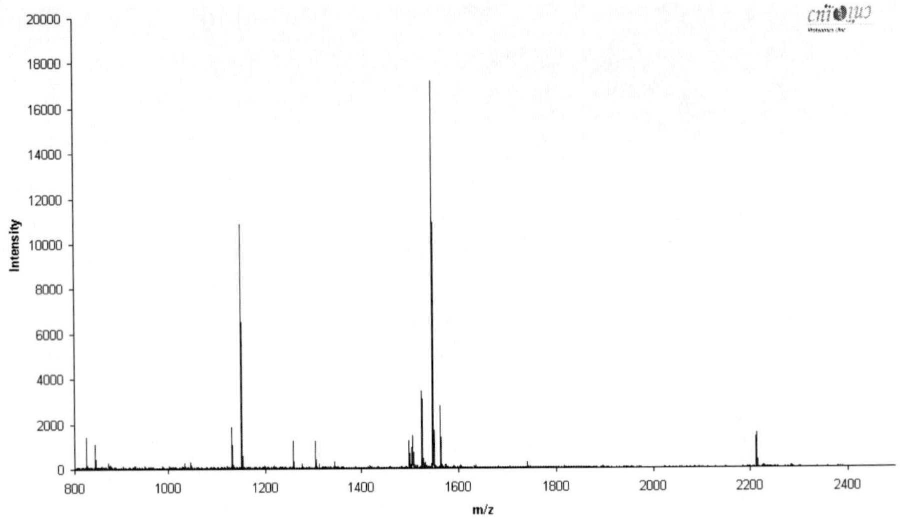

(a) MALDI spectrum

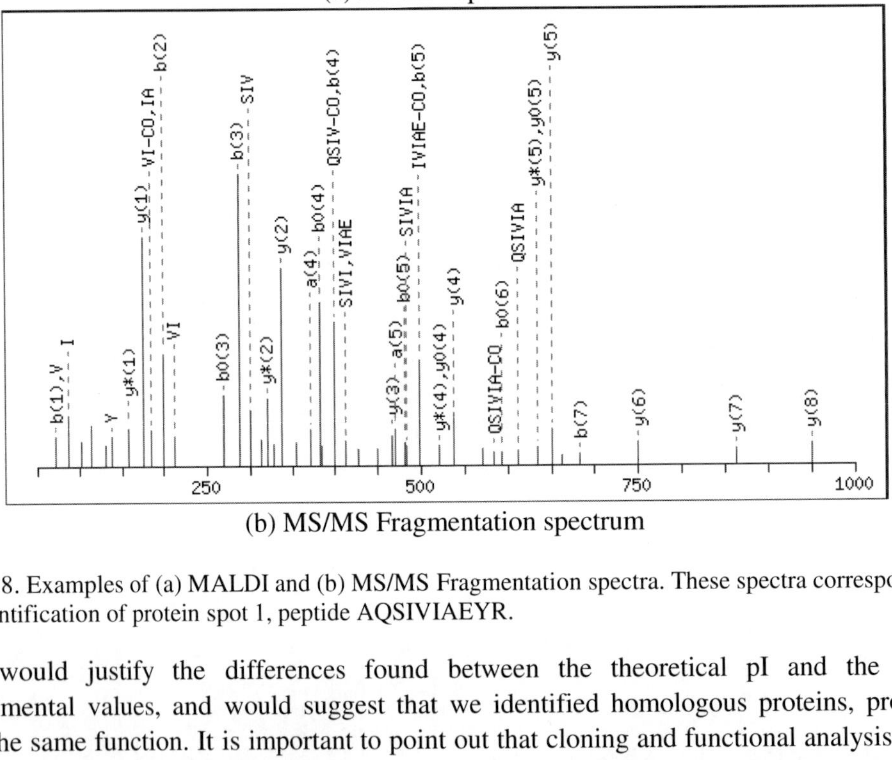

(b) MS/MS Fragmentation spectrum

Figure 8. Examples of (a) MALDI and (b) MS/MS Fragmentation spectra. These spectra correspond to the identification of protein spot 1, peptide AQSIVIAEYR.

This would justify the differences found between the theoretical pI and the actual experimental values, and would suggest that we identified homologous proteins, probably with the same function. It is important to point out that cloning and functional analysis of the corresponding genes should be performed to confirm this hypothesis. Therefore, we could only hypothesize about the benefits that the overexpression of these proteins might have for MTBE-degrading cells.

Of the four selected proteins, L10 is one of the 34 proteins found in the large 30S ribosomal subunit of bacteria. A complex formed by L10 and L7/L12 is able to bind specifically to the leader sequence of the L10 operon mRNA and prevent translation (Petersen, 1990). L10 overexpression could therefore account for a decrease in protein translation. This is to be expected because cells tended to decrease the translation rate (Wool, 1996) in stressful situations such as incubation in the presence of MTBE.

Table 2. Identified protein spots of A. xylosoxidans MCM1/1 in response to MTBE

Spot n°	Spot intensity (Average)		Fold change	*p*	Protein name	Species	NCBI ID	Swissprot entry n°	Peptides identified and MASCOT score	Theoretical		Observed pI
	Control	MTBE								Mr (kDa)	pI	
1	3526 ± 2252	11739 ± 2075	3.3	0.018	50s ribosomal protein L10	*Bordetella pertussis Tohama I*	gi33591286	Q7W0S1	AQSIVIAEYR 202	18.5	9.93	9.7
2	10434 ± 3398	19390 ± 2901	1.9	0.031	Aminoacid-binding periplasmic protein	*Bordetella parapertussis 12822*	gi33594953	Q7W1V8	FTALQSGEVDVLTR 121	36.2	6.64	7.6
3	7008 ± 5501	19265 ± 6716	2.7	0.037	Conserved hypothetical protein	*Bordetella avium 197N*	gi115422182	Q2KUN0	AWDANVAAGNSPPR 116	12.6	5.01	4.2
4	6976 ± 2844	12682 ± 2309	1.8	0.026	ATP synthase delta chain	*Bordetella avium 197N*	gi115424274	Q2KU33	VAVGDQVLDTSVKAQLAR 152	19.2	4.79	3.9

Both the transporter and chaperone activities of the amino acid-binding periplasmic protein (Fath & Kolter, 1993; Richarme & Caldas, 1997) could be advantageous for the cell in the presence of MTBE: they might protect the cell against toxicity, and prevent protein misfolding due to MTBE- or MTBE metabolite-promoted modifications. Efflux of organic solvents by induced transporters has been observed in other strains like *Pseudomonas putida* in response to aromatic and aliphatic solvents and alcohols (Kieboom et al., 1998; Rojas et al., 2001). Moreover, Krayl et al. (2003) has given a plausible explanation of the overexpression of a solvent efflux pump.

The overexpression of an ATP synthase might prove to be beneficial in situations in which ATP production is down-regulated (Krayl et al. [2003], MTBE lowered ATP content) or hydrolyzed to allow active transport to/from the cell of organic compounds that could cause irreversible lesions to cell membranes (Segura et al., 1999).

Endoribonuclease L inhibits protein synthesis by cleavage of mRNA (Bisbal & Silverman, 2007). Therefore, up-regulation of this protein would again lead to a down-regulation of protein synthesis. Pandey & Rath (2004) observed that the expression of recombinant human RNase L degraded RNA and inhibited cell growth in *Escherichia coli*.

CONCLUSION

We have demonstrated that resting cells of the isolated *A. xylosoxidans* MCM1/1, E. *cloacae* MCM2/1, *O. anthropi* MCM5/1 and *E. dermatitidis* MCM3/4 were capable of biodegrading MTBE in a simple mineral solution. These microorganisms achieve a high rate of MTBE biodegradation in a short period of time. Interestingly, in these conditions bacteria do not need any nutrient requirements to maintain cell viability. This avoids any growth problems that may arise when they are applied directly to the soil for bioremediation, thus competing with other possible autochthonous bacteria. These four microorganisms have considerable potential as biocatalysts for in situ bioremediation. Moreover, when *A. xylosoxidans* MCM1/1 cells were immobilized in Na-alginate beads the biodegradation rate was also good. This procedure may stabilize and protect the cells in a bioremediation process. Oxygenase activity was observed in *A. xylosoxidans* MCM1/1 which might indicate the participation of cytochrome P450 in the biodegradation process. However, further studies are needed to demonstrate this.

On the other hand, little is known about the molecular mechanism during the biodegradation process. This is a complex topic, which is complicated even further in the case of environmental bacteria as here. To fill this gap in knowledge, we provide information about the proteins that were overexpressed during the biodegradation of MTBE by *A. xylosoxidans* MCM1/1. The presence of these proteins may help bacteria to adapt to the stressed situation when MTBE is present and contribute to rapid MTBE biodegradation, which means that this microorganism is a good biocatalyst. We have contributed knowledge about the mechanism that may enable bacterial cells to survive in an initially hostile environment. Once they had survived, cells adapted and began to metabolize MTBE, so this is a significant tool for pollutant biodegradation.

ACKNOWLEDGMENTS

This study was supported by a project grant from the Spanish Ministerio de Medio Ambiente.

REFERENCES

Alcalde, M., Farinas, E. T., Arnold, F. H., 2004. Colorimetric highthroughput assay for alkene epoxidation catalyzed by cytochrome P450 BM-3 variant 139-3. *J. Biomol. Screen* 9, 141-146.

Andreoni, V., Cavalca, L., Rao, M. A., Nocerino, G., Bernasconi, S., Dell'Amico, E. et al., 2004. Bacterial communities and enzyme activities of PAHs polluted soils. *Chemosphere* 57(5), 401-412.

Awong-Taylor, J., Craven, K., Griffiths, L., Bass, C., Muscarella, M., 2007. Comparison of biochemical and molecular methods for the identification of bacterial isolates associated with failed loggerhead sea turtle eggs. *J. Appl. Microbiol.* 104(5), 1244-51.

Baek, K., Yang, J. S., Yang, J. W., 2003. Immobilization behavior of methyl *tert*-butyl ether by cyclodextrins. *J. Hazard. Mater.* 105, 169-177.

Barberà, M. J., Mateo, E., Monkaityte R., Constantí, M., 2011. Biodegradation of methyl tert-butyl ether by newly identified soil microorganisms in a simple mineral solution. *World J. Microbiol. Biotechnol.* 27, 813-821.

Barceló, D., 2007. Fuel oxygenates. Springer, Berlin.

Bartos, P., Balazs, M., Kiss, I., Bihari, Z., Kelemen, O., Mecs, I., 2008. Toxic effect of methyl *tert*-butyl ether on growth of soil isolate *Pseudomonas veronii* T1/1. *World J. Microbiol. Biotechnol.* 24, 875-878.

Bisbal, C., Silverman, R. H., 2007. Diverse functions of RNase L and implications in pathology. *Biochimie* 89(6-7), 789-798.

Bosshard, P. P., Zbinden, R., Abels, S., Böddinghaus, B., Altwegg, M., Böttger, E. C., 2006. 16S rRNA gene sequencing versus the API 20 NE system and the VITEK 2 ID-GNB card for identification of nonfermenting Gram-negative bacteria in the clinical laboratory. *J. Clin. Microbiol.* 44(4), 1359-66.

Bradford, M. M., 1976. A rapid and sensitive method for the quantitation of microgram quantities of protein utilizing de principle of protein-dye binding. *Anal. Biochem.* 72, 248-254.

Chen, D., Chen, J., Zhong W., Cheng, Z., 2008. Degradation of methyl *tert*-butyl ether by gel immobilized *Methylibium petroleiphilum* PM1. *Bioresource Technol.* 99, 4702-4708.

Chen, D., Chen, J., Zhong, W., 2009. Enhancement of methyl *tert*-butyl ether degradation by the addition of readily metabolizable organic substrates. *J. Hazard. Mater.* 167, 860-865.

Cheng, Z., Chen, J., Chen, D., Zhang, L., 2011. Biodegradation of methyl *tert*-butyl ether in a bioreactor using immobilized *Methylibium petroleiphilum* PM1 cells. *Water Air Soil Poll.* 214(1-4), 59-72.

Constantí, M., Giralt, J., Bordons, A., 1996. Degradation and desulfurization of dibenzothiophene sulfone and other sulfur compounds by *Agrobacterium* MC501 and a mixed culture. *Enzyme Microb. Technol.* 19, 214-219.

Dursun, A. Y., Aksu, Z., 2002. Effect of internal diffusivity of ferrous(II) cyanide complex ions in Ca-alginate immobilized *Pseudomonas fluorescens* gel beads on the biodegradation rate. *Process Biochem.* 37, 747–752.

Eixarch, H., Constantí, M., 2010. Biodegradation of MTBE by *Achromobacter xylosoxidans* MCM1/1 induces synthesis of proteins that may be related to cell survival. *Process Biochem.* 45, 794-798.

Fath, M. J., Kolter, R., 1993. ABC transporters: bacterial exporters. *Micro Rev.* 57, 995-1017.

François, A., Mathis, H., Godefroy, D., Piveteau, P., Fayolle, F., Monot, F., 2002. Biodegradation of methyl *tert*-butyl ether and other fuel oxygenates by a new strain, *Mycobacterium austroafricanum* IFP 2012. *Appl. Environ. Microbiol.* 68(6), 2754-2762.

Guindon, S., Gascuel, O., 2003. A simple, fast, and accurate algorithm to estimate large phylogenies by maximum likelihood. *Syst. Biol.* 52(5), 696-704.

Hanson, J. R., Ackerman, C. E., Scow, K. M., 1999. Biodegradation of methyl *tert*-butyl ether by a bacterial pure culture. *Appl. Environ. Microbiol.* 65, 4788-4792.

Hardison, L. K., Curry, S. S., Ciuffetti, L. M., Hyman, M. R., 1997. Metabolism of diethyl ether and cometabolism of methyl *tert*-butyl ether by a filamentous fungus, a *Graphium sp*. *Appl. Environ. Microbiol.* 63, 3059-3067.

Iwashita, S., Callahan, T. P., Haydu, J., Wood, T. K., 2004. Mesophilic aerobic degradation of a metal lubricant by a biological consortium. *Appl. Microbiol. Biotechnol.* 65, 620-626.

Kennes, C., Veiga, M. C., 2004. Fungal biocatalysts in the biofiltration of VOC-polluted air. *J. Biotechnol.* 113, 305-31.

Kieboom, J., Dennis, J. J., Jylstra, G. J., Bont, J. A. M., 1998. Active efflux of organic solvents by *Pseudomonas putida* S12 is induced by solvents. *J. Bacteriol.* 180, 6769-6772.

Krayl, M., Behndorf, D., Loffhagen, N., Babel, W., 2003. Use of proteomics and physiological characteristics to elucidate ecotoxic effects of methyl *tert*-butyl ether in *Pseudomonas putida* KT2440. *Proteomics* 3, 1544-1552.

Lacayo-Romero, M., Quillaguaman, J., van Bavel, B., Mattiasson, B., 2005. A toxaphene-degrading bacterium related to *Enterobacter cloacae* strain D1 isolated from aged contaminated soil in Nicaragua. *Syst. Appl. Microbiol.* 28(7), 632-639.

Lee, E. H., Cho, K. S., 2009. Effect of substrate interaction on the degradation of methyl *tert*-butyl ether, benzene, toluene, ethylbenzene and xylene by *Rhodococcus sp. J. Hazard. Mater.* 167 (1-3), 669-674.

Lyew, D., Guiot, S. R., Monot, F., Fayolle-Guichard, F., 2007. Comparison of different support materials for their capacity to immobilize *Mycobacterium austroafricanum* IFP 2012 and to adsorb MTBE. *Enzyme Microb. Technol.* 40, 1524-1530.

Maciel, H., Mathis, H., Ferreira, N. L., Lyew, D., Guiot, S., Monot, F., Greer, C. W., Fayolle-Guichard, F., 2008. Use of *Mycobacterium austroafricanum* IFP 2012 in a MTBE-degrading bioreactor. *J. Mol. Microbiol. Biotechnol.* 15, 190-198.

Magaña-Reyes, M., Morales, M., Revah, S., 2005. Methyl *tert*-butyl ether and tert-butyl alcohol degradation by *Fusarium solani*. *Biotechnol. Lett.* 27, 1797-1801.

Maidak, B. L., Cole, J. R., Parker, Jr C. T., 1999. A new version of the RDP (Ribosomal Database Project). *Nucleic Acids Res.* 27, 171-173.

Mo, K., Lora, C. O., Wanken, A. E., Javanmardian, M., Yang, X., Kulpa, C. P., 1997. Biodegradation of methyl *tert*-butyl ether by pure bacterial cultures. *Appl. Microbiol. Biotechnol.* 47(1), 69-72.

Muller, R. H., Rohwerder, T., Harms, H., 2008. Degradation of fuel oxygenates and their main intermediates by *Aquincola tertiaricarbonis* L108. *Microbiology*, 154, 1414-1421.

Muñoz-Castellanos, L. N., Torres-Muñoz, J. V., Keer-Rendon, A., Manzanares-Papayanopoulos, L. I., Nevarez-Moorillon, G. V., 2006. Aerobic biodegradation of methyl *tert*-butyl ether (MTBE) by pure bacterial cultures isolated from contaminated soil. *World J. Microbiol. Biotechnol.* 22(8), 851-855.

Olaniran, A. O., Mfumo, N. H., Pillay, D., Pillay, B., 2006. Synergistic utilization of dichloroethylene as sole carbon source bacterial consortia isolated from contaminated sites in Africa. *Biotechnol. Bioprocess Eng.* 11(3), 205-210.

Pandey, M., Rath, P.C., 2004. Expression of interferon-inducible recombinant human Rnase L causes RNA degradation and inhibition of cell growth in *Escherichia coli*. *Biochem. Biophys. Res. Commun.* 317(2), 586-97.

Pannier, A., Oehm, C., Fischer, A. R., Werner, P., Soltmann, U., Böttcher, H., 2010. Biodegradation of fuel oxygenates by sol-gel immobilized bacteria *Aquincola tertiaricarbonis* L108. *Enzyme Microb. Technol.* 47, 291-296.

Perkins, D. N., Pappin, D. J., Creasy, D. M., Cottrell, J. S., 1999. Probability-based protein identification by searching sequence databases using mass spectrometry data. *Electrophoresis* 20, 3551-3567.

Petersen, C., 1990. *Escherichia coli* ribosomal protein L10 is rapidly degraded when synthesized in excess of ribosomal protein L7/L12. *J. Bacteriol.* 172, 431-436.

Pimentel-Gonzalez, D., Revah, S., Campos-Montiel, R., Monroy-Hermosillo, O., Vernon-Carter, E. J., 2008. A laboratory study of the biodegradation of MTBE solubilised in water by a microbial consortium entrapped in a water-in-oil-in-water double emulsion. *Process Biochem.* 43, 1239-1243.

Purswani, J., Pozo, C., Rodríguez-Diaz, M., Gonzalez-Lopez, J., 2008. Selection and identification of bacterial strains with Methyl *tert*-butyl ether, Ethyl *tert*-butyl ether, and *tert*-amyl methyl ether degrading capacities. *Environ. Toxicol. Chem.* 27, 2296-2303.

Quan, C. S., Fan, S. D., Ohta, Y., 2003. Immobilization of *Candida krusei* cells producing phytase in alginate gel beads: an application of the preparation of myo-inositol phosphates. *Appl. Microbiol. Biotechnol.* 62, 41–47.

Richarme, G., Caldas, T. D., 1997. Chaperone properties of the bacterial periplasmic substrate-binding proteins. *J. Biol. Chem.* 272, 15607-15612.

Rittmann, B. E., 2004. Definition, objectives, and evaluation of natural attenuation. *Biodegradation* 15, 349-357.

Rojas, A., Duque, E., Mosqueda, G., Goleen, G. et al., 2001. Three efflux pumps are required to provide efficient tolerance to toluene in *Pseudomonas putida* DOT-T1E. *J. Bacteriol.* 183, 3967-3973.

Salanitro, J. P., Diaz, L. Q., WillIams, M. P., Wisniewski, H. L., 1994. Isolation of a bacterial culture that degrades methyl *t*-butyl ether. *Appl. Environ. Microbiol.* 60, 2593–2596.

Segura, A., Duque, E., Mosqueda, G., Ramos, J. L., Junker, F., 1999. Multiple responses of Gram-negative bacteria to organic solvents. *Environ. Microbiol.* 1, 191-198.

Sequillace, P. J., Pankow, J. F., Korte, N. F., Zogorski, J. S., 1997. Review of the environmental behavior and fate of methyl *tert*-butyl ether. *Environ. Toxicol. Chem.* 16, 1836–1844.

Shevchenko, A., Tomas, H., Vorm, O., Havlis, J., Olsen, J. V., Mann, M., 2006. In-gel digestion or mass spectrometric characterization of proteins and proteomes. *Nat. Protoc.* 1, 2856-60.

Smith, C. A., Hyman, M. R., 2004. Oxidation of methyl *tert*-butyl ether by alkane hydroxylase in dicyclopropylketone-induced and n-octane-grown *Pseudomonas putida* GPo1. *Appl. Environ. Microbiol.* 70, 4544-4550.

Stamatakis, A., 2006. RAxML-VI-HPC: Maximum likelihood-based phylogenetic analyses with thousands of taxa and mixed models. *Bioinformatics* 22(21), 2688-2690.

Streger, S. H., Vainberg, S., Dong, H. L., Hatzinger, P. B., 2002. Enhancing transport of *Hydrogenophaga flava* ENV735 for bioaugmentation of aquifers contaminated with methyl *tert*-butyl ether. *Appl. Environ. Microbiol.* 68, 5571-5579.

Suckau, D., Resemann, A., Schuerenberg, M., Hufnagel, P., Franzen, J., Holle, A., 2003. A novel MALDI LIFT-TOF/TOF mass spectrometer for proteomics. *Anal. Bioanal. Chem.* 376, 952-965.

Urlacher, V. B., Schmid, R. D., 2006. Recent advances in oxygenase-catalyzed biotransformations. *Curr. Opin. Chem. Biol.* 10(2), 156-161.

von Wintzingerode, F., Böcker, S., Schlötelburg, C., Chiu, N. H. L. et al., 2002. Base-specific fragmentation of amplified 16S rRNA genes analyzed by mass spectrometry: a tool for rapid bacterial identification. *PNAS* 99, 7039-44.

White, T. J., Bruns, T., Lee, S., Taylor, J. W., 1990. Amplification and direct sequencing of fungal ribosomal RNA genes for phylogenetics. In: PCR Protocols: a Guide to Methods and Applications (pp. 315-322). New York : Academic Press.

Woese, C. R., 1987. Bacterial evolution. *Microbiol. Rev.* 51, 221-271.

Wool, I. G., 1996. Extraribosomal functions of ribosomal proteins. *TIBS* 21, 164-165.

Yeh, C. K., Novak, J. T., 1994. Anaerobic biodegradation of gasoline oxygenates in soils. *Water Environ. Res.* 66, 744–752.

Zaitsev, G. M., Uotila, J. S., Häggblom, M. M., 2007. Biodegradation of methyl *tert*-butyl ether by cold-adapted mixed and pure bacterial cultures. *Appl. Microbiol. Biotechnol.* 74(5), 1092-1102.

Zhong, W., Li, Y., Sun, K., Jin, J., Li, X., Zhang, F., Chen, J., 2011. Aerobic degradation of methyl *tert*-butyl ether in a closed symbiotic system containing a mixed culture of *Chlorella ellipsoidea* and *Methylibium petroleiphilum* PM1. *J. Hazard. Mater.* 185, 1249-1255.

In: Bioremediation: Processes, Challenges and Future Prospects ISBN: 978-1-62948-513-3
Editors: Jesús Bernardino Velázquez-Fernández et al.

Chapter 4

BIOREMEDIATION AS A USEFUL BIOTECHNOLOGICAL STRATEGY FOR THE TREATMENT OF PHENOLICS: ADVANCES, CHALLENGES AND FUTURE PROSPECTS[†]

Cintia E. Paisio, Ana L. Wevar Oller, Sabrina G. Ibañez, Melina A. Talano, Paola S. González, María I. Medina and Elizabeth Agostini*

Dpto. de Biología Molecular, Facultad de Ciencias Exactas, Físico- Químicas y Naturales, Universidad Nacional de Río Cuarto, Río Cuarto, Córdoba, Argentina

ABSTRACT

Bioremediation is an attractive biotechnological tool for the treatment of water and effluents containing toxic pollutants, including phenolics, which are very hazardous pollutants due to their high toxicity and environmental persistence. Several bacterial strains are used for biodegradation or biotransformation of these compounds, as an eco-friendly technology to treat contaminated environments. Most of these bacteria can utilize phenolics as sole carbon and energy source and the metabolic pathways implicated in this process are well established.

Immobilization of bacterial cells on different support matrices represents a very important advance in the application of bioremediation. Furthermore, the use of this strategy and the design of new types of bioreactors would allow to remediate great volumes of highly contaminated water and effluents. Due to the continuous release of industrial wastewater the application and improvement of different strategies could be very useful.

This chapter highlights the more recent applications of bacteria for phenol bioremediation, new advances for the improvement of this technology as well as the less

---

[†] All authors contributed equally to this book chapter.

[*] Corresponding author: eagostini@exa.unrc.edu.ar.

explored aspects that must be deepened to assess the feasibility of phenolics bioremediation from highly contaminated water and effluents.

Keywords: Phenol, bacteria, bioremediation, immobilization, bioreactors, phenolic wastewaters

INTRODUCTION

Phenol and its derivatives represent a large group of chemical compounds consisting of a hydroxyl functional group (–OH) attached to an aromatic hydrocarbon group, with a ring structure like that of benzene. They are considered as priority pollutants because of their adverse effects on animal and human health, widespread diffusion and persistence in the environment. Their origin may be natural or anthropogenic; they occur naturally in water and soil as degradation products of lignin, one of the main components of plant cell walls (Michałowicz & Duda, 2007). However, human activities are responsible for the main introduction of phenolics into the environment. In fact, phenolic compounds have broad spectrum of domestic, agricultural and industrial usage and they are widely employed as the components of dyes, polymers, drugs and other organic substances. Industrial phenol production is estimated to be over three million tons per year, being used mostly in petrochemical industry, synthesis of resins, pharmaceuticals, perfumes, as well as intermediates in the preparation of other chemicals (e.g., plastics, drugs, explosives, pesticides and detergents), solvents or lubricating oils (Iurascu et al., 2009). In this sense, phenol is considered as an important degradation product of substituted phenols (Basu, Oleszkiewicz & Sparling, 1996; Khodadoust et al., 1997). Moreover, chlorophenols may be formed as by-products during the desinfection of water by chlorination and as a result of the natural chlorination of organic material. Phenols are also released through automobile exhaust, fireplaces, cigarette smoke, and gases from incinerators. Although they are not release directly into water, their transference to aquatic systems may occur and high levels of these compounds have been found in the rain (Canadian Council of Ministers of the Environment, 1999; Schummer et al., 2009) and, also, in fog and cloud (Harrison et al., 2005).

Measured values of phenolics in the environment show a wide range of variability depending on the procedence. For instance, phenol concentrations from petroleum refinery effluents vary between 50 and 2000 mg L^{-1} for distillation units while in waste solutions generated from coal conversion processes phenols concentrations of 200-600 mg L^{-1} are usually detected (Nayak & Singh, 2007). Moreover, it can be assumed that phenols levels will increase over time if more stringent regulations are not taken to prevent its discharge.

Phenol is included in the class 2 water hazardous pollutants list in several countries (Iurascu et al., 2009). A phenol concentration of 1 mg L^{-1} or greater, affects aquatic life and may represent a risk to human health. Therefore, in most cases the accepted limit for effluents is less than 0.5 mg L^{-1}. In this sense, the United States Enviromental Protection Agency (US EPA) and the World Health Organization (WHO) have established a limit concentration of 1 μg L^{-1} (Srivastava et al., 2006) while the European Community defined a limit of 5 μg L^{-1} for phenolic compounds in drinking water (Schummer et al., 2009). However, phenols are frequently found in the environment in higher concentrations than those established by Regulatory Organizations, as it was already mentioned.

The evaluation of phenol toxicity has become an area of active research. It had been studied on selected microorganisms (e.g. protozoa, yeast and bacteria), algae, duckweed and numerous invertebrates and vertebrates and varies widely with the organism tested, dissolved oxygen content and water temperature. It was observed that phenol induced teratogenic effects (Paisio et al., 2009) and depending on the organism tested, the acute toxicity estimated by the Media Lethal Concentration (LC_{50}) value, varied from 6.5 to 1840 mg L^{-1} (Bernardini et al., 1996; Iurascu et al., 2009; Paisio et al., 2009).

Human consumption of phenol contaminated water can cause severe pain, blood changes, liver injury, muscular effects and even death (Flocco et al., 2002; Aksu, 2005). In addition, chronic toxic effects on human include vomiting, difficulty in swallowing, anorexia, liver and kidney damage, headache and other mental disturbances (Srivastava et al., 2006). It also induce hemolysis in human erythrocytes *in vitro (*Bukowska & Kowalska, 2004). The determination of urinary concentrations of phenols has been considered to be quite useful for the estimation of exposure, and its relationship with human diseases as well as laboratory abnormalities. Widespread human exposure to phenols has been documented recently and some of them, which are potential endocrine disruptors, have demonstrated adverse effects on male reproduction in animal and *in vitro* studies. The associations between human exposures to phenols on reproductive health are largely unknown. However, Chen et al. (2013) provide the first evidence that exposure to alkylphenols is associated with idiopathic male infertility.

Due to the environmental problems realted with the presence of phenolics in natural waters and soils, attemps have been focused on developing effective technologies for the removal of such compounds. Treatment systems generally only focus on the removal of regulated phenols such as phenol, 2,4-dichlorophenol (2,4-DCP) and pentachlorophenol (PCP) while the occurrence of un-regulated phenols in wastewater has not received sufficient attention. Nevertheless, many un-regulated phenols such as 4-t-octylphenol and 4-chloro-3-methylphenol are potentially hazardous to aquatic environments. Thus, the removal of both regulated and un-regulated phenolics is of considerable importance (Zhong, Wang & Xu, 2012).

Numerous conventional physico-chemical methods based on the principles of adsorption, precipitation and coagulation, chemical oxidation, sedimentation, filtration, osmosis and ion exchange have been employed for the dephenolization of industrial wastewaters. Although various of these methods have demonstrated to be effective, most of them have several disadvantages such as high cost, incomplete purification and formation of hazardous by-products; which limited their use. Thus, it has become necessary to look for environmentally friendly alternative technologies, such as biological methods, which are generally considered as safe and least disruptive treatments. These methods are based on the use of living organisms such as microorganisms like bacteria and fungi (bioremediation); algae (phycoremediation); terrestrial and aquatic plants (phytoremediation) and earthworms (vermiremediation) to manage or remediate polluted soils and water. All of them are well recognized as biotechnological tools to degrade or transformate contaminants into non- or less-toxic compounds. Generally, biological methods are cost-effective, environmentally sustainable and also socially acceptable. The present chapter focusses in bacterial bioremediation technologies for the treatment of water and effluents containing phenolics. In particular, phenol is taken as a representative toxic contaminant belonging to this class of compounds.

BACTERIAL PHENOL BIOREMEDIATION

Biodegradation of Phenol

Bioremediation is an innovative technology that involves the use of different microorganisms to attenuate, immobilize, remove, degrade or breakdown xenobiotic compounds in less toxic or non-toxic ones (Sharma & Gupta, 2012). This technology includes different processes such as a) contaminant transformation, b) degradation to simpler molecules, c) mineralization into inorganic molecules such as CO_2, H_2O, H_2, NH_3, etc, d) sorption on cell surfaces and e) intracellular accumulation, among others (Krastanov, Alexieva & Yemendzhiev, 2013). Some of these processes are mediated by enzymatic reactions that generate simpler compounds that enter into normal catabolic cell cycles (Fetzner, 2012).

Since phenol is widespread in nature, bacteria capable of utilizing this compound as a carbon and energy source can be found in many different habitats. Several bacterial genera have been described as phenol degraders like *Rhodococcus, Arthrobacter, Acinetobacter, Bacillus* and *Pseudomonas,* among others (Sridevi et al., 2012). However, not all of them have developed adequate mechanisms to metabolize high phenol concentrations that can be found in anthropogenic sources such as industrial effluents. Moreover, there are many man-made phenol derivatives for which microorganism have not evolved degradation mechanisms, thus, these compounds persist in the environment as xenobiotics. The great potential of microorganisms to remediate phenolic compounds has motivated a number of studies focused on the elucidation of degradation process.

Phenol Degradation Pathways under Aerobic and Anaerobic Conditions

Phenol metabolism can occur under aerobic (O_2 as electron acceptor) and/or anaerobic conditions (NO_3^-, SO_4^{-2}, CO_2 as final electron acceptors) although most research has been carried out on aerobic phenol degradation (Ruiz et al., 2001). Under aerobic conditions, O_2 is a very important factor for hydroxylation and aromatic ring cleavage. In a first step, the enzyme phenol hydroxylase (a monoxygenase), which requires reduced pyrimidine nucleotide ($NADH + H^+$) adds a second hydroxyl group at the *ortho* position to the pre-existing hydroxyl group, to form catechol (1,2-dihydroxybenzene) (Figure 1). Regarding phenol hydroxylases, it has been described that they can be monocomponent or multicomponent flavoproteins. In some species of *Bacillus*, *Pseudomonas*, *Acinetobacter* and *Alcaligenes*, phenol hydroxylases are monocomponent (Kim & Oriel, 1995), however, in *Acinetobacter radioresistens* and *Pseudomonas* sp. CF600, phenol hydroxylases are multicomponent (Shingler, 1996; El-Sayed, Ismaeil & El-Beih, 2009). The formed catechol can be then degraded by dioxygenases through two different pathways, *ortho* or *meta* cleavage, depending on the involved microorganism. In the *ortho*- or β-ketoadipate pathway, the aromatic ring is cleaved between hydroxyl groups to form *cis,cis*-muconate, by a catechol 1,2-dioxygenase (Harwood & Parales, 1996; Tuan et al., 2011). This resulting compound is then metabolized to form acetyl-

CoA and succinyl-CoA, which are finally degraded in the Krebs cycle (El Sayed et al., 2003; Agarry & Solomon, 2008). Diverse bacteria are capable to degrade phenol by *ortho-* pathway, like *Acinetobacter calcoaceticus*, *Pseudomonas* species, *Arthrobacter*, *Sphingomonas*, *Geobacillus* and *Rhodococcus* (Paller, Hommel & Kleber, 1995; Guzik et al., 2011). Regarding *meta-* pathway, the ring cleavage is catalyzed by the enzyme catechol 2,3-dioxygenase producing 2-hydroxymuconic semialdehyde. This product is finally transformed, after several reactions, in acetaldehyde and pyruvate, which are intermediates of the Krebs cycle (Agarry, Solomon & Layokun, 2008) (Figure 1). Leonard & Lindely (1998) described that *Pseudomonas putida*, *P. cepacia*, *P. picketti* and *Alcaligenes eutrophus* metabolized phenol by *meta-* pathway. Comparing both pathways, o*rtho-* pathway seems to be more efficient to convert carbon into biomass than *meta-* pathway, although phenol biodegradation process occurs at higher rate in bacteria that use *meta*-pathway (Kiesel & Muller, 2002). It is important to note that some bacterial strains were capable to degrade phenol through both pathways as it was described for *Comamonas* sp. and *Cupriavidus* sp. (Dong et al., 2008).

Regarding to phenol anaerobic degradation, the knowledge is limited. It has been established that anaerobic degradation includes phenol carboxilation in the *para* position to form 4-hydroxybenzoate, mediated by the enzyme 4-hydroxybenzoate carboxylase (Figure 2).

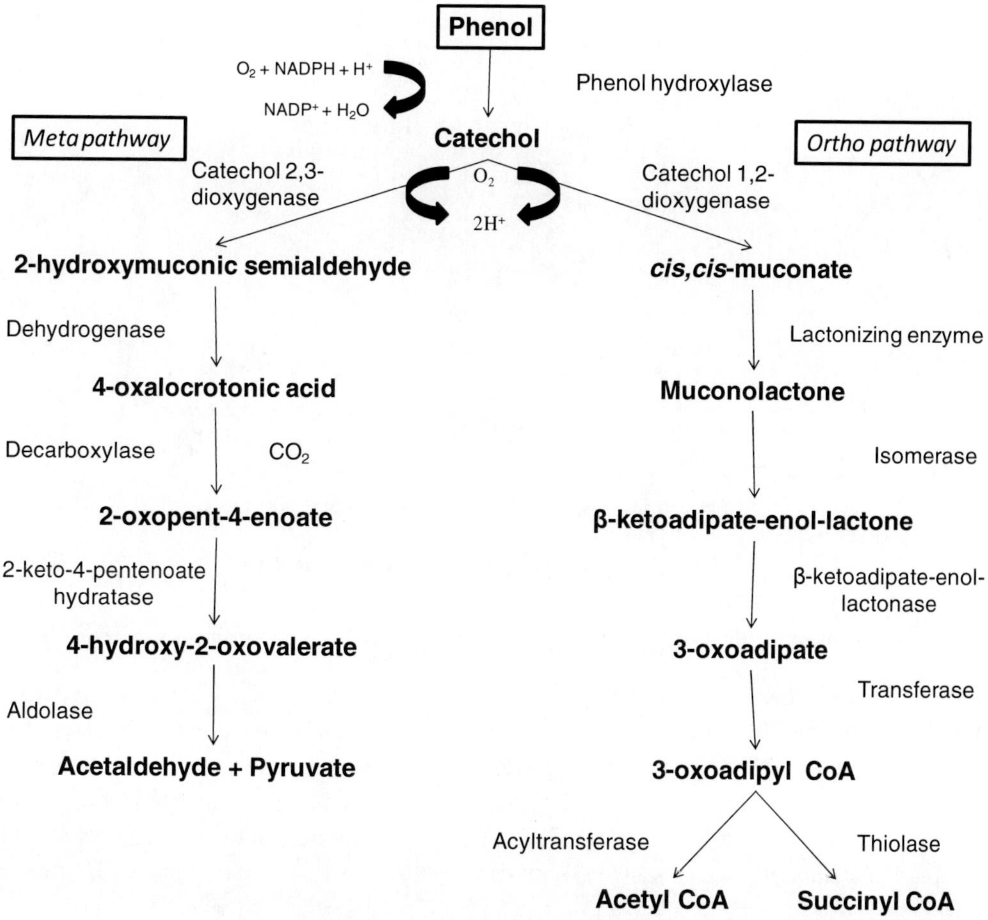

Figure 1. Aerobic phenol biodegradation pathways proposed for bacterial strains.

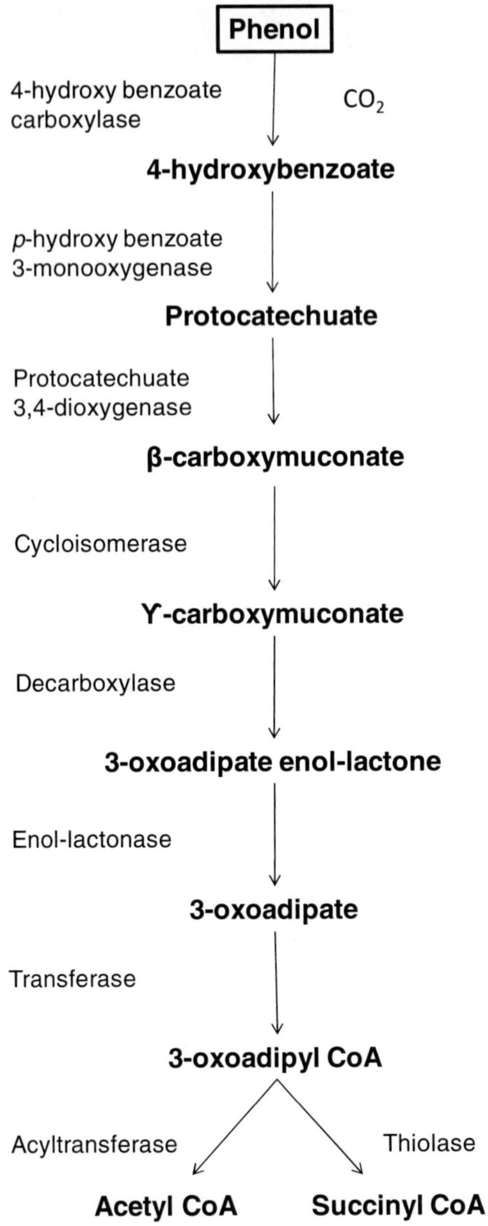

Figure 2. Anaerobic phenol biodegradation pathway proposed for bacterial strains.

Besides phenol, other aromatic compounds are also carboxylated as the first step of the anaerobic pathway. It was demonstrated for denitrifying *Paracoccus*, as well as for a methanogenic consortium capable to degrade catechol, *ortho*-cresol and *ortho* halogenated phenols, that they used *para* carboxylation, followed by dehydroxylation. Some microorganisms capable to degrade phenol under anaerobic conditions are *Desulphobacterium phenolicum* and *Thauera aromatic* (Basha, Rajendran & Thangavelu, 2010).

As it was previously described, both phenol degradation pathways have been thoroughly studied. Furthermore, the elucidation of the involved enzymes has recently allowed their use as molecular tools to estimate the diversity of phenol degrading bacterial communities. In this context, the coding gene for phenol hydroxylase and catechol 1,2 -dioxygenase enzymes have been applied with this purpouse (Basile & Erijman, 2008; Sandhu, Halverson & Beattie, 2009; El Azhari et al., 2010).

Phenol Bioremediation: In Vitro Experiments

In this section the focus will be put on *in vitro* experiments carried out to test optimal conditions for phenol removal from synthetic solutions that are generally performed before the scale-up in bioreactors.

Table 1. Recently studied bacteria for phenol biodegradation

Microorganisms	Isolation source	Maximum phenol concentration used (mg L^{-1})	Reference
Acinetobacter sp. *Comamonas* sp. *Cupriavidus* sp. *Pseudomonas* sp.	Natural soil	180-900	Dong et al., 2008
Acinetobacter sp. XA05	Activated sludge	1000	Liu, Zhang &Wang, 2009
Rhodococcus coprophilus	Semiarid soil	800	Nagamani & Lowry, 2009
Rhodococcus opacus 1G	ND	750	Shumkova et al., 2009
Rhodococcus sp. *UKM-P*	Oil contaminated soil	500	Nor Suhaila et al., 2010
Consortium of 6 bacterial strains (*Bacillus* sp., *Arthrobacter* sp., *Halomonas* sp., *Pseudomonas* sp.)	Soil from saline environments	300	Gayathri & Vasudevan, 2010
Citrobacter freundi *Proteus mirabilis*	Oil contaminated soil	100	Mohite, Pawar & Morankar, 2011
Pseudomonas sp. *a3*	Activated sludge	100	Wu et al., 2012
Acinetobacter sp. EBR01 *Acinetobacter* sp. EBR02 *Cobetia marina* EBR04	Intestine of marine creatures	100	Kobayashi, Maki & Nakamura, 2012
Ralstonia pickettii	Petroleum refinery oil sludge	200	Al-Zuhair & El-Naas, 2012
Rhodococcus sp. *CS1*	Tannery effluent	1000	Paisio et al., 2012
Planococcus sp. *S5*	Activated sludge	380	Hupert-Kocurek, Guzik & Wojcieszyńska, 2012
Acinetobacter sp. RTE1.4	Industrial effluent	600	Paisio et al., 2013
Rhodococcus sp. *UKMP-5M*	Oil contaminated soil	1000	Nor Suhaila et al., 2013
Rhodococcus sp. *AQ5NOL 2*	Contaminated soil	500	Arif et al., 2013
Bacillus cereus WJ1	Contaminated wastewater	600	Zhang et al., 2013

Ref. ND Not determined.

Several bacterial species such as *Bacillus* sp., *Pseudomonas* sp., *Acinetobacter* sp. and *Achromobacter* sp. have been widely studied for phenol bioremediation (Nair, Jayachandran & Shashidhar, 2008). This pollutant was also degraded by *Actinobacillus* species (Khleifat & Khaled, 2007) as well as *Alcaligenes* sp. (Nair & Shashidhar, 2004) and *Rhodococcus erythropolis* (Prieto et al., 2002). Considering the available data in the literature it is clear that *Pseudomonas* has been the most widely genus applied for the degradation of phenolic compounds as a single carbon and energy source (Hinteregger et al., 1992; Kwon & Yeon, 2009). For example, it has been described that *P. putida* strains exhibit high degradation activity toward significant concentrations of phenol or its derivatives (Farrell & Quilty, 2002; Kargi & Eker, 2004). In particular, the strain *P. putida* MTCC 1194 degraded 1000 mg L^{-1} phenol after 162 h (Kumar & Kumar, 2005).

Despite the significant amount of information gathered during the last years, phenol biodegradation is still an important challenge, which stimulates researchers to search new and more effective microbial species. In Table 1 a compilation of the most recent *in vitro* studies related to bacterial phenol degradation is presented. It is noteworthy, from this Table, that the focus of these works was the isolation and characterization of native phenol degrading bacterial strains from highly polluted sites and the optimization of culture conditions to maximize the biodegradation process (pH, temperature, initial phenol concentrations, agitation and presence of different nutrients, among others). For instance, Zhang et al. (2013) isolated and identified a *B. cereus* strain from a phenol-contaminated wastewater that exhibited optimum degradation performance at pH 7.0 and 30 °C. Moreover, the supplement ation with low glucose concentration (100 mg L^{-1}) was helpful for accelerating the biodegradation. In addition, Arif et al. (2013) isolated and identified a *Rhodococcus* sp.strain AQ5NOL 2 which was capable not only to degradate high phenol concentrations but also to grow on diesel and several other aromatic compounds. This strain exhibited a broad optimum temperature to grow on phenol however it was susceptible to several heavy metals, which could suppress phenol biodegradation. Thus, the isolation of phenol-degrading microbes should take into account heavy metal´s tolerance, as phenolic wastes often contain metallic pollutants (Arif et al., 2013 and references therein). On the other hand, several studies have contributed to elucidate possible degradation pathways and key enzymes involved in the process, like phenol hydroxilase and catechol dioxygenases (Hupert-Kocurek, Guzik & Wojcieszyńska, 2012; Arif et al., 2013). Nevertheless, there are other enzymes that could be implicated in phenol degradation such as laccases, tyrosinases and peroxidases, which have been extensively studied in plants and fungi and less described in prokariotic cells. In this context, *Pseudomonas putida* F6 has shown to possess a tyrosinase enzyme capable of transforming phenol to catechol. Furthermore, the coexistence of this tyrosinase with a laccase has been demonstrated in this strain and also in the marine bacterium *Marinomonas mediterranea* MMB-1, *Streptomyces* spp. and *Sinorhizobium meliloti* (McMahona et al., 2007 and references therein).

It is important to note that frequently 1000 mg L^{-1} appears to be the maximum concentration of phenol used in the *in vitro* experiments. At higher phenol concentration microorganisms could suffer from substrate inhibition, by which the growth is inhibited and hence phenol is not degraded (Prieto et al., 2002). For this reason, new developments to improve biodegradation are being performed including the use of bacterial consortia, immobilized cells on different natural or synthetic materials and co-metabolism of phenol with another substrate, among others.

Bacterial Immobilization As a Useful Strategy for an Efficient Phenol Biodegradation

Immobilization of microorganisms on various supports has gained much interest in recent years due to its potential for biodegradation efficiency improvement (Banerjee & Ghoshal, 2011). This technology gives some advantages to biodegradation process over free-cells such as: greater active biomass availability, greater resistance to high toxic compound concentrations, increase of catalytic activity and the alternative of reusing the immobilized cells for consecutive biodegradation cycles (Ahamad & Kunhi, 2011; Robatjazi et al., 2013). In particular, for phenol biodegradation, the immobilization of bacterial biomass is an effective technique to protect bacteria from high concentrations of this compound. With this purpose, cells immobilized in different support matrices such as polyacrylamide, polyvinyl alcohol (PVA), agar, agarose, polysulphone, polyacrylnitrile, calcium alginate (Ca-alginate) and others have been used with different results (Junter & Jouenne, 2004).

There are many evidences about the reduction of the toxicity of high phenol concentrations (considered as bactericides for free cells) and hence a higher degradative ability when bacteria are immobilized on Ca-alginate (Santos et al., 2003; Gouda, 2007; Wang et al., 2012). This increase in the efficiency was explained not only by a higher reaction ability due to the high density of immobilized cells, but also by a protective effect of the polymer giving a more favorable microenvironment for the catalytic reaction (Tallur, Megadi & Ninnekar, 2009). Two Ca-alginate immobilized *Bacillus cereus* strains showed higher phenol degradation efficiency than free cells at high phenol concentrations (1500-2000 mg L^{-1}), indicating the improved tolerance of the immobilized cells toward phenol toxicity. Besides, these immobilized cells were able to degrade more than 50% of 2000 mg L^{-1} phenol within 26 and 36 d (Banerjee & Ghoshal, 2011). However, there are some reports showing better rates and efficiencies for phenol degradation using Ca-alginate immobilized cells. For example, Ali, Namane & Hellal (2013) found that *P. aeruginosa* cells immobilized in Ca-alginate were able to completely degrade 900 and 1200 mg L^{-1} of phenol after 80 and 290 h, respectively. In Table 2 different reports about the use of Ca-alginate immobilized cells for phenol removal are shown, however, the comparison is difficult because the bacterial density in beads is determinant for the efficiencies found and this data is often absent or missed in the description.

Other essential factors for practical application of Ca-alginate immobilized cells are the optimization of the polymer percentage in the beads, reusability and stability during long-term storage. The increase of Ca-alginate percentage (2-5 %) in the beads affects negatively the degradation rate of contaminants, which could be explained by a lower diffusion of oxygen, nutrients and even the substrate into the beads, as it was demonstrated by Sreenivasulu et al. (2012) and Konsoula & Liakopoulou-Kyriakides (2006). Many authors evaluate the stability of the bacterial immobilized cells after long periods of time, varying from months to years, in order to propose the strategy as a cost effective and economically viable process for bioremediation.

Different polymer types and combinations for cell immobilization were used for phenol removal assays and the results were compared. Mollaci et al. (2010) found that *Pseudomonas* cells immobilized on either Ca-alginate beads, combined beads of PVA-alginate or alginate-chitosan-alginate (ACA) capsules resulted in remarkable reduction (65%) in phenol degradation time compared to free cells. Among them, cells immobilized on PVA-alginate and ACA provided the best performance for high phenol concentrations, producing complete degradation of 2000 mg L^{-1} phenol after 100 and 110 h, respectively (Mollaei et al. 2010). Furthermore, the effect of activated carbon (AC) and clay addition to alginate beads was studied and the performance for phenol biodegradation of the immobilized cells was compared (Massalha, Shaviv & Sabbah, 2010). These authors concluded that this combination contributed to a better mineralization of high phenol concentrations, although they found a notorious limitation of diffusion.

Table 2. Examples of immobilized bacterial cells in different support matrices used for biodegradation of phenols

Immobilized bacterial species	Immobilization matrix	Phenol concentration (mg L^{-1})	Time required for degradation	References
Bacillus cereus strains AKG1 and AKG2	Ca-alginate	2000	26-36 days (50%)	Banerjee & Ghoshal, 2011
Pseudomonas aeruginosa	Ca-alginate	900 1200	80 h (100%) 290 h (100%)	Ali, Namane & Hellal, 2013
Acinetobacter and Sphingomonas	PVA	800	35 h (100%)	Liu, Zhang & Wang, 2009
Acinetobacter sp. strain PD12	PVA	500	9 h (100%)	Wang et al., 2007
Pseudomonas sp. SA01	Alginate Pectin PVA-alginate Glycerol-alginate ACA	2000	PVA-alginate 100 h (100%) ACA 110 h (100%)	Mollaei et al., 2010
Enriched microorganisms from a compost of agricultural wastes	AC-alginate AC-clay-alginate	1630	38 h (90%) 44 h (100%)	Massalha, Shaviv & Sabbah, 2010

Different studies have used PVA for cell immobilization. In this sense, mixed cells of *Acinetobacter* and *Sphingomonas* immobilized in PVA exhibited higher degradation rate than free cells for phenol concentrations from 500 to 1000 mg L^{-1}, even in a broader range of pH and temperatures (Liu, Zhang & Wang, 2009). However, there is a main drawback with the entrapment immobilization technique using PVA because it includes a step during which bacteria are exposed to sub-freezing temperatures during the cross linking stage. In this sense, although some bacterial strains like *P. putida* are not affected by this step, Al-Zuhair & El-Naas (2011) modified some steps of the process avoiding chilling temperatures aimed to applicate PVA-immobilization to other bacterial genera. Actually, there is an increasing search for low-cost, more efficient and easier to handle support matrices (Li et al., 2008). Thus, different residual products have been used to immobilize cells, such as tubings, packed columns, granulated activated carbon, coconut shells, etc. (Carvalho et al., 2001; Chang et al., 2003; Amuda & Ibrahim, 2006; Eker & Kargi, 2008). In this context, Robledo Ortíz et al. (2010) proposed and thoroughly studied the adhesion property of bacteria to fiber-polymer foamed composites produced from waste materials like recycled polyethylene and agave fiber from *Agave tequilana*, a plant used for tequila production. For phenol biodegradation, *P.*

putida cells have been immobilized in activated pumice particles (Pazarlioglu & Telefoncu, 2005). Del Castillo et al. (2012) found that several (chloro) phenol degrading bacteria were able to colonize the surface of cork particles in the presence of phenol the authors proposed that bacterial immobilization on these particles would be a valuable tool to perform the self-remediation of wastewaters derived of cork processing.

Other cost effective and environmental friendly matrices have been proved for bacterial immobilization such as linen shells, an abundantly available agro-waste, or a natural fibrous material obtained from the mature dried fruits of *Luffa cylindrica* L., which could be ideal alternatives to the expensive solid matrices for the development of bioremediation processes and other biotechnological applications (Krishnani et al., 2006; Behera, Mohanty & Ray, 2012). Possibly, in the following years, new alternative matrices derived from residual materials would become useful for immobilization process, contributing with the disposal of great volumes of wastes.

Another important aspect that is gaining attention within the scientific community for an efficient bioremediation process is the design and improvement of different types of bioreactors with free and immobilized cells, as it will be described in the following section.

Use of Bioreactors for Phenol Biodegradation

Current technology for biodegradation of toxic compounds, including phenols, involves the use of bioreactors, in batch and continous processes, using either free or immobilized cells. Bioreactors are generally defined as devices in which biological and/or biochemical processes are developed under closely monitored and tightly controlled environmental and operating conditions (e.g. pH, temperature, pressure, nutrient supply and agitation). The high degree of reproducibility, low contamination risk, control and automatization introduced by bioreactors for specific experimental bioprocesses has been key for their transfer to large-scale applications, including its use for wastewater treatments (Martin, Wendt & Heberer, 2004). In the case of wastewater containing phenol, its treatment has been focused on employing and exploring new types of bioreactors with high performance for practical utilization (El-Naas, Al-Zuhair & Makhlouf, 2010b). These include the use of hollow fiber membrane contactors, fluidized bed bioreactor, microbial fuel cells system and fixed-biofilm process. Several novel bioreactors have been developed for biotreatment applications including rotating rope bioreactor, two phase partitioning bioreactors (TPPBs) and foam emulsion bioreactor (Li & Wang, 2008; El-Naas, Al-Zuhair & Makhlouf, 2010b). It is important to note that although biotreatment schemes using bioreactors are based on aerobic, anaerobic and combined aerobic-anaerobic processes (Field & Sierra-Alvarez, 2008), the focus in this section will be put on the aerobic ones.

It is noteworthy that the overall efficiency of the bacterial biomass in the biodegradation of phenol may be affected by many factors such as phenol concentration, temperature, the presence of other nutrients or other pollutants and bacterial density (ElNaas, Al-Muhtaseb & Makhlouf, 2010a and references therein). In the case of *P. putida* it was demonstrated that biodegradation capabilities were highly affected by temperature, pH, initial phenol concentration and the biomass density in a bubble column bioreactor El-Naas, Al-Muhtaseb & Makhlouf (2009). Furthermore, optimization of aeration is another essential aspect in order to maximize phenol removal efficiency in a bioreactor (Safont, Vitas & Peñas, 2012). Paisio et al. (2012) also evaluated the effect of aeration and agitation on growth and phenol removal efficiency of *Rhodococcus* sp. CS1 in a stirred tank bioreactor. In this study, when aeration of 1 vvm and 3 vvm were used, complete phenol removal was observed without significant differences between two agitation rates used (200 and 400 rpm). In contrast, at 600 rpm of agitation and 3 vvm of aeration, only 38% of phenol removal and a lower biomass were observed, compared with aeration of 1 vvm. This result could indicate that high aeration and agitation produced stress in bacteria and probably may result in cell rupture, which inhibited their metabolism and growth and consequently decreased phenol removal efficiency.

Due to the advantages of immobilized bacterial cells, they have been used in combination with bioreactors. In this sense, Aneez Ahamad & Mohammad Kunhi (2011) compared the performance of free and immobilized cells of *Pseudomonas* CP4 in a fluidized bed bioreactor. In batch runs, with an aeration rate of 1 vvm, at 30°C and pH 7.0 agar immobilized cells degraded up to 3000 mg L^{-1} of phenol as compared to 1500 mg L^{-1} by Ca-alginate immobilized cells whereas free cells could degrade only 1000 mg L^{-1}. In a continuous process with Ca-alginate immobilized cells a degradation rate of 200 mg L^{-1} h^{-1} was obtained while agar-entrapped cells were better since they could withstand and degrade up to 4000 mg L^{-1} phenol with a maximum degradation rate of 400 mg L^{-1} h^{-1} (Aneez Ahamad & Mohammad Kunhi, 2011). In another study, Al-Zuhair & El-Nass (2011), using two types of bioreactors namely bubble column and spouted bed bioreactor, demonstrated that *P. putida* immobilized on PVA particles could remove high phenol concentrations and that bacteria remained active for a period of 72 h, even without the addition of nutrients.

Another alternative to enhance the biodegradation of the pollutant in bioreactors can be called a cooperative phenol removal system between biological and physical treatment. In this process, adsorption plays a key role in decreasing phenol concentration during adaptation period at which microorganisms cannot degrade phenol, avoiding substrate inhibition. After the adaptation period, the microorganism starts to degrade the remaining phenol concentration (Kwon & Yeom, 2009). For example, Li & Wang (2008) enhanced phenol biodegradation using a hollow-fiber membrane, by the addition of granular activated carbon (GAC) (hybrid bioreactor). In batch biotransformation experiments, complete removal of 1000 mg L^{-1} phenol (concentration at which free cells cannot grow) was accomplished within 18 h in the hybrid bioreactor, comparing with 23 h in the GAC free bioreactor. At continuous running, the GAC bioreactor showed its superiority over the GAC free bioreactor during start-up and elevated loading phase. More than 90% of phenol was transformed in the GAC bioreactor when the phenol loading was < 24 mg h^{-1}. The authors concluded that the better bioreactor performance may be due to the enhanced mass transportation and adsorption capacity with the incorporation of GAC. Kwon, Jung & Yeom (2009) found that a porous polymer bead of PVA and Xanthan gum was the best entrapment for phenol degradation by *P. fluorescens* KNU417. Besides, when AC (1%) was co-immobilized with microorganisms in the bead, the start-up period was shortened by 40 h and the removal efficiency of phenol during the period was increased by 28% than beads containing only microorganisms. On the other hand, Zhao et al. (2009) used organic modified montmorillonite (OMMT) as delivery agents for phenol to microorganisms in TPPBs (OMMT-PSF) and polyurethane foam immobilized microorganisms (PUF-immobilized microorganisms) as biocatalysts. Phenol biodegradation rates of batch fermentation were examined, concluding that the maximum volumetric consumption rate of phenol decreased in the order: immobilized microorganisms with OMMT-PSF capsules (342.4 mg L^{-1} h^{-1}) > immobilized microorganisms without OMMT-PSF capsules (300 mg L^{-1} h^{-1}) > free microorganisms with OMMT-PSF capsules (208.4 mg L^{-1} h^{-1}) > free microorganisms without OMMT-PSF capsules (125.8 mg L^{-1} h^{-1}). These results demonstrated that the use of immobilized microorganisms and OMMT-PSF capsules in TPPB offers improved degradation of phenol.

Finally, the choice of the most suitable bioreactor will be determined by many factors including the scale, the method of aeration, the mixing mechanism and the resistance of the cells to the shear stress generated in the bioreactor. At last, all of these factors will determine the biodegradation capabilities. In addition, bioreactor-based technology allows the management of high volumes and degradation of toxic compounds commonly found in industrial effluents.

Biodegradation of Phenolic Compounds from Industrial Wastewaters

Industrial effluents can contain different organic and/or inorganic contaminants, therefore their bioremediation represent an important challenge. As it was previously mentioned, biological treatments using both anaerobic and aerobic systems have been recognized as effective methods for the degradation of highly polluted industrial wastewaters, including those containing different phenolic compounds (McNamara et al., 2008; Maszenan, Liu & Ng, 2011). Biotreatments can be performed at different scales such as Erlenmeyers flasks, bioreactors of variable volume as well as pilot-scale experiments. Most of the studies in this field employ chemical or biological oxygen demand (COD/BOD) determinations as indicators of bioremediation. In addition, the removal of phenolic compounds is also evaluated in some effluents such as those derived from olive oil extraction, paper factories, tanneries, crude oil refineries, palm oil and phenolic resin producing industries and coke and coal gasification processes. In Table 3 some recent research in this area are shown.

Table 3. A brief summary of recent research studies on bacterial wastewater bioremediation

Wastewater	Phenols concentration (mg L^{-1})	Microorganisms	Reference
Olive mill extraction	Phenol (1200)	Mixed culture	Omer, 2012
	Phenols (4300)	Actived sludge	Gonçalves et al., 2012
Paper factory	Phenols (364)	Mixed culture	Chandra & Singh, 2012
	Chlorophenol (82)/PCP (100)	*Pseudomonas stutzeri* CL7	Karn, Chakrabarty & Reddy, 2010
Tannery	Phenols (17.5)	*Rhodococcus* sp. CS1	Paisio et al., 2012
Petroleum refinery	Phenol (637)	*Pseudomonas* sp. PD39	Ren et al., 2008
Phenolic resin manufacture	Phenol (250-4000)	Nitrifying and heterotrophic bacteria	Eiroa et al., 2008
POME	Phenols (33.6)	*Lactobacillus plantarum* SF5	Limkhuansuwan & Chaiprasert, 2010
	Phenol (100-1000)	*Thermoanaerobacterium*-rich sludge	Mamimin et al., 2012
Coal gasification	Phenols (342-487)	Actived sludge	Li et al., 2011

Some authors have focused on olive mill wastewater biotreatment. This is important considering that olive oil extraction is one of the activities that generate great quantities of highly toxic phenolic wastes (DellaGreca et al., 2001). Omer (2012) indicated that the phenol content (1200 mg L^{-1}) of olive mill wastewater gradually decreased reaching the maximum removal (71.9 and 71.4%) after 25 days of treatment using a bacterial mixture constituted by *Azotobacter vinelandii*, *P. putida* and *P. fluorescens*, grown at 50 and 30% of effluent dilution, respectively. In addition, Gonçalves et al. (2012) also observed remarkable phenols removal efficiencies (up 60%) in up-flow anaerobic reactors, inoculated with active sludge, which were intermittently feeded with raw olive mill wastewater containing 4300 mg $L^{-1}$of phenols. Khoufi, Aloui & Sayadi (2009) described a novel process developed at pilot scale for the treatment of olive mill wastewater, which combines electro-Fenton, anaerobic digestion and ultrafiltration. Application of these procedures allowed to obtain high removal efficiencies of phenolic compounds (95% of initial concentration of 1150 mg L^{-1}). This work remarks the need of integrating physico-chemical treatments and biological processes for the degradation of recalcitrant compounds, as it was previously indicated by other researchers (Oller, Malato & Sánchez-Pérez, 2011).

The efficiency of bacteria for phenolic compounds removal from paper factory effluents has been extensively studied. Recently, Chandra & Singh (2012) demonstrated that a mixed culture constituted by *Pseudochrobactrum glaciale* IITRP1, *Providencia rettgeri* IITRP2 and *Pantoea* sp RCT2 degraded 61% and 90% of total phenols and chlorophenol from pulp paper mill effluent containing initial concentrations of 364 and 82 mg L^{-1}, respectively, within 216 h. Other authors have also evaluated PCP degradation in these effluents. Regarding this, Singh et al. (2008) studied a mixed culture of two bacterial strains, *Bacillus* sp. and *Serratia marcescens,* which was able to degrade up to 94% of the PCP present in a pulp paper mill effluent at initial concentration of 1440 mg L^{-1} after 168 h. Moreover, Karn, Chakrabarty & Reddy (2010) demonstrated that *P. stutzeri* strain CL7 was able to remove 66.8% of PCP from the secondary sludge of pulp and paper mill supplemented with 100 mg L^{-1} of PCP after two weeks. Nair, Jayachandran & Shashidhar (2007) used free and immobilized cells of *Alcaligenes* sp. d(2) in a packed bed reactor to treat a paper factory effluent containing 9.41 mg L^{-1} of phenol. Both cells removed a maximum of 99% of phenol after 20 h of treatment using a batch process. In the continuous mode of operation the strain was able to achieve 99% phenol removal. A combined chemical/biological process for the treatment of paper mill effluent (activated sludge-ozonation processes) was described by Assalin, Dos Santos Almeida & Durán (2009). The combined activated sludge-O_3/pH 10 treatment was able to remove around 70% of total phenols (initial concentration of 10-30 mg L^{-1}).

Tannery wastewater usually contains high concentrations of organic matter, including phenols and other chemicals. These effluents cannot be released into the environment without pre-treatment because of their high toxicity (Bajza & Vrcek, 2001). Thus, in recent years the efficiency of several bacterial strains for treating tannery wastewater has been studied. For instance, Paisio et al. (2012) observed that *Rhodococcus* sp. CS1 strain was able to grow and completely degrade phenols from tannery effluents containing 17.5 mg L^{-1} of total phenols, after 9 h of incubation. Srivastava, Ahmad & Thakur (2007) studied PCP biodegradation in a sequential bioreactor and determined that tannery effluents treated initially by bacterial consortium followed by the fungus *Aspergillus niger* FK1 removed 67% of PCP, whereas in another set of bioreactor in which effluents were treated initially by fungi followed by bacteria could remove 58% of PCP. Chandra et al. (2011) evaluated the tannery wastewater bioremediation in aeration lagoons at a common effluent treatment plant from Uttar Pradesh, India. Most of the organic pollutants detected in the tannery wastewaters were diminished and 77% reduction of phenolics was obtained by bacterial treatment.

Although reports on petroleum refinery effluents biotreatment are relatively abundant in the literature, (Diya'uddeen, Wan Daud & Abdul Aziz, 2011; Vennila & Kannan, 2011) only few of them describe the biodegradation of phenols. This is an important feature to take into account because high phenolics contents have been found in these wastewaters (up to 2000 mg L^{-1}) Ojumu et al. (2005) studied the ability of *P. aeruginosa* and *P. fluorescens* strains for phenol biodegradation from refinery effluent, in a batch reactor. Phenol (30 mg L^{-1}) was degraded completely by both bacterial species after 60 and 84 h, respectively. Subsequently, Bako et al. (2008) studied mixed cultures of *P. aeruginosa* and *Penicillium janthinellum*, they removed 50-100% phenol (1.48-3.10 mg L^{-1}) from effluents of a refinery of crude oil after two weeks of incubation. Better results were achieved by Ren et al. (2008) using *Pseudomonas* sp. PD39, that was capable to remove 637 mg L^{-1} of phenol from wastewaters of a petroleum chemical plant only in 72 h.

Similarly to the petroleum refinery effluents, few research studies showing the bacterial phenolics biodegradation from wastewaters derived from phenolic resin industries have been published. Among them, Eiroa et al. (2008) showed that phenol (250-4000 mg L^{-1}) from these wastewater was completely removed at all concentrations combining nitrifying bacteria, contained in an anoxic reactor, with heterotrophic bacteria of an aerobic reactor. These results are promising taking into account the high phenol concentrations detected and removed from these wastewaters.

Another understudied area is the biotreatment of phenols contained in palm oil mill effluent (POME). POME is a highly polluted wastewater exhibing high COD, BOD and phenol concentration. Limkhuansuwan & Chaiprasert (2010) demonstrated that *Lactobacillus plantarum* SF5 removed 34% of phenolic compounds (33.6 mg L^{-1}) contained in POME. More recently, Mamimin et al. (2012), used a *Thermoanaerobacterium*-rich sludge for hydrogen production and phenol removal from POME in the presence of phenol concentrations from 100 to 1000 mg L^{-1}, obtaining high removal efficiencies (up to 92%).

Conventional biological processes do not always provide satisfactory results for industrial wastewater treatment, since many of the organic substances produced by chemical industries are toxic or resistant to biological treatment (Oller, Malato & Sánchez-Pérez, 2011). In this context, some inhibitory effects of coke and coal gasification wastewaters on different microorganisms were observed. For example, Cordova-Rosa et al. (2009) observed that an indigenous bacterial consortium as well as a pure culture of *Acinetobacter calcoaceticus var. anitratus* growing in coke gasification wastewater, were inhibited and no phenol biodegradation was observed after 10 days of incubation. Similarly, direct biological treatment of coke factory wastewaters by a consortium of *Chlorella vulgaris* strain and *Alcaligenes* sp. was not possible due to the toxicity of organic compounds. However, complete phenol degradation was achieved when the effluent was pre-treated with activated carbon adsorption and UV(A-B)-irradiation, demonstrating the importance of combined physicochemical and biological tools (Essam et al., 2006). On the other hand, Li et al. (2011) also obtained promising results using a laboratory-scale moving bed biofilm reactor (MBBR) to study the biodegradation of coal gasification wastewater by an acclimated activated sludge. Maximum removal efficiencies of 89% were obtained for phenols after 12 d.

With the advent of new manufacturing products, the releasing of different industrial effluents will continuously increase, making necessary to develop suitable and efficient strategies for waste management. In this context, bioremediation could play a key role to deal with these hazardous wastewaters, leading to enhanced public acceptance and compliance with environmental legislation.

CONCLUSION

As it was highlighted in this chapter, the application of bacteria for phenol bioremediation is a promising technology that is being continuously improved. Several bacterial genera have demonstrated to be efficient for phenol removal from wastewaters. Moreover, mixed bacterial cultures sometimes offer more advantages because they can better withstand different environmental conditions. Although an extensive knowledge is now available in this regard, the appropriate and continuous selection of more efficient microorganisms is still an area of great interest.

In order to implement an efficient bioremediation technology, a scientific and well formulated strategy must be carried out taking into consideration several aspects that influence the process. In this sense, immobilization is an effective methodology to protect bacteria from high concentrations of toxic compounds, and also allow reusing and stability during long-term storage. These properties are essential factors for practical application of immobilized cells for bioremediation. The search of new and low-cost matrices would be useful to make this process more economical- and environmentally viable. Moreover, combination of immobilization with the design and use of proper bioreactors would confer several advantages to optimize and improve phenol removal of great volumes of contaminated solutions and/or effluents. Nevertheless, additional basic research, such as toxicity studies of the formed metabolites is of concern and need further investigation to apply this technology in a safe way. In addition, the use of the currently emerging -omic approaches (genomics, metagenomics, proteomics, metabolomics, etc) applied in a novel and imaginative way could allow to understand interesting aspects of the complex biodegradation pathways and, therefore, they can be used to enhance phenol remediation applicability.

It is expected that a combined approach, integrating recent findings and using different strategies simultaneously, could be used to successfully improve the efficiency of phenol bioremediation, allowing the application of this technology in a large scale which is one of the most important challenges from an environmental point of view.

REFERENCES

Agarry, S. E., Salomon, B. O., 2008. Kinetics of batch microbial degradation of phenols by indigenous *Pseudomonas fluorescens*. *Int. J. Environ. Sci. Technol.* 5, 223-232.

Agarry, S. E., Solomon, B. O., Layokun, S. K., 2008. Kinetics of batch microbial degradation of phenols by indigenous binary mixed culture of *Pseudomonas aeruginosa* and *Pseudomonas fluorescence*. *African J. Biotechnol.* 7, 2417-2423.

Aksu, Z., 2005. Application of biosorption for the removal of organic pollutants: a review. *Process Biochem.* 40, 997-1026.

Ali, O., Namane, A., Hellal, A., 2013. Use and recycling of Ca-alginate biocatalyst for removal of phenol from wastewater. *J. Indust. Eng. Chem.* DOI 10.1016/j.jiec.2012.12.045 (*in press*).

Al-Zuhair S., El-Naas M. H., 2011. Immobilization of *Pseudomonas putida* in PVA gel particles for the biodegradation of phenol at high concentrations. *Biochem. Eng. J.* 56, 46-50.

Al-Zuhair, S., El-Naas, M. H., 2012. Phenol biodegradation by *Ralstonia pickettii* extracted from petroleum refinery oil sludge. *Chem. Eng. Commun.* 199(9), 1194-1204.

Amuda, O. S., Ibrahim, A. O., 2006. Industrial wastewater treatment using natural material as adsorbent. *African J. Biotechnol.* 5(16), 1483-1487.

Aneez Ahamad, P. Y., Mohammad Kunhi, A.A., 2011. Enhanced degradation of phenol by *Pseudomonas* sp. CP4 entrapped in agar and calcium alginate beads in batch and continuous processes. *Biodegradation* 22, 253-265.

Arif, N. M., Ahmad, S. A., Syed, M. A., Shukor, M. Y., 2013. Isolation and characterization of a phenol-degrading *Rhodococcus* sp. strain AQ5NOL 2 KCTC 11961BP. *J. Basic Microbiol.* 53(1), 9-19.

Assalin, M. R., Dos Santos Almeida, E., Durán, N., 2009. Combined system of activated sludge and ozonation for the treatment of kraft E1 effluent. *Int. J. Environ. Res. Public Health* 6(3), 1145-1154.

Bajza, Z., Vrcek, I. V., 2001. Water quality analysis of mixtures obtained from tannery waste effluents. *Ecotoxicol. Environ. Saf.* 50(1), 15-18.

Bako, S. P., Chukwunonso, D., Adamu, A. K., 2008. Bio-remediation of refinery effluents by strains of *Pseudomonas aeruginosa* and *Penicillium janthinellum*. *Appl. Ecol. Environ. Res.* 6(3), 49-60.

Banerjee, A., Ghoshal, A. K., 2011. Phenol degradation performance by isolated *Bacillus cereus* immobilized in alginate. *Int. Biodet. Biodeg.* 65, 1052-1060.

Basha, M. K., Rajendran, A., Thangavelu, V., 2010. Recent advances in the biodegradation of phenol: A review. *Asian J. Exp. Biol. Sci.* 1(2), 219-234.

Basile, L. A., Erijman, L., 2008. Quantitative assessment of phenol hydroxylase diversity in bioreactors using functional gene analysis. *Appl. Microbiol. Biotechnol.* 78, 863-872.

Basu, S. K., Oleszkiewicz, J. A., Sparling, R., 1996. Dehalogenation of 2-chlorophenol (2-CP) in anaerobic batch cultures. *Water Res.* 30, 315-322.

Behera, S., Mohanty, R. C., Ray, R. C., 2012. Ethanol fermentation of sugarcane molasses by *Zymomonas mobilis* MTCC 92 immobilized in *Luffa cylindrical* L. sponge discs and Ca-alginate matrices. *Braz. J. Microbiol.* 43(4), 1499-1507.

Bernadini, G., Spinelli, O., Presutti, C., Vismara, C., Bolzacchini, E., Orlando, M., Settimi, R., 1996. Evaluation of the developmental toxicity of pesticide MCPA and its contaminants phenol and chlorocresol. *Environ. Toxicol. Chem.* 15, 754-760.

Bukowska, B., Kowalska, S., 2004. Phenol and catechol induce prehemolytic and haemolytic changes in human erythrocytes. *Toxicol. Lett.* 152, 73-84.

Canadian Council of Ministers of the Environment. 1999. Canadian water quality guidelines for the protection of aquatic life: Phenols-Mono- and dihydric phenols. In: Canadian Environmental Quality Guidelines, 1999, Canadian Council of Ministers of the Environment, Winnipeg. ISBN 1-896997-34-1.

Carvalho, M. F., Vasconcelos, I., Bull, A. T., Castro, P. M. L., 2001. A GAC biofilm reactor for the continuous degradation of 4-chlorophenol: treatment efficiency and microbial analysis. *Appl. Microbiol. Biotechnol.* 57, 419-426.

Chandra, R., Bharagava, R. N., Kapley, A., Purohit, H. J., 2011. Bacterial diversity, organic pollutants and their metabolites in two aeration lagoons of common effluent treatment plant (CETP) during the degradation and detoxification of tannery wastewater. *Biores. Technol.* 102, 2333-2341.

Chandra, R., Singh, R., 2012. Decolourisation and detoxification of rayon grade pulp paper mill effluent by mixed bacterial culture isolated from pulp paper mill effluent polluted site. *Biochem. Eng. J.* 61, 49-58.

Chang, C. C., Tseng, S. K., Chang, C. C., Ho, C. M., 2003. Reductive dechlorination of 2-chlorophenol in a hydrogenotrophic, gas permeable, silicone membrane bioreactor. *Biores. Technol.* 90, 323-328.

Chen, M., Tang, R., Fu, G., Xu, B., Zhu, P., Qiao, S., Chen, X., Xu, B., Qin, Y., Lu, C., Hang, B., Xia, Y., Wang, X., 2013. Association of exposure to phenols and idiopathic male infertility. *J. Hazard. Mater.* (250-251), 115-121.

Cordova-Rosa, S. M., Dams, R. I., Cordova-Rosa, E. V., Radetski, M. R., Corrêa, A. X. R., Radetski, C. M., 2009. Remediation of phenol-contaminated soil by a bacterial consortium and *Acinetobacter calcoaceticus* isolated from an industrial wastewater treatment plant. *J. Hazard. Mater.* 164(1), 61-66.

Del Castillo, I., Hernández, P., Lafuente, A., Rodríguez-Llorente, I. D., Caviedes, M. A., Pajuelo E., 2012. Self-bioremediation of cork-processing wastewaters by (chloro)phenol-degrading bacteria immobilised onto residual cork particles. *Water Res.* 46, 1723-1734.

DellaGreca, M., Monaco, P., Pinto G., Pollio A., Previtera L., Temussi F., 2001. Phytotoxicity of low-molecular-weight phenols from olive mill waste waters. *Bull. Environ. Contam. Toxicol.* 67(3), 352-359.

Diya'uddeen, B. H., Wan Daud, W. M. A., Abdul Aziz, A. R., 2011. Treatment technologies for petroleum refinery effluents: A review. *Process Saf. Environ. Protec.* 89(2), 95-105.

Dong, X., Hong, Q., He, L., Jiang X., Li, S., 2008. Characterization of phenol-degrading bacterial strains isolated from natural soil. *Int. Biodet. Biodeg.* 62(3), 257-262.

Eiroa, M., Vilar, A., Kennes, C., Veiga, M. C., 2008. Effect of phenol on the biological treatment of wastewaters from a resin producing industry. *Biores. Technol.* 99(9), 3507-3512.

Eker, S., Kargi, F., 2008. COD, 2,4,6-trichlorophenol (TCP) and toxicity removal from synthetic wastewater in rotating perforated-tubes biofilm reactor. *J. Hazard. Mater.* 159(2-3), 306-312.

El Azhari, N., Devers-Lamrani, M.C., Guillaume, R., Nadine, M., Laurent, F., 2010. Molecular analysis of the catechol-degrading bacterial community in a coal wasteland heavily contaminated with PAHs. *J. Hazard. Mater.* 177(1-3), 593-601.

El-Naas, M. H., Al-Muhtaseb, S. A., Makhlouf, S., 2010a. Batch degradation of phenol in a spouted bed bioreactor system. *J. Indust. Eng. Chem.* 16, 267-272.

El-Naas, M. H., Al-Zuhair, S., Makhlouf, S., 2010b. Continuous biodegradation of phenol in a spouted bed bioreactor (SBBR). *Chem. Eng. J.* 160, 565-570.

El-Naas, M. H., Al-Muhtaseb, S. A., Makhlouf, S., 2009. Biodegradation of phenol by *Pseudomonas putida* immobilized in polyvinyl alcohol (PVA) gel. *J. Hazard. Mater.* 164, 720-725.

El-Sayed, W., Ibrahim, M., Abu-Shady, M., El-Beih, F., Ohmura, N., Saiki, H., Ando, A., 2003. Isolation and characterization of phenol-degrading bacteria from a coking plant. *Biosci. Biotechnol. Biochem.* 67(9), 2026-2029.

Essam, T., Amin, M. A., El Tayeb, O., Mattiasson, B., Guieysse, B., 2006. Biological treatment of industrial wastes in a photobioreactor. *Water Sci. Technol.* 53(11), 117-125.

Farrell, A., Quilty, B., 2002. The enhancement of 2-chlorophenol degradation by a mixed microbial community when augmented with *Pseudomonas putida* CP1. *Water Res.* 36, 2443-2450.

Fetzner, S., 2012. Biotechnology. Vol X. Biodegradation of xenobiotics. Encyclopedia of life support systems (EOLLS). UNESCO. http://www.eolss.net/Sample-Chapters/C17/E6-58-09-08.pdf.

Field, J. A., Sierra-Alvarez, R., 2008. Microbial degradation of chlorinated phenols. *Rev. Environ. Sci. Biotechnol.* 7, 211-241.

Flocco, C. G., Lo Balbo, A., Carranza, M. P., Giulietti, A. M., 2002. Removal of phenol by alfalfa plants (*Medicago sativa* L.) grown in hydroponics and its effect on some physiological parameters. *Acta Biotechnol.* 22, 43-54.

Gayathri, V. K., Vasudevan, N., 2010. Enrichement of phenol degrading moderately halophilic bacterial consortium from saline. *J. Biorem. Biodeg.* 1(1), 1-6.

Gonçalves, M. R., Costa, J. C., Marques, I. P., Alves, M. M., 2012. Strategies for lipids and phenolics degradation in the anaerobic treatment of olive mill wastewater. *Water Res.* 46(6), 1684-1692.

Gouda, M. K., 2007. Immobilization of *Rhodococcus* sp. DG for efficient degradation of phenol. *Fresen. Environ. Bull.* 16(12b), 1655-1661.

Guzik, U., Greń, I., Hupert-Kocurek, K., Wojcieszyńka, D., 2011. Catechol 1,2-dioxygenase from the new aromatic compounds-degrading *Pseudomonas putida* strain N6. *Int. Biodet. Biodeg.* 65, 504-512.

Harrison, M. A. J., Barra, S., Borghesi, D., Vione, D., Arsene, C., Olariu, R.I., 2005. Nitrated phenols in the atmosphere: a review. *Atmospheric Environ.* 39, 231-248.

Harwood, C. S., Parales, R. E., 1996. The β-ketoadipate pathway and the biology of self-identity. *Ann. Rev. Microbiol.* 50, 553-590.

Hinteregger, C. R., Leinter, M., Loidl, A., Streichshier, F., 1992. Degradation of phenol and phenolic compounds by *Pseudomonas* EKII. *Appl. Microbiol. Biotechnol.* 37, 252-295.

Hupert-Kocurek, K., Guzik, U., Wojcieszyńska, D., 2012. Characterization of catechol 2,3-dioxygenase from *Planococcus* sp. strain S5 induced by high phenol concentration. *Acta Biochim. Pol.* 59(3), 345-351.

Iurascu, B., Siminiceanu, I., Vione, D., Vicente, M.A., Gil, A., 2009. Phenol degradation in water through a heterogeneous photo-fenton process catalyzed by Fe- treated laponite. *Water Res.* 43, 1313-1322.

Kargi, F., Eker, S., 2004. Kinetics of 2,4-dichlorophenol degradation by *Pseudomonas putida* CP1 in batch culture. *Int. Biodet. Biodeg.* 55, 25-28.

Karn, S. K., Chakrabarty, S. K., Reddy, M. S., 2010. Pentachlorophenol degradation by *Pseudomonas stutzeri* CL7 in the secondary sludge of pulp and paper mill. *J. Environ. Sci.* 22(10), 1608-1612.

Kheleifat K. M., 2007. Biodegradation of phenol by *Actinobacillus* sp. Mathematical interpretation and effect of some growth conditions. *Bioremediation J.* 11,103-112.

Khodadoust, A. P., Wanger, J. A., Suidan, M. T., Brenner, R. C., 1997. Anaerobic treatment of PCP in fluidized-bed GAC bioreactors. *Water Res.* 31, 1776-1786.

Khoufi, S., Aloui, F., Sayadi, S., 2009. Pilot scale hybrid process for olive mill wastewater treatment and reuse. *Chem. Eng. Process.* 48(2), 643-650.

Kim, I. C., Oriel, P. J., 1995. Characterization of the *Bacillus stearothermophilus* BR219 phenol hydroxylase gene. *Appl. Environ. Microbiol.* 61(4), 1252-1256.

Kobayashi, F., Maki, T., Nakamura, Y., 2012. Biodegradation of phenol in seawater using bacteria isolated from the intestinal contents of marine creatures. *Int. Biodet. Biodeg.* 69, 113-118.

Krastanov, A., Alexieva, Z., Yemendzhiev, H., 2013. Microbial degradation of phenol and phenolic derivatives. *Eng. Life Sci.* 13, 76-87.

Krishnani, K. K., Parimala, V., Gupta, B. P., Azad, I. S., Meng, X., Abraham, M., 2006. Bagasse-assisted bioremediation of ammonia from shrimp farm wastewater. *Water Environ. Res.* 78, 938-950.

Kumar, A., Kumar, S., 2005. Biodegradation kinetics of phenol and catechol using *Pseudomonas putida* MTCC 1194. *Biochem. Eng. J.* 22, 151-159.

Kwon, K. H., Jung, K. Y., Yeom, S. Y., 2009. Comparison between entrapment methods for phenol removal and operation of bioreactor packed with co-entrapped activated carbon and *Pseudomonas fluorescence* KNU417. *Bioprocess Biosyst. Eng.* 32, 249-256.

Kwon, K. H., Yeom, S. H., 2009. Optimal microbial adaptation routes for the rapid degradation of high concentration of phenol. *Bioprocess Biosyst. Eng.* 32, 435-442.

Leonard, D., Lindely, N. D., 1998. Growth of *Ralstonia eutropha* on inhibitory concentrations of phenol-diminished growth can be attributed to hydrophobic perturbation of phenol hydroxylase activity. *Enz. Microbiol. Technol.* 25, 271-277.

Li, H., Han, H., Du, M., Wang, W., 2011. Removal of phenols, thiocyanate and ammonium from coal gasification wastewater using moving bed biofilm reactor. *Biores. Technol.* 102(7), 4667-4673.

Li, H., Liu, T., Li, Z., Deng, L., 2008. Low-cost supports used to immobilize fungi and reliable technique for removal hexavalent chromium in wastewater. *Biores. Technol.* 99, 2234-2241.

Li, Y., Wang, C., 2008. Phenol biodegradation in hybrid hollow-fiber membrane bioreactors. *World J. Biotechnol.* 24, 1843-1849.

Limkhuansuwan, V., Chaiprasert, P., 2010. Decolorization of molasses melanoidins and palm oil mill effluent phenolic compounds by fermentative lactic acid bacteria. *J. Environ. Sci.* 22(8), 1209-1217.

Liu, Y. J., Zhang, A. N., Wang, X. C., 2009. Biodegradation of phenol by using free and immobilized cells of *Acinetobacter sp.* XA05 and *Sphingomonas sp.* FG03. *Biochem. Eng. J.* 44(2-3), 187-192.

Mamimin, C., Thongdumyu, P., Hniman, A., Prasertsan, P., Imai, T., O-Thong, S., 2012. Simultaneous thermophilic hydrogen production and phenol removal from palm oil mill effluent by *Thermoanaerobacterium*-rich sludge. *Int. J. Hydrogen Energy* 37(20), 15598-15606.

Martin, I., Wendt, D., Heberer, M., 2004. The role of bioreactors in tissue engineering. *Trends Biotechnol.* 22, 81-86.

Massalha, N., Shaviv, A., Sabbah, I., 2010. Modeling the effect of immobilization of microorganisms on the rate of biodegradation of phenol under inhibitory conditions. *Water Res.* 44, 5252-5259.

Maszenan, A. M., Liu, Y., Ng, W. J., 2011. Bioremediation of wastewaters with recalcitrant organic compounds and metals by aerobic granules. *Biotechnol. Adv.* 29(1), 111-123.

McMahona, A., Doylea, E. M., Broosksa S., O´Connorb K. E., 2007. Biochemical characterisation of the coexisting tyrosinase and laccase in the soil bacterium *Pseudomonas putida* F6. *Enz. Microb. Technol.* 40, 1435-1441.

McNamara, C. J., Anastasiou, C. C., O'Flaherty, V., Mitchell, R., 2008. Bioremediation of olive mill wastewater. *Int. Biodet. Biodeg.* 61, 127-134.

Michałowicz, J. & Duda W., 2007. Phenols – Sources and Toxicity. *Polish J. Environ. Stud.* 16(3), 347-362.Mohite, B. V., Pawar, S. P., Morankar, A., 2011. Isolation, selection and biodegradation profile of phenol degrading bacteria from oil contaminated soil. *Bull. Environ. Contam. Toxicol.* 7(2), 143-146.

Mollaei, M., Abdollahpour, S., Atashgahi, S., Abbasi, H., Masoomi, F., Rad, I., Lotfi, A. S., Zahiri, H. S., Vali, H., Noghabi, K. A., 2010. Enhanced phenol degradation by *Pseudomonas sp.* SA01: gaining insight into the novel single and hybrid immobilizations. *J. Hazard. Mater.* 175(1-3), 284-292.

Nagamani, A., Lowry, M., 2009. Phenol biodegradation by *Rhodococcus coprophilus* isolated from semi arid soil samples of Pali, Rajasthan. *Int. J. Appl. Environ. Sci.* 4(3), 295-302.

Nair, I. C., Jayachandran, K., Shashidhar, S., 2008. Biodegradation of phenol. *African J. Biotechnol.* 7(25), 4951-4958.

Nair, I. C., Jayachandran, K., Shashidhar, S., 2007. Treatment of paper factory effluent using a phenol degrading *Alcaligenes* sp. under free and immobilized conditions. *Biores. Technol.* 98(3), 714-716.

Nair, I. C & Shashidhar S., 2004. Microbial degradation of phenol by species of Alcaligenes isolated from a tropical soil. *Soil Sci.* T.53(4), 47-51.

Nayak, P. S., Singh, B. K., 2007. Removal of phenol from aqueous solutions by sorption on low cost clay. *Desalination* 207(1-3), 71-79.

Nor Suhaila, Y., Ariff, A., Rosfarizan, M., Abdul Latif, I., Ahmad, S. A., Norazah, M. N., Shukor, M. Y. A., 2010. Optimization of parameters for phenol degradation by *Rhodococcus* UKM-P in shake flask culture. *World Congress on Engineering 2010.* 1, 601-604.

Nor Suhaila, Y., Ramanan, R. N., Rosfarizan, M., Abdul Latif, I., Ariff A. B., 2013. Optimization of parameters for improvement of phenol degradation by *Rhodococcus* UKMP-5M using response surface methodology. *Annu. Microbiol. 2012*, 1-9.

Ojumu, T. V., Bello, O. O., Sonibare, J. A., Solomon, B. O., 2005. Evaluation of microbial systems for bioremediation of petroleum refinery effluents in Nigeria. *African J. Biotechnol.* 4(1), 31-35.

Oller, I., Malato, S., Sánchez-Pérez, J. A., 2011. Combination of advanced oxidation processes and biological treatments for wastewater decontamination-A review. *Sci. Total Environ.* 409(20), 4141-4166.

Omer, A. M., 2012. Production of organic biofertilizer from olive mill waste water. *Australian J. Basic Appl. Sci.* 6(10), 654-663.

Paisio, C. E., Agostini, E., González, P. S., Bertuzzi, M. L., 2009. Lethal and teratogenic effects of phenol on *Buffo arenarum* embryos. *J. Hazard. Mater.* 167, 64-68.

Paisio, C. E., Talano, M. A., González P. S., Busto, V. D., Rodríguez Talou, J., Agostini E., 2012. Isolation and characterization of a *Rhodococcus* strain with phenol-degrading ability and its potential use for tannery effluent biotreatment. *Environ. Sci. Pollut. Res.* 19, 3430-3439.

Paisio, C. E., Talano, M. A., González, P. S., Pajuelo-Domínguez, E., Agostini, E., 2013. Characterization of a phenol-degrading bacterium isolated from an industrial effluent and its potential application for bioremediation. *Environ. Technol.* 34(4), 485-493.

Paller, G., Hommel, R. K., Kleber, H. P., 1995. Phenol degradation by *Acinetobacter calcoaceticus* NCIB 8250. *J. Basic Microbiol.* 35, 325-335.

Pazarlioglu, N. K., Telefoncu, A., 2005. Biodegradation of phenol by *Pseudomonas putida* immobilized on activated pumice particles. *J. Process Biochem.* 40, 1807-1814.

Prieto, M., Hidalgo, A., Rodríguez-Fernández C., Serra J., Llama, M., 2002. Biodegradation of phenol in synthetic and industrial wastewater by *Rhodococcus erythropolis* UPV-1 immobilized in an air-stirred reactor with clarifier. *Appl. Microbiol. Biotechnol.* 58(6), 853-860.

Ren, H. S., Wang, Y., Zhao, H. B., Cai, B. L., 2008. Isolation and identification of phenol-degrading strains and the application in biotreatment of phenol-containing wastewater. *Huan Jing Ke Xue* 29(2), 482-487.

Robatjazi, S. M., Shojaosadati, S. A., Khalilzadeh, R., Farahani, E. V., Zeinoddini, M., 2013. Continuous biodegradation of parathion by immobilized *Sphingomonas* sp. in magnetically fixed-bed bioreactors and evaluation of the enzyme stability of immobilized bacteria. *Biotechnol. Lett.* 35(1), 67-73.

Robledo-Ortíz, J. R., Ramírez-Arreola, D. E., Gómez, C., González-Reynoso, O., González-Nuñez, R., 2010. Bacterial immobilization by adhesion onto agave-fiber/polymer foamed composites. *Biores. Technol.* 101, 1293-1299.

Ruíz, N., Ruíz, J., Castañón, J., Hernández, E., Cristiani, E., Galíndez, J., 2001. Phenol biodegradation using a repeated batch culture of *Candida tropicalis* in a multistage bubble column. *Revista Latinoam. Microbiol.* 43, 19-25.

Safont B., Vitas A. I., Peñas F. J., 2012. Isolation and characterization of phenol degrading bacteria immobilized onto cyclodextrin-hydrogel particles within a draft tuve spouted bed bioreactor. *Biochem. Eng. J.* 64, 69-75.

Sandhu, A., Halverson, L. J., Beattie, G. A., 2009. Identifcation and genetic characterization of phenol-degrading bacteria from leaf microbial community. *Microb. Ecol.* 57, 276-285.

Santos, V. L., Heilbuth, N. M., Braga, D. T., Monteiro, A. S., Linardi, V. R., 2003. Phenol degradation by a *Graphium* sp. FIB4 isolated from industrial effluents. *J. Basic Microbiol.* 43(3), 238-248.

Schummer, C., Groff, C., Al Chami, J., Jaber, F., Millet, M., 2009. Analysis of phenols and nitrophenols in rainwater collected simultaneously on an urban and rural site in east of France. *Sci. Total Environ.* 407, 5637-5643.

Sharma, N., Gupta, V. C., 2012. Comparative biodegradation analysis of phenol from paper and pulp industrial effluent by free and immobilized cells of *Aspergillus niger*. Int. Conference on Environmental Science and Technology IPCBEE vol. 30. IACSIT Press, Singapore.

Shingler, V., 1996. Molecular and regulatory checkpoints in phenol degradation by *Pseudomonas* sp. CP600. In: Nakazawa, T., Furukawa, K., Haas, D., and Silver, S., (eds.). Molecular biology of *Pseudomonads*. *Am. Soc. Microbiol.*Washington, D. C. pp. 153-164.

Shumkova, E. S., Solianikova, I. P., Plotnikova, E. G., Golovleva, L. A., 2009. Phenol degradation by *Rhodococcus opacus* strain 1G. *Prikl Biokhim Mikrobiol*. 45(1), 51-57.

Singh, S., Chandra, R., Patel, D. K., Reddy, M. M. K., Rai, V., 2008. Investigation of the biotransformation of pentachlorophenol and pulp paper mill effluent decolorisation by the bacterial strains in a mixed culture. *Biores. Technol.* 99(13), 5703-5709.

Sreenivasulu, C., Megharaj, M., Venkateswarlu, K., Naidu, R., 2012. Degradation of p-nitrophenol by immobilized cells of *Bacillus* spp. isolated from soil. *Int. Biodet. Biodeg.* 68, 24-27.

Sridevi, V., Chandana Lakshmi, M. V. V., Manasa, M., Sravani, M., 2012. Metabolic pathways for the biodegradation of phenol. *Int. J. Eng. Sci. Adv. Technol.* 2(3), 695-705.

Srivastava, S., Ahmad, A. H., Thakur, I. S., 2007. Removal of chromium and pentachlorophenol from tannery effluents. *Biores. Technol.* 98, 1128-1132.

Srivastava, V. C., Swamy, M. M., Mall, I. D., Prasad, B., Mishra, I. M., 2006. Adsorptive removal of phenol by bagasse fly ash and activated carbon: equilibrium, kinetics and thermodynamics. *Colloids Surf. A: Physicochem. Eng. Asp.* 272, 89-104.

Tallur, N., Megadi, V. B., Ninnekar, H. Z., 2009. Biodegradation of *p*-cresol by immobilized cells of *Bacillus* sp. strain PHN 1. *Biodegradation* 20(1), 79-83.

Tuan, N. N., Hsieh, H., Lin, Y., Huang, S., 2011. Analysis of bacterial degradation pathways for long-chain alkylphenols involving phenol hydroxylase, alkylphenol monooxygenase and cathecol dioxygenase genes. *Biores. Technol.* 102, 4232-4240.

Vennila, R., Kannan, V., 2011. Bioremediation of petroleum refinery effluent by *Planococcus halophilus. African J. Biotechnol.* 10(44), 8829-8833.

Wang, Y., Tian, Y., Han, B., Zhao, H. B., Bi, J. N., Cai, B. L., 2007. Biodegradation of phenol by free and immobilized *Acinetobacter* sp. strain PD12. *J. Environ. Sci.* 19, 222-225.

Wang, Z., Xu, Y., Wang, H., Zhao, J., Gao, D., Li, F., Xing, B., 2012. Biodegradation of crude oil in contaminated soils by free and immobilized microorganisms. *Pedosphere* 22(5), 717-725.

Wu, Z., Liu, Y., Liu, H., Xia, Y., Shen, W., Hong, Q., Li, S., Yao, H., 2012. Characterization of the nitrobenzene-degrading strain *Pseudomonas* sp. a3 and use of its immobilized cells in the treatment of mixed aromatics wastewater. *World J. Microbiol. Biotechnol.* 28(8), 2679-2687.

Zhao, G., Zhou, L., Li, Y., Liu, X., Ren, X., Liu, X., 2009. Enhancement of phenol degradation using immobilized microorganisms and organic modified montmorillonite in a two-phase partitioning bioreactor. *J. Hazard. Mater.* 169, 402-410.

Zhong, W., Wang, D., Xu, X., 2012. Phenol removal efficiencies of sewage treatment processes and ecological risks associated with phenols in effluents. *J. Hazard. Mater.* (217-218), 286-292.

In: Bioremediation: Processes, Challenges and Future Prospects ISBN: 978-1-62948-513-3
Editors: Jesús Bernardino Velázquez-Fernández et al.

Chapter 5

COLORIMETRY AS A PETROLEUM TOXICITY AND BIODEGRADATION SCREENING TECHNIQUE

Ederio Dino Bidoia, Renato Nallin Montagnolli and Paulo Renato Matos Lopes
UNESP – Universidade Estadual Paulista, Instituto de Biociências Rio Claro, São Paulo, Brazil

ABSTRACT

Treatment of hydrocarbon-contaminated environments is still a great challenge today. Oily residues present in refinery separation systems waste or even the accumulation in the storage tanks has been an increasing issue. There are many places contaminated with oil that may affect public health or the balance of local ecosystems, compromising the future economic development of affected regions. Several factors control biodegradation intensity of oil and its toxicity to organisms in the environment. Waste oil have a wide variety of hazardous components with high toxicity. Petroleum components such as nitrogen and sulfur are highly toxic when metabolized, primarily in form of ammonia and hydrogen sulfide. Biodegradation is a dynamic process: its environment conditions, microbiota and metabolites are all variables as the process advances. The conditions for biodegradation processes depends on many factors such as electron acceptors like oxygen, nitrate, sulfate or carbon dioxide that allow biodegradation to be complete.

On this basis, it is possible to develop toxicity and biodegradation colorimetric assays that provide important information on the effects of environmental pollution on organisms. In contrast to other chemical analyzes, redox dyes can easily detect the effects of multiple contaminants, byproducts and metabolites during bioremediation processes. Resazurin is an indicator that changes from blue to red during bacterial respiration. The cell growth inhibition may indicate toxicity and it can be determined by methods based on the reduction of redox indicators. The same principle can be used in the analysis of biodegradation of oil in microbial community using other redox indicator, 2,6-dichlorophenol-indophenol, whose color change when reduced. When the indicator is oxidized its color is blue, and when reduced is colorless. The color change is due to the change in the structure of the molecule window to receive electrons resulting from reactions in biodegradation processes.

Thus, using the colorimetry is possible to analyze different microcosms according to their capacity of biodegradation and their response to toxicity of various petroleum products. All the measurements take into account the time required for complete discoloration as an indicator of biodegradation of the oil or inhibition of cells for toxicity. Toxicity of the oils varies with time and may cause growth inhibition and mortality in organisms.

Furthermore, different profiles for different discoloration are observed to occur in each substance degraded. In addition, comparison with the phyto-toxicity data is also important to provide conclusive data on the biodegradation generated oil metabolites affecting the growth of plants in soil. The biotransformation of oils can cause an increase toxicity of these substances and intermediates toxic oils may be formed over the test duration and thereby increasing the toxicity of the oil. The assessment of acute toxicity have shown that the volatile aromatic hydrocarbons as compounds that influence on toxicity. Added to this, low molecular weight hydrocarbons as present in gasoline and diesel exhibit intense acute toxic effect, mainly due to their high solubility and presence of more volatile molecules with ability to penetrate cells and alter and damage cellular structures.

Keywords: DCPIP, rezazurin, phyto-toxicity, *Bacillus subtilis*

PETROLEUM BIODEGRADATION

Microorganisms can be great tools for pollutants removal in soil, water and sediment due to its numerous advantages over other processes in substances processing and degradation (Nano, Borroni and Rota, 2003; Morelli et al., 2005; Demnerova et al., 2005). Thus, potential bioremediation microorganisms must be screened and evaluated when applied to oil contaminated water restoration.

Microbial biotransformation is considered a major environmental technique in treating hydrocarbon pollution in both terrestrial and aquatic ecosystems (Churchill, Harper and Churchill, 1999; Kleinsteuber et al., 2006). This suggests that the biodegradative contribution of the indigenous microorganisms is often significant (Margesin and Schinner, 2001; Wang and Bartha, 1990). Biodegradation has long been one of the main fields of research in microbiology, which includes biodegradation of organic and inorganic substances.

Bioremediation exploits the ability of some microorganisms to degrade organic contaminants and has been established as an efficient, cost-effective, and environmentally friendly treatment (Norris, 1994). This process depends on environment conditions and the microbial community structure and, in unfavorable conditions, is generally achieved via bioaugmentation, biostimulation or both (Lin, Pan and Cheng, 2010).

Large catastrophic accidents involving oil tankers have attracted public attention to the fate of petroleum hydrocarbons in marine environments. In response to this concern, research on petroleum hydrocarbons biodegradation in natural environments has been enhanced. Therefore, in order to eliminate or minimize the effects of such contaminations, bioremediation shows up as an effective alternative, which can be used by potentially biodegrading microorganisms (Atlas, 1995). Microorganisms are capable of performing the mineralization of organic chemicals, transforming substances ultimately into CO_2, water and biomass (Cunha and Leite, 2000). This transformation capacity is of main importance in bioremediation processes, which relies on biodegradation.

Due to hydrocarbons being widespread pollutants, their hydrophobicity causes a low bioavailability and therefore a particular persistence against bioremediation measures (Anderson et al., 1995; Azadpour, Powell and Mattews, 1997). However, numerous terrestrial and aquatic species of microorganisms possess the ability to degrade these hydrocarbons. In order to increase the bioavailability of these otherwise hardly accessible substrates, microorganisms use strategies like increasing the hydrophobicity of their surfaces or the production of biosurfactants (Niescher et al., 2006).

Long chain hydrocarbons, specifically cyclic alkanes present in petrol derived oils, are difficult for bacteria to degrade (Perry, 1984). In order to efficiently promote biodegradation, many hydrocarbon degrading bacteria were investigated after isolation in an oil polluted environment (Pritchard et al., 1992; Korda et al., 1997; Dua et al., 2002; Aislabie, Saul and Foght, 2006). Referring to automotive lubricants, biodegradation studies demonstrated that used oil showed considerable assimilation by microbial metabolism when compared to new fluids such as mineral, semi-synthetic and synthetic ones. Automotive lubricant processing in engines modifies its physical-chemical properties due to high pressure and temperature that promote breaks in the hydrocarbon chains. Thus, shorter carbon molecules resulting from biodegradation processes may facilitate microorganism degradation (Montagnolli, Lopes and Bidoia, 2009).

It is important to mention that, even abiotically, product composition begins to change immediately because of various biochemical processes in the environment. The characteristics of discharged oil pollutants may be changed by physical, chemical and biological mechanisms after large periods in altered environmental conditions. Mixed-function oxidase systems present a great catalytic versatility and are also widely distributed in the environment, taking part in the biotransformation of a wide variety of both exogenous and endogenous organic compounds (Bonaventura and Johnson, 1997). Biodegradation involves a series of monooxygenation reactions that are catalyzed, including heteroatom oxygenation and dealkylation, aliphatic hydroxylation, desaturation, oxidative group migration and many other modes of inactivation (May and Katoposis, 1990).

The activation of natural degradation potentials in environmental media is the challenge for environmental research that addresses remediation methods. Notwithstanding the widespread application, in situ bioremediation is a site-specific process and feasibility studies are required before full-scale remediation can be applied successfully (Balba, Al-Awadhi and Al-Daher, 1998). Moreover, degradation rates and in particular residual concentrations derived from preliminary studies in the laboratory can be used to predict achievable rates and concentrations in the field (Höhener et al., 2003).

Hence, feasibility studies are a prerequisite for any planned strategy in bioremediation contaminated environments in order to identify limitations towards biodegradation and to predict remediation performance and thereby rule out technologies that may be inappropriate for the clean-up of the site of concern (Aichberger et al., 2005; Alisi et al., 2009). Additionally, the environmental behavior and chemical-physical properties of lubricant oils are the basis for new fluid developments and for ecological treatment strategies (Eisentraeger et al., 2002). According to Torstensson (1988) and Aichberger et al. (2005), preliminary assessments account for three critical prerequisites in bioremediation, namely: the availability of chemicals for the biodegrading microorganisms; the quantity of these microorganisms; and their activity level. Other factors are also important in moderating and driving biodegradation, such as temperature, pH, aeration, organic matter and nutrients.

Viable methods in biodegradation data generation during biotechnological process application are fundamental in the elaboration of original references about the biodegradability of certain substances (Battersby, 1997). There are many techniques capable of precisely evaluating biodegradation processes, including colorimetric methods. The isolation, characterization and profile of specific bacteria for petrol derived oils biodegradation capacity studies are important when deciding the correct bioremediation strategy.

TOXICITY

Petroleum exploration, transportation and consumption result in accidental spills that release hydrocarbons into the environment (Okoh and Trejo-Hernandez, 2006; Oluwole, Makinde and Philips, 2005). During refining, large residues amounts are released. Their components are not only toxic to local animals and plants, but have a long term carcinogenic potential for local human population. Some petroleum components are highly toxic, including commonly found compounds in form of ammonia and hydrogen sulfide (Atlas and Buyukgungor, 2008).

In aqueous media, the sulfhydric acid (H_2S) exists in equilibrium as hydrosulfide (HS^-) and sulfide ions (S^{2-}). Sulfide has a high oxygen demand, which contributes to the depletion of oxygen and overall water quality. Degradation of water quality by refinery residues accumulation should therefore promote a mobilization in agencies and regulators worldwide to establish an accurate concentration limit of these substances in the environment (Poulton et al., 2002).

Discharge of organic matter from crude oil containing hydrocarbons results in excess consumption of oxygen by bacteria in an attempt to oxidize the pollutants. The microorganisms in the biodegradation process consume oxygen faster than it dissolves air to water. Thus, the dissolved oxygen in aquatic environments falls below 2 $mg.L^{-1}$, the minimum required to sustain aerobic life in those environments (Attiogbe, Glover-Amengor and Nyadziehe, 2007). The lack of oxygen affects the more complex life forms.

Due to hydrocarbons persistence, contamination is likely to last long. This places a hazard to the whole ecosystem. The macroscopic effects of crude oil toxicity for humans includes liver necrosis and dissociation of liver cells (Khan and Ryan, 1991), also birds and other animals from areas polluted with oil have hydrocarbons adhered to their gastrointestinal tract. This leads to decreased absorption of nutrients, causing the death of many animals. Among other symptoms of contamination of crude oil include capillary rupture and hemorrhage, necrosis, renal anemia and hepatocellular dissociation.

Aromatic crude oil have the strongest effects on cellular levels. There are reports of intense adverse effects on local microbiota by inhibited the growth in presence of such substances (Sikkema, Debont and Poolman, 1995).

Waste generated in refineries are toxic due to the presence of high concentrations of polycyclic aromatic hydrocarbons (PAH). Refinery waste oils have a wide variety of hazardous substances with high toxicity. When reaching water bodies, refinery waste oils derease algae growth, which affects the whole food chain (Pardeshi and Patil, 2008).

Phenolic compounds pose threats to the environment due to their extreme toxicity (Kavitha and Palanivelu, 2004), stability, and persistence bioaccumulation. In general they are carcinogenic and cause considerable damage to human health when they reach bodies of water (Yang et al., 2008).

Even though directly unrelated to toxicity, oily residues may cause structural damages, since they tend to aggregate the pipe walls of sewage, causing bad odors, and promote chemical and microbiological corrosion where they adhere. They also adhere to the pipe and eventually blocking filters, eventually leading to further pathogenicity issues (Xu and Zhu, 2004).

COLORIMETRIC ANALYSIS

Treatment strategies to combat the hazardous effects of oil pollution are needed. Therefore, many techniques are currently capable of precisely evaluating biodegradation processes.

Colorimetric methods are referred to as a low cost and rapid procedure in detecting microbial metabolism occurrence, both in aerobic or anaerobic instances. These methods represent a rapid catalytic microbial based method in which the natural co-substrate (which can be oxygen, sulfates or nitrates) is substituted by a synthetic mediator (Pasco et al., 2000). Colorimetry is characterized by high levels of microorganisms and mediators, facilitating the fast reaction. Mediators basically act as chemical electron carriers that are able to pass through the cell walls of microorganisms, moving between intracellular and extracellular spaces. Once inside the microbial cell, an interaction between mediators and their bioelectrochemical counterparts, such as NAD, NADP, flavoproteins, iron–sulfur proteins, quinones and cytochrome, happens by their interaction with the metabolic process and current redox reactions (Wilkinson, Klar and Applegarth, 2006). Microbial cells catalyze mediator reduction with certain ease when in a very favorable environment. The reduction activity is a sensitive indicator directly coupled to respiration via the electron transport chain, and thus reflects the metabolic status of the cell (Zhao et al., 2007). One of these indicators, 2,6 dichlorophenol-indophenol (DCPIP), which will be further discussed within this chapter, is a lypofilic mediator and electron acceptor, capable of probing microbial metabolism.

Means of determining when reducing conditions exist are observed from such from redox indicators.

Generally, it is observed that oxidized forms are colored and reduced forms are colorless. Due to this fact, such color change can be monitored with a spectrophotometer, evaluating changes in redox indicator absorbance value. The redox half reaction of an indicator is described by Equation 1. As the oxidized indicator reacts with a reductant, the absorbance decreases and this way the reducing power of the sample can be estimated.

$$\mathrm{Ind_{ox}} + \mathrm{ne^-} + \mathrm{mH^+} \rightleftharpoons \mathrm{Ind_{red}}, \tag{1}$$

where: Ind_{ox} is the oxidized form of the indicator, Ind_{red} is the reduced form, ne^- is the number of electrons transferred, and m is the number of protons transferred - pH dependant (Bishop, 1972).

Feasibility studies are a prerequisite for any planned strategy in bioremediation of contaminated environments to predict remediation performance and thereby rule out technologies for the cleanup. The characterization environmental behavior of contaminants is important when deciding the right bioremediation strategy.

This technique is aimed at a sensitive biocatalyst and mediators choice, capable of detecting the metabolic perturbations caused by various substrates, including toxic ones (Bentley et al., 2001) or biodegrading processes.

It is well known that mediators fundamentally work by interacting with the metabolic pathways of the microbial cell. Even though the facts regarding interaction of microbial metabolic pathways with electron transfer mediators is quite an important study, with some important practical usage, few publications are devoted to finding the solution to such an important question as the search for the interaction mechanism between microbial and mediator, and also the origin of the electrons. Thus, more studies are necessary in order to elucidate completely this subject. According to Heiskanen et al. (2004), once the interaction mechanism between microbial and mediator is made clear, the mediator will assume the function of an electrochemical probe, as the functionalized mediators would be useful for a better comprehension of the effects of physiochemical reactions occurring within microorganisms in the redox state.

This chapter will discuss two main redox reactions applied to crude oil and petroleum derivatives such as gasoline, diesel, motor oil and phenol. These techniques allowed a brief, yet responsive and reliable analysis of short term biodegradation and toxicity in different compounds. In addition, biodegradation products underwent standard germination toxicity analysis.

2,6 Dichlorophenol-Indophenol (DCPIP)

DCPIP is an enzyme-catalyzed redox electron acceptor that is blue in its oxidized form and colorless in its reduced form (Dawson et al., 1986). Its loss of color is monitored at a wavelength of 600 nm. A peak in the absorbance is observed at 600 nm, as reported by Yoshida et al. (2001). Some studies affirm that DCPIP is unstable in darkness or when exposed to light for a long time (1992). Moreover, the color of the DCPIP indicator in the presence of light is reasonably stable and can be used reliably for some time, especially when adding $NAHCO_3$ to the dye solution (2006).

Hydrocarbon oxidation processes by microorganisms involve redox reactions, in which electrons are transferred to electron acceptors, such as O_2 nitrates and sulfate (Hutchins et al., 1991), as seen in Figure 1. The microbial sulfate-reducing redox is one of the most reducing redox environmental systems. In terms of formal reduction potentials, sulfate reduction to sulfide occurs at about 220 mV at pH 7 (Brock et al., 1994). A low redox level is achieved in an environment where conditions are completely anaerobic and the terminal electron acceptors (TA) such as NO_3^-, Mn(IV) and Fe(III) have been fully depleted (Chapelle, 1993). Sulfate-reducers use sulfate as the TA and commonly lactate, acetate, or H_2 as the electron donor in the process of creating ATP, and producing sulfide (Brock et al., 1994).

Thus, by incorporating an electron acceptor such as DCPIP, it is possible to ascertain the ability of a microorganism to utilize a hydrocarbon substrate by simply observing the color change, in this case, from blue (oxidized) to colorless (reduced).

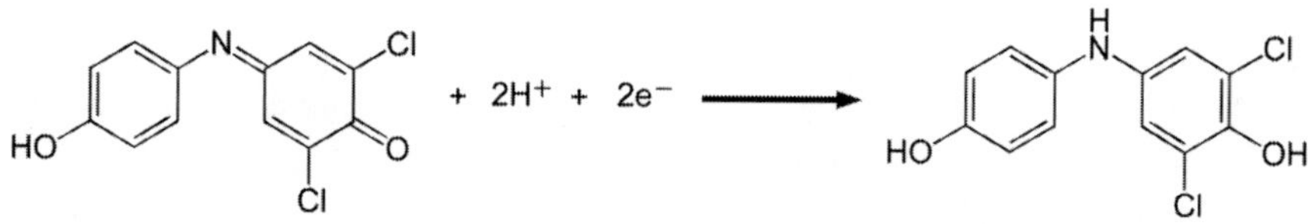

Figure 1. DCPIP reduction reaction.

Table 1. Colorimetric assays content

	DCPIP	BH	B. subtilis inoculum	Oil
Control 0 (C0)	400 µL	7.5 mL	-	-
Control 1 (C1)	400 µL	7.5 mL	100 µL	-
Media 1 (M1)	400 µL	7.5 mL	-	50 µL of Crude Oil
Media 2 (M2)	400 µL	7.5 mL	-	50 µL of Diesel Oil
Media 3 (M3)	400 µL	7.5 mL	-	50 µL of Motor Oil
Media 4 (M4)	400 µL	7.5 mL	-	50 µL of Gasoline
Media 5 (M5)	400 µL	7.5 mL	-	50 µL of Phenol
Inoculum 1 (I1)	400 µL	7.5 mL	100 µL	50 µL of Crude Oil
Inoculum 2 (I2)	400 µL	7.5 mL	100 µL	50 µL of Diesel Oil
Inoculum 3 (I3)	400 µL	7.5 mL	100 µL	50 µL of Motor Oil
Inoculum 4 (I4)	400 µL	7.5 mL	100 µL	50 µL of Gasoline
Inoculum 5 (I5)	400 µL	7.5 mL	100 µL	50 µL of Phenol

In this section, a rapid colorimetric assay will be discussed to exemplify DCPIP application as an effective biodegradation indicator. To study petroleum derivatives biodegradation processes, colorimetric assays were setup with *Bacillus subtilis* CCT 2576. The cells were transferred from storage culture tubes and streaked onto a Petri dish with PCA medium. Then, microbial cultures reactivated after 48 h were inoculated to 50 mL of BH medium at 35°C for biomass growth. No agitation or darkness conditions were applied. The BH medium contents are, (in $g.L^{-1}$: $MgSO_4$, 0.2; $CaCl_2$, 0.02; KH_2PO_4, 1.0; K_2HPO_4, 1.0; NH_4NO_3, 1.0; $FeCl_3$, 0.05; according to Difco (1984).

After 48 h in BH medium, *B. subtilis* culture was inoculated to tubes along with DCPIP indicator and the oil under analysis.

Colorimetry tests were conducted in test tubes with lids, to keep CO_2 saturation. The tubes were transparent and suitable for reading in Hach DR-2500 spectrophotometer. The quantities and substances included in colorimetric assays followed concentrations in Table 1. Tube contents and assays were performed in triplicates.

After assembled, each tube underwent vortex agitation for 10 seconds and then stored in an incubator at 35°C at 180 rpm. Control assays (C0-1) evaluated interactions of the assay components and DCPIP indicator. C0 contained only DCPIP indicator and BH medium. On the other hand, C1 presented DCPIP indicator, BH medium and *B. subtilis* inoculum; and was done to determinate inoculum influences in DCPIP through time. Another series of blank assays, named "Media" contained DCPIP indicator, BH medium and different types of oils (M1-5) according to Table 1. Finally, Inoculum assays were included DCPIP indicator, BH medium, *B. subitilis* inoculum and the different oils (I1-5).

The absorbance of all assays was measured by Hach® DR 2500 spectrophotometer. Data was collected three times per day, totalizing 320 h. Then, absorbance values were compared between different assays. Quantitative measurements were made using the spectrophotometer Odyssey Hach DR-2500 in length from 600 nm to all tubes. The tubes were left without agitation for 10 min before the start of measurement and not abruptly moved to avoid any turbulence of the liquid, which could alter the results obtained absorbance.

The colorimetric analysis successfully demonstrated biodegradation. It occurred in all the oil containing tubes, whereby DCPIP color loss appeared as a general pattern. Results are shown in Figure 2.

In the assays containing oil without *B. subtilis* inoculum (M1-5), total biodegradation times for diesel (M2) and motor oil (M3) were 172 h and 280 h, respectively. No significant biodegradation was observed in crude oil (M1) and gasoline (M4) media controls. There was no decolorization in phenol assays in both cases (M5 and I5). Both petroleum and gasoline did not degrade without inoculum in media control assays. For the Inoculum assays, the DCPIP color changes occurred slowly for crude oil (I1) and motor oil (I3) assays. A faster biodegradation time was observed for diesel (I2) and gasoline (I4), which ocurred in 80 h and 88 h, respectively.

The control assays (C0 and C1) sustained their blue color, whereas the oil containing assays passed through the biodegradation process. The biodegradation time necessary for the blue color to completely vanish was different between different types of oils and allowed biodegradation performance quantification.

According to observation results in Figure 2, *B. subtilis* inoculum assays demonstrated a faster biodegradation compared to assays without the bacteria. Even though no biomass was inoculated, biodegradation still occurred in M1-5 assays probably due to microorganism previously present in oils. Moreover, total biodegradation time observed generally followed an ascendant order: gasoline, diesel, crude oil and motor oil.

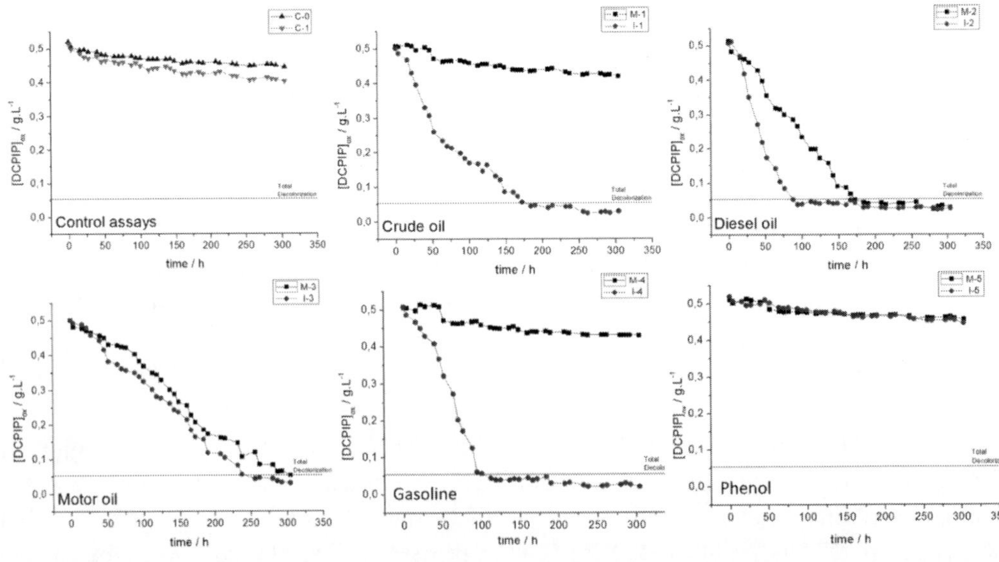

Figure 2. DCPIP concentration quantified at 600 nm absorbance control assays - without oils; and in oil containing assays without (M) and with (I) *Bacillus subtilis*.

A minor decrease in DCPIP concentration was noticed even in control assays. This decrease indicates an overall pattern of natural disintegration of DCPIP chemical structure through time. Therefore, it cannot be due to *Bacillus subtilis* inoculum, since the DCPIP concentration decreases equally in both C0 and C1. Even though the DCPIP concentration tends to naturally decrease, tubes containing oils tend to reduce original DCPIP concentration even further because of biodegradation.

At the outset, the last measurement point was plotted after complete loss of original color. So, the DCPIP was considered reduced as a result of biodegradation when the absorbance value led to a DCPIP concentration below 0.030 $g.L^{-1}$. Values below this concentration could not be measured by this spectrophotometric method.

Control assays DCPIP quantification demonstrated a discoloring tendency through time (Figure 2) but it was not as significant as the DCPIP concentration reduction attributable to oil biodegradation, especially when B. subtilis was added. In microbial biodegradation assays where a DCPIP absorbance decreases occured, a positive interaction of the bacteria in oil degradation can be assumed, since all "Inoculum" assays yielded a faster biodegradation.

Other studies also succesfully applied DCPIP to conduct a survey of petroleum-degrading bacteria to evaluate distribution of naturally occurring petroleum-degrading aerobic bacteria. Some genera, such as *Pseudomonas*, *Mycobacterium*, *Klebsiella*, *Acinetobacter*, *Micrococcus*, and *Nocardia* were found to be the most common petroleum degraders. Other heterotrophic bacteria included several species of *Escherichia*, *Klebsiella*, non-oil-degrading *Pseudomonas*, *Vibrio*, *Streptococcus*, *Staphylococcus* and *Bacillus*. Following preliminary selection, five strains, showing best growth in medium with oil fraction as the sole carbon source, were chosen for estimation of the efficiency of crude oil biodegradation using DCPIP as one of the bioindicators (Roy et al., 2002).

The use of this method proved to be reliable enough in experiments to quantify biodegradation kinetics, taking into account the time taken for decolorization of the blue DCPIP to the colorless form. According to Roy et al. (2002), the strains, which decolorized the DCPIP in the quickest time, were chosen as the best oil degraders, but the absorbance values in 600 nm could also be used to precisely collect data regarding the biodegradation process through time.

This indicator was also applied to (Cormack and Fraile, 1997) research, where enzyme induction activity and the study of the mycelium of Aspergillus ochraceus in a kerosene biodegradation were also evaluated with DCPIP. The subject extract was used as an enzyme source for investigation of the induction of NADPH-DCPIP reductase, aminopyrine N-demethylase and kerosene degradation activity, therefore evaluating biodegradation.

Rezazurin

Using the same colorimetric principle and setup as in DCPIP redox reactions, it was possible to apply another redox indicator named rezasurin to determine acute toxicity of certain substances to a *Bacillus subtilis* community. This method is also based on the reduction of a redox dye by bacterial respiration, but, different from DCPIP, it is aimed to provide an immediate toxicity resposnse from a microrganisms to certain substances.

The test organism used in this assay was the same *B. subtilis* used in the previous biodegradation assays with DCPIP.

The percentage of inhibition of cellular toxicity was determined by following the suggested methodology by commercial kit "ToxTrack Toxicity Test" manufactured by the Hach Company using a Hach Spectrophotometer Odyssey - DR/2500 model (Method 10017). However, along with the test tubes containing sample, resazurin and inoculum, there was also a chemical catalyst added to decrease reaction time, allowing rapid analysis of acute toxicity from each substance.

When the resazurin is reduced, its color changes from blue to pink. However, toxic substances can prevent the reduction rate of the dye by inhibiting cell. Thus, variations in coloration of the mixture by reduction of resazurin were measured in a spectrophotometer at 603 nm after 60 min. Through these readings percentage inhibition (% I) for the different samples were obtained according to Equation 2.

$$\%I = [1 - \frac{\Delta A_a}{\Delta A_c}] \times 100 \quad (2)$$

where: ΔAa= change in absorbance of the sample during testing time; ΔAc = change in absorbance of control during testing time

As an example, we tested the toxicity of various petroleum products to *B. subtilis* through the "ToxTrack Toxicity Test". The results are shown in Figure 3.

The results were expressed as inhibition percentage (% I) as a relative measure, since they do not represent an absolute value or toxicity index. According to method 10.017 (Hach Company), the results below 10% are not entirely reliable, but become quite consistent when estimates are made repeats similar results. In such cases, the occurrence of many results between -10% and 10%, should only be regarded as mildly toxic to the microorganism tested. Very negative results (below -10%) in turn should be considered toxic.

According to the results shown in Figure 3, gasoline and phenol are toxic to *B. subtilis*. Phenol is the most toxic compound, with the highest inhibition percentage.

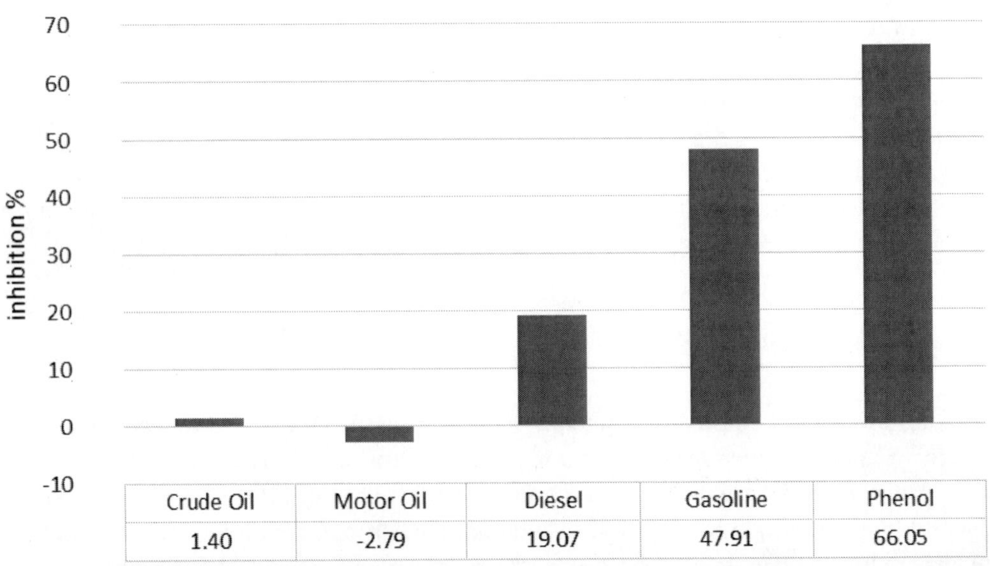

Figure 3. Toxicity index based on inhibition percentage of *B. subtilis* metabolism in various substances.

Diesel is mildly toxic. Motor oil and crude oil, on the other hand, are not toxic to the microbial comunity. Low toxicity to *B. subtilis*, may be tied to high biodegradation observed in DCPIP colorimetric assays.

The findings of this assay corroborate other methodologies for assessing acute toxicity. Other studies have shown that aromatic hydrocarbons – including phenols - are compounds with a high toxicity (Neff, 2002; Gamo, Oka and Nakanishi, 2003). Added to this, low molecular weight hydrocarbons present in gasoline and diesel exhibit intense acute toxic effect, mainly due to their high solubility and presence of more volatile molecules with ability to penetrate cells and alter cellular structures (Farida and Das 2005).

Phyto-Toxicity Assays

It was demonstrated that by using colorimetry allows assessing different microcosms capacity of biodegradation and their response to toxicity of various petroleum products. All the measurements take into account the time required for complete discoloration as an indicator of biodegradation of the oil or inhibition of cells for toxicity. However, toxicity of the oils may varies with time and may cause growth inhibition and mortality in organisms. Thus, toxicity tests were conducted in *Lactuca sativa* seeds, by evaluating germination and growth inhibition ratio when exposed to respirometric media before and after a 120 days biodegradation time span.

In plants, toxicity occur when certain compounds are absorbed or accumulated into the embryo tissue and they cause damage and affect growth (Ayers and Westcot, 1999). Germination is a biological phenomenon when embryo grows and develops, with subsequent disruption of tegument by radicle. From the physiological point of view, germination means metabolic activity beginning (Nassif, Vieira and Fernandes, 1998), provided that the conditions necessary for internal and environmental organisms that can develop. Any change in these parameters by toxic substances can affect the development of plant tissue (Ayers and Westcot, 1999).

Phyto-toxicity tests conducted in this study followed standardized methodologies by Morales (2004), Dutka (1989) and Wang (1987) with relatively low cost and simple methods. The acute toxicity tests determined intensity of adverse effects from petroleum products in plant tissues, thus allowing toxicity comparison of both substances in lettuce seedlings. The seeds were all purchased from the same supplier lot and free of any previous chemical treatment. Each seed was placed in petri dish with filter paper. Into each petri dish, 4 ml of an aqueous solution containing a 5% slurry soil m/v and 1% oil v/v mixture was added to soak the filter paper. The vairous petroleum products were tested, before and after being biodegraded in this solution. The seeds were then added to the filter paper soaked petri dishes. Twenty lettuce seeds were placed on each petri dish assay. The experiments were conducted in duplicate for each substance tested.

The plates were covered with plastic wrap in order not to lose moisture, and covered with dark paper. The plates were incubated in a climatic chamber (BOD incubator) at 22 ± 1 °C in the dark for 120 h.The counting of germinated seeds was performed after 120 h, being considered germinated seeds any 2 mm radicle protrusion. Data obtained from tests were used to calculate the percentage of germination (% G) given by Labouriau and Agudo (1987) accordingly.

Regarding toxicity using lettuce seeds, production of toxic metabolites may have occured during biodegradation of crude oil, motor oil and gasoline was observed at different degrees in Figure 4. This may be indicative growth inhibitors appearance after 140 days. It was observed that a higher percentage of germination occurred in all the seeds that were exposed to substances before the occurrence of biodegradation process. The germination of lettuce during 120 h exposure to the respirometric media before and after biodegradation was analyzed statistically by the Tukey test at a significance level of 5%. It was found that the averages and variances of samples from groups "before biodegradation" and "after biodegradation" differ statistically from control and each other, indicating that the variables which seeds were exposed influenced the rate of germination of lettuce.

Crude oil presented less than 75% germination percentage before biodegradation. No germination occurred when phenol was present. After 140 days biodegradation, diesel oil kept its germination percentage close to the initial time value, whereas no germination occurred in gasoline. Both crude oil and motor oil provided a less germination inducing environment after biodegraded. These toxicity assays results may indicated plant growth inhibition from intermediate metabolites with low molecular weight substances with intense acute toxic effect. There was a possible formation of secondary metabolites toxic to plant cells.

The findings in this toxicity assay corroborate other reported studies where petroleum derivated hydrocarbons containing aromatic hydrocarbons and phenols increase toxicity (Neff 2002). Added to this, low molecular weight and volatile hydrocarbons as present specially in gasoline caused intense acute toxic effect to vegetable cells. According to Haigh (1995) biodegradable substances are not necessarily less persistent and toxic. Biodegradation products can also affect germination and growth of plants. These results were also observed in corn and oats (Salanitro et al., 1997). This is justified by the fact that the oils after biodegradation by microorganisms have their size of their carbon chain and reduced oil viscosity becomes smaller. This can lead to interfacial area between cells and biodegradation products increase thus causing toxic responses.

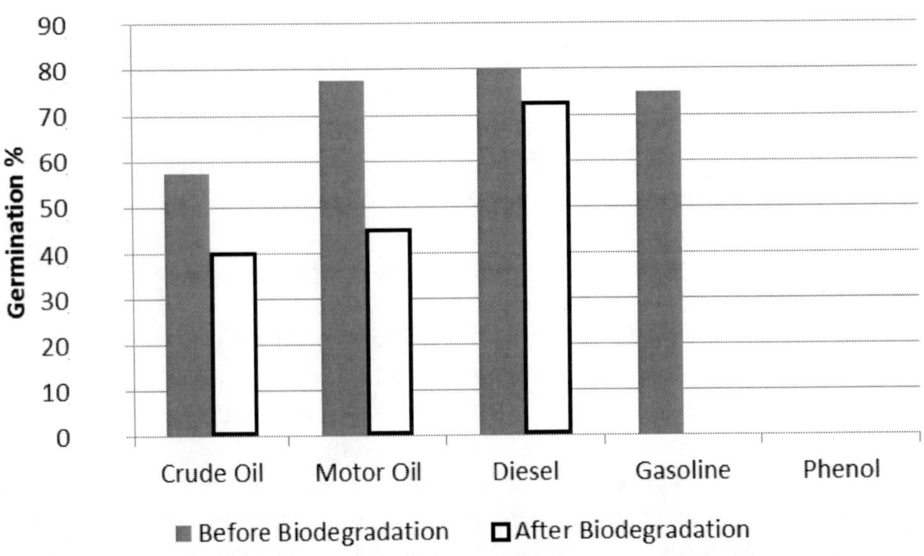

Figure 4. Germination percentage of lettuce seeds in acute toxicity tests in different substances before and after biodegraded.

Phenol degrading microorganisms are crucial for growth and germination of plants in environments contaminated with oil. However, in the experimental setup no phenolitic conditions were achieved. The high toxicity indicated that aromatic rings oxidizing bacteria were not present in the microbial consortia in order to decreasing overall toxicity of phenol in the assays (Huesemann, 1995). In fact, the biodegradation process produced an increased toxicity of all analysed compounds. Biodegradation produces intermediate compounds which can be present more toxic substance source. Thus, the products generated in may interfere with germination and hence the development of the root and hypocotyl of plants.

Regarding other compounds toxicity, it has been reported that gasoline biodegradation generate ca. 20 metabolites and byproducts. During biodegradation intermediates such as tert-butyl alcohol (TBA), tert-butyl format (TBF) and lactate is formed (Martienssen et al. 2006), and the presence of volatile components such as benzene, toluene, ethylbenzene and xylene (BTEX). A gasoline additive ether, methyl tert-butyl ether (MTBE) was found at high concentrations (87 $mg.L^{-1}$) in gasoline spills and is considered a serious environmental contaminant (Schmidt et al. 2002). The potential toxicity of these compounds and byproducts is low for humans, but was able to inhibit 100% germination of lettuce (Figure 4) and can be quite carcinogenic for animals, particularly MTBE, as a studie by Belpoggi, Soffritti and Maltoni (1995) suggests.

According Dwyner and Moore (1974) some parameters influencing the toxicity are known, for example, the fact that larger carbon chains presenting a lower toxicity than oil derivatives with low molecular weight, whose presence may cause an aromatics high toxicity. Longer chain derivatives include motor oil. The results of this study indicate that compounds with a higher long hydrocarbons chains rate have a lower toxicity after biodegraded. This is justified by an inert effect of heavier molecules, which are unable to affect tested organisms as much as low weight molecules.

Toxicity bioassays provide important information on the effects of environmental pollution on organisms. In contrast to chemical analysis, they can detect the effects of multiple contaminants, byproducts and metabolites during bioremediation processes (Maila and Cloete, 2005). The composition of the oil is very complex and its biological properties vary widely. Just as different biodegradation profiles occur for different substances, the toxicity varies with many factors. Thus the biological effect toxicity of these substances can vary widely due to each situation.

CONCLUSION

The methods discussed by the assays in this chapter allowed a quick initial assessment to any experimental setup regarding toxicity and biodegradation of soil, water and sediments contaminated with petroleum products. In conclusion, this rapid and simple colorimetric methodology applied to oils biodegradation promotes some discussion leading to a better handling of contaminated water medium by oils and their respective biodegrading profile. Thus, it promotes the development of new techniques in bioremediation.

Microorganisms represent an reliable way to biodegrade petrol derived substances. This ability is highly exploitable during bioremediation processes to ultimately free the environment from subsequent pollutants derived from these hydrocarbons.

The development of efficient techniques in order to obtain biodegradation data is a fundamental tool when proposing different strategies for bioremediation of polluted areas. Colorimetric technique proved itself to be a rapid, easy and low cost method when detecting microbial metabolism from carbon and other nutrients sources, including hydrocarbons. To ascertain microbial ability to utilize hydrocarbon substrates by simply observing the color changes, in which the quickest decolorization time represents the best oil degradation, could be an excellent tool for biodegradation and toxicity studies.

Further studies in the applicability of this method and other different redox indicators are recommended in order to reach a wide and precise selection of mediators capable of detecting the metabolic perturbations caused by many types of substrates, even though colorimetry usage has already been successfully reported for a wide range of situations.

REFERENCES

Aichberger, H., Hasinger, M., Braun, R., Loibner, A. P., 2005. Potential of preliminary test methods to predict biodegradation performance of petroleum hydrocarbons in soil. *Biodegradation*. 16, 115–125.

Aislabie, J., Saul, D. J., Foght, J. M., 2006. Bioremediation of hydrocarbon-contaminated polar soils. *Extremophiles*. 10, 171–179.

Alisi, C., Musella, R., Tasso, F., Ubaldi, C., Manzo, S., Cremisini, C., Sprocati, A. R., 2009. Bioremediation of diesel oil in a co-contaminated soil by bioaugmentation with a microbial formula tailored with native strains selected for heavy metals resistance. *Sci. Total Environ*. 407, 3024–3032.

Anderson, A. J., Tucker, B., Radtke, C., Kown, S. I., 1995. Suppression of bioremediation by Phanerochaete chrysosporium by soil factors. *J. Hazard. Mater.* 41, 251–265.

Atlas, L., Büyükgüngör, H., 2008. Sulfide removal in petroleum refinery wastewater by chemical precipitation. *J. Hazard. Mater.* 153, 462–469.

Atlas, R. M., 1995. Bioremediation of petroleum pollutants. *Int. Biodeterior. Biodegradation*. 35, 317-327.

Attiogbe, F. K., Glover-Amengor, M., Nyadziehe, K. T., 2007. Correlating biochemical and chemical oxygen demand of effluents—a case study of selected industries in Kumasi, Ghana. *West Africa Journal of Applied Ecology*. 11, 110–118.

Ayers, R. S., Westcot, D. W. A., 1999. Qualidade da água na agricultura, UFPB, Campina Grande.

Azadpour, A., Powell, P. D., Matthews, J. 1997. Use of lignin degrading fungi in bioremediation. *Remediation Autumn*. 5, 25–49.

Balba, M. T., Al-Awadhi, N., Al-Daher, R., 1998. Bioremediation of oil-contaminated soil: microbiological methods for feasibility assessment and field evaluation. *J. Microbiol. Methods*. 32, 155-164.

Battersby, N. S., 1997. The ISO Headspace CO2 biodegradation test. *Chemosphere*. 34, 1813–1822.

Belpoggi, F., Soffritti, M., Maltoni, C., 1995. Methyl-tertiary-butyl ether (Mtbe)—a gasoline additive—causes testicular and lymphohaematopoietic cancers in rats. *Toxicol. Ind. Health*. 11, 119–149.

Bentley, A., Atkinson, A., Jezek, J., Rawson, D. M., 2001. Whole cell biosensors - electrochemical and optical approaches to ecotoxicity testing. *Toxicol. In Vitro*. 15, 469-475.

Bishop, E., 1972. *Indicators*. 1st ed. Oxford, Pergamon Press.

Bonaventura, C., Johnson, F. M., 1997. Healthy environments for healthy people, bioremediation today and tomorrow. *Environ. Health Perspect*. 105, 5-20.

Brock, T. D., Madigan, M. T., Martinko, J. M., Parker, J., 1994. *Biology of Microorganisms*. 7th ed. Englewood Cliffs, Prentice-Hall.

Chapelle, F. H., 1984. *Groundwater Microbiology and Geochemistry*. 1st ed. New York, Wiley.

Churchill, S. A., Harper, J. P., Churchill, P. F., 1999. Isolation and characterization of a Mycobacterium species capable of degrading three- and four-ring aromatic and aliphatic hydrocarbons. *Appl. Environ. Microbiol.* 65, 549–552.

Cormack, W. P. M., Fraile, E. R., 1997. Characterization of a hydrocarbon degrading psychrotrophic Antarctic bacterium. *Antarctic Science*. 9, 150-155.

Cunha, C. D., Leite, S. G. F., 2003. Gasoline biodegradation in different soil microcosms. *Braz. J. Microbiol.* 31, 45-49.

Dawson, R. M. C., Elliott, D. C., Elliott, W. H., Jones, K. M., 1986. Artificial and natural substrates: D. substrates for phosphatases. In: Dawson, R. M. C., ed. *Data for Biochemical Research*. New York, Oxford Science.

Demnerova, K., Mackova, M., Spevakova, V., Beranova, K., Kochankova, L., Lovecka, P., Ryslava, E., Macek, T., 2005. Two approaches to biological decontamination of groundwater and soil polluted by aromatics characterization of microbial populations. *Int. Microbiol.* 8, 205–211.

Difco., 1984. *Difco Manual*. 10th ed. Detroit, Difco Laboratories.

Dua, M., Johri, A. K., Singh, A., Sethunathan, N., 2002. Biotechnology and bioremediation: successes and limitations. *Appl. Environ. Microbiol.* 59, 143–152.

Dutka, B., 1989. Short-term root elongation toxicity bioassay - Methods for toxicological analysis of waters, wastewaters and sediments, National Water Research Institute (NWRI), Canada.

Eisentraeger, A., Schmidt, M., Murrenhoff, H., Dott, W., Hahn, S., 2002. Biodegradability testing of synthetic ester lubricants – effects of additives and usage. *Chemosphere*. 48, 89-96.

Gamo, M., Oka, T., Nakanishi, J., 2003. Ranking the risk of 12 major environmental pollutants that occur in Japan. *Chemosphere*. 53, 277-284.

Haigh, D. S., 1995. Fate and effects of synthetic lubricants in soil: biodegradation and effect on crops in field studies. *Sci. Total Environ.* 168, 71-83.

Heiskanen, A., Yakovleva, J., Spegel, C., Taboryski, R., Koudelka-Hep, M., Emneus, J., Ruzgas, T., 2004. Amperometric monitoring of redox activity in living cells: comparison of menadione and menadione sodium bisulfite as electron transfer mediators. *Electrochem. Commun*. 6, 219-224.

Höhener, P., Duwig, C., Pasteris, G., Kaufmann, K., Dakhel, N., Harms, H., 2003. Biodegradation of petroleum hydrocarbon vapors: laboratory studies on rates and kinetics in unsaturated alluvial sand. *J. Contam. Hydrol.* 1917, 1–23.

Huesemann, M. H., 1995. Predictive model for estimating the extent of petroleum hydrocarbon biodegradation in contaminated soils. *Env. Sci. Technol*. 29, 7–18.

Hutchins, S. R., Sewell, G. W., Kovacs, D. A., Smith, G. A., 1991. Biodegradation of aromatic hydrocarbons by aquifer microorganisms under denitrifying conditions. *Environ. Sci. Technol.* 25, 68-76.

Kavitha, V., Palanivelu, K., 2004. The role of ferrous ion in Fenton and photo-Fenton processes for the degradation of phenol. *Chemosphere*. 55, 1235–1243.

Khan, R. A., Ryan, P., 1991. Long Term Effects of Crude Oil on Common Murres (Uria aalge) Following Rehabilitation. *Bull. Environ. Contam. Toxicol.* 46, 216-222.

Kleinsteuber, S., Riis, V., Fetzer, I., Harms, H., Müller, S., 2006. Population dynamics within a microbial consortium during growth on diesel fuel in saline environments. *Appl. Environ. Microbiol.* 72, 3531-3542.

Konidari, C. N., Tzouware, S. M., Bowman, L. E., Karayannis, M. I., 1992. Kinetic and mechanistic study of the reaction of 2,6-dichlorophenol-indophenol and cysteine. *Talanta*. 39, 863-868.

Korda, A., Santas, P., Tenente, A., Santas, R., 1997. Petroleum hydrocarbon bioremediation: sampling and analytical techniques, in situ treatments and commercial microorganisms currently used. *Appl. Environ. Microbiol*. 48, 677–686.

Labouriau, L. G., Agudo, M., 1987. On the physiology of seed germination in Salvia hispanica L. I. temperature effects. *Anais da Academia Brasileira de Ciências*, 59, 37-56.

Lin, T., Pan, P., Cheng, S., 2010. Ex-situ bioremediation of oil-contaminated soil. *J. Hazard. Mater.* 176, 27–34.

Maila, M. P., Cloete, T. E., 2005. The use of biological activities to monitor the removal of fuel contaminants-perspective for monitoring hydrocarbon contamination: a review. *Int. Biodeterior. Biodegradation.* 55, 1–15.

Margesin, R., Schinner, F., 2001. Bioremediation (natural attenuation and biostimulation) of diesel-oil-contaminated soil in an alpine glacier skiing area. *Appl. Environ. Microbiol.* 67, 3127-3133.

Martienssen, M., Fabritius, H., Kukla, S., Balcke, G. U., Hasselwander, E., Schirmer, M., 2006. Determination of naturally occurring MTBE biodegradation by analysing metabolites and biodegradation by-products. *J. Contam. Hydrol.* 87, 37–53.

May, S. W., Katoposis, A. G., 1990. Hydrocarbon monooxygenase system of Pseudomonas oleovolans. *Meth. Enzymol*. 188, 3-9.

Montagnolli, R. N., Lopes, P. R. M., Bidoia, E. D., 2009. Applied models to biodegradation kinetics of lubricant and vegetable oils in wastewater. *Int. Biodeterior. Biodegradation*. 63, 297–305.

Morales, C. G., 2004. Ensayos Toxicológicos y Métodos de Evaluación de Calidad de Agua: estandarización, intercalibración, resultados y aplicaciones, IMTA, México.

Morelli, I. S., Del Panno, M. T., De Antoni, G. L., Painceira, M. T., 2005. Laboratory study on the bioremediation of petrochemical sludge contaminated soil. *Int. Biodeter. Biodegr.* 55, 271–278.

Nano, G., Borroni, A., Rota, R., 2003. Combined slurry and solid-phase bioremediation of diesel contamined soils. *J. Hazard. Mater.* 100, 79–94.

Nassif, S. M. L., Vieira, I. G., Fernandes, G. D., 1998. Informativo Sementes do Instituto de Pesquisas e Estudos Florestais, Escola Superior de Agricultura Luís de Queiroz (ESALQ), Universidade de São Paulo (USP), Piracicaba.

Neff, J. M., 2002. Bioaccumulation in marine organisms. *Effect of contaminants from oil well produced water*. 1st ed, Elsevier, London.

Norris, R. D., 1994. *Handbook of Bioremediation.* 1st ed. Boca Raton, CRC Press.

Okoh, A. I., Trejo-Hernandez, M. R., 2006. Remediation of petroleum hydrocarbon polluted systems: exploiting the bioremediation strategies. *Afr. J. Biotechnol.* 5, 2520–2525.

Oluwole, O. S., Makinde, S. C. O., Philips, D. A., 2005. The impact of spent engine oil pollution on the growth of Celosia argentea. Lagos State University (LASU), Ojo.

Pardeshi, S. K., Patil, A. B., 2008. A simple route for photocatalytic degradation of phenol in aqueous zinc oxide suspension using solar energy. *Solar Energy.* 82, 700–705.

Pasco, N., Baronian, K. H., Jeffries, C., Hay, J., 2000. Biochemical mediator demand—a novel rapid alternative for measuring biochemical oxygen demand. *J. Microbiol. Biotechnol.* 53, 613-618.

Perry, J. J., 1984. Microbial metabolism of cyclic alkanes. In: Atlas, R. M. *Petroleum Microbiology.* New York, Macmillan Publishing Co.

Poulton, S. W., Krom, M. D., Rijn, J. V., Raiswell, R., 2002. The use of hydrous iron (III) oxides for the removal of hydrogen sulphide in aqueous systems. *Water Res.* 36, 825–834.

Pritchard, P. H., Mueller, I. J. G., Rogers, J. C., Kremer, F. V., Glaser, J. A., 1992. Oil spill bioremediation: experiences, lessons and results from the Exxon Valdez oil spill in Alaska. *Biodegradation.* 3, 315–335.

Roy, S., Hens, D., Biswas, D., Kumar, R., 2002. Survey of petroleumdegrading bacteria in coastal waters of Sunderban Biosphere Reserve. *World J. Microbiol. Biotechnol.* 18, 575-581.

Salanitro, J. P., Dorn, P. B., Huesemann, M. H., Moore, K. O., Rhodes, L. M., Jackson, R., Vipoad, T. E., Western, M. M. ve Wisniewski, H. L., 1997. Crude oil hydrocarbon bioremediation and soil ecotoxicity assessment. *Env. Sci. Technol.* 31, 1769-1776.

Schmidt, T. C., Morgenroth, E., Schirmer, M., Effenberger, M., Haderlein, S. B., 2002. *Oxygenates in Gasoline: Environmental Aspects.* ACS, Washington, 2002.

Sikkema, J., Debont, J. A. M., Poolman, B., 1995. Mechanisms of Membrane Toxicity of Hydrocarbons. *Microbiol. Rev.* 59, 201–222.

Torstensson, L., 1988. Microbial decomposition of herbicides in the soil. *Outlook in Agriculture.* 17, 120-124.

Wang, W., 1987. Root elongation method for toxicity testing of organic and inorganic pollutants. *Environ. Toxicol. Chem.* 6, 409-414.

Wang, X., Bartha, R., 1990. Effects of bioremediation on residues, activity and toxicity in soil contaminated by fuel spills. *Soil Biol. Biochem.* 22, 501-505.

Wilkinson, S., Klar, J., Applegarth, S. P., 2006. Optimizing biofuel cell performance using a targeted mixed mediator combination. *Electroanalysis.* 18, 2001-2007.

Xu, X., Zhu, X., 2004. Treatment of refectory oily wastewater by electro-coagulation process. *Chemosphere.* 56, 889–894.

Yang, S., Zhu, W., Wang, J., Chen, Z., 2008. Catalytic wet air oxidation of phenol over CeO2–TiO2 catalyst in the batch reactor and the packed-bed reactor. *J. Hazard. Mater.* 153, 1248–1253.

Yari, A., Kargosha, K., 2006. Simple photometric determination of free cyanide ion in aqueous solution with 2,6-dichlorophenolindophenol. *Eur. J. Chem.* 4, 329-337.

Yoshida, N., Hoashi, J., Morita, T., McNiven, S. J., Nakamura, H., Karube, I., 2001. Improvement of a mediator-type biochemical oxygen demand sensor for on-site measurement. *J. Biotechnol.* 88, 269-275.

Zhao, J., Wang, M., Yang, Z., Wang, Z., Wang, H., Yang, Z., 2007. The different behaviors of three oxidative mediators in probing the redox activities of the yeast Saccharomyces cerevisiae. *Anal. Chim. Acta*. 597, 67-74.

In: Bioremediation: Processes, Challenges and Future Prospects ISBN: 978-1-62948-513-3
Editors: Jesús Bernardino Velázquez-Fernández et al.

Chapter 6

DEVELOPMENTS IN BIOREMEDIATION STRATEGIES: ENHANCING BIODEGRADATION USING SURFACTANT AND MICRO-BUBBLES

Aditya Veer, Wei Tao and Naresh Singhal
Department of Civil and Environmental Engineering,
University of Auckland, New Zealand

ABSTRACT

Bioremediation has emerged as a promising technology for restoring contaminated soils. Its advantage over traditional methods is that the resulting end products are usually less toxic and further biodegradable. Bioremediation involves the use of living organisms, especially micro-organisms, to degrade complex organic pollutants into simpler substrates. However, the process can be limited by factors such as the bioavailability of contaminant and dissolved oxygen. A range of recalcitrant organic pollutants can potentially be biodegraded. However, in practice this degradation potential is not realised for hydrophobic compounds as their bioavailability is limited by their sorption to soil. This limitation can be overcome by enhancing the solubility of such compounds. Surfactants are known to solubilise hydrophobic compounds by micellar encapsulation. As synthetic surfactants could potentially be toxic, the use of biosurfactants is preferred for enhancing the effectiveness of bacterial degradation by increasing the bioavailability of hydrophobic organic compounds like polycyclic aromatic hydrocarbons (PAHs). In addition to bioavailability, other parameters that affect the effectiveness of remediation are the availability of electron receptors and microbes with appropriate enzymatic capability to degrade the contaminant. Microbubble suspensions can serve as a delivery vehicle for oxygen and bacteria to simultaneously achieve biostimulation and bioaugmentation in the contaminated environment. For successful implementation of these strategies *in-situ*, soil transport characteristics of the pollutant and degrader organisms need to be studied individually. Studying the transport behaviour of micro-bubble suspension will give us further insight into its application as a bioremediation strategy. This chapter discusses the use of surfactants in solution and foam to enhance bioremediation and to model the interactions between bacteria, pollutant, and surfactant to illustrate the benefits of using this approach.

Keywords: Bioremediation, PAH, biosurfactant, microbubbles, transport models

1. INTRODUCTION

Human activities have led to widespread pollution of the environment. Anthropogenic organic pollutants such as polycyclic aromatic hydrocarbons, polychlorinated biphenyls and pesticides in soil have become the focus of ongoing research due to potential threat to both wildlife and human resulting from their toxicity, resistance to degradation, and a tendency to bioaccumulate. Some of the commonly detected pollutants in soil and groundwater include chlorinated hydrocarbons, petroleum hydrocarbons, and organic compounds such as trichloroethylene, benzene, toluene, and PAHs (Grathwohl, 2011). Soil contaminants needing most attention include heavy metals (lead, arsenic, cadmium and mercury) and organic compounds (hydrocarbons, trichloroethylene and petroleum products). Residues of explosives, radionuclides, and pesticides (the latter especially near agricultural land) comprise the other common contaminants. This chapter focuses on organic compounds as the main pollutant from a perspective of their treatment via bioremediation.

Organic pollutants comprise numerous compounds from a wide range of sources. The increasing use of petroleum products and their derivatives has resulted in the accumulation of PAHs in the natural environment. These compounds are structurally composed of two or more fused aromatic hydrocarbon rings and can natural sources, such as forest fires, as well as anthropogenic causes, such as house fires, heat and energy power stations, vehicle traffic, waste incineration and industrial plants (Hao, Wan, Song, Jiang, & Peng, 2007; Elliot, Singhal, & Swift, 2011). Petroleum products comprise of a complex mix of alkanes and aromatic compounds. The long chain alkane molecules are also recalcitrant and hydrophobic, and some complex chains have very low water solubility (Baek & Kim, 2009).

Ingestion of some organic compounds can affect the central nervous system, ranging from altering the cognitive behaviour to loss of consciousness. PAH toxicity depends on the chemical structure of the compound, with isomers of a molecule displaying different toxic properties. Several PAHs are known to be carcinogenic, teratogenic and mutagenic (Luch, 2005). Due to their hydrophobic nature, they are usually found trapped in soil and sediments. They are also suspended in air attached to particulate matter. Their aversion from aqueous liquid entraps them in low permeable hydrophobic regions of soil and increases their life span in environment. Due to absence of functional groups, alkanes are not as susceptible to biodegradation in the environment and tend to persist in the environment. The heavier alkanes are classified as harmful and dangerous; in fact, hexane in straight-chain isomer configuration acts as a neurotoxin.

The drive to increase agricultural output has led to greater worldwide use of pesticides. The commonly used chemical pesticides have varying chemical structures to make them effective against weeds and pest organisms. Their purpose makes them toxic to local ecology if left unchecked in the environment. Considerable dispersal and losses of chemical occur during and after application onto land (Miller, 2004). Molecules on the surface are also prone to volatilisation or transformation by photolysis. Adsorption onto subsurface soil and leaching into groundwater spreads the contaminants further from the source and also leads to their accumulation in soil biota.

Storm water runoff carries the pesticide chemicals to rivers and lakes, compromising the quality of potable water (Katagi, 2008). Exposure to pesticides incites a variety of health effects, ranging from skin irritation to toxic effect on nervous system and cancer (Bassil, Vakil, Sanborn, Cole, Kaur, & Kerr, 2007; Sanborn, Kerr, Sanin, Cole, & Bassil, 2007).

Physical and chemical treatment methods have traditionally been used to treat a variety of environmental pollutants. However, use of these techniques for treatment of pollutants in soil can be expensive and result in formation of toxic by-products. Bioremediation can potentially be a cost-effective method for treating contaminated soils. The increased interest is on part due to reduced operational and capital costs.

Furthermore, the bioremediation may result in formation of non-toxic end products as aerobic environments promotes oxidative degradation leading to production of carbon dioxide and water as end products, while reductive transformation in anoxic environment promotes their conversion to organic acids or methane.

2. Bioremediation

Bioremediation has emerged as a promising treatment approach for restoring contaminated environments. It involves the degradation of organic pollutants into non-harmful end-products by naturally occurring or cultured microorganisms. Bioremediation can be undertaken via two different modes: *ex-situ* and *in-situ* (Table 1). In *ex-situ* treatment, polluted soil is excavated and transported to other locations for treatment. Common treatment methods include land farming, composting, biopiles and bioreactors. Due to controlled location, more deliberate techniques and different approaches can be combined and used for treatment, thereby increasing the effectiveness of biodegradation. Some of the by-products are soil forming materials and offers choice for disposal. However, this is not always the case, and thus creates issues with disposal. Operational costs are also increased on account of soil transportation and landscape reconfiguration.

On the other hand, *in-situ* bioremediation is designed to carry out in-place treatment, without causing damage to the soil ecosystem. It also avoids potential threat to human health and environment during the transport of contaminated material. Bioremediation does have a common limiting factor of treatment time, as the degradation pathways for microbes are extremely complex.

There are several factors that can lower the effectiveness of traditional bioremediation approaches. Bioremediation process is highly dependent on specific environmental factors. Temperature, pH, moisture content, presence of electron acceptors, local microbial flora, contaminant type and other such parameters require specific biodegradation techniques to be developed for specific treatment sites.

The treatment process can be difficult to monitor without invasive techniques and treatment may require a long time. It is limited by the bioavailability of pollutants, as many contaminants are hydrophobic and aggregate in impermeable spaces and soil pores where it is difficult for bacteria to get in surface-contact with. The distribution of degrader bacteria also plays a major role in the efficiency of this technique. The shortage of oxygen in subsurface can promote anaerobic degradation. (Philp, Bamforth, Singleton, & Atlas, 2005; Vidali, 2001).

Table 1. Summary of Bioremediation techniques

Technique	Examples	Benefits	Limitation	Limiting factors
Ex situ	Land farming Composting Biopiles Bioreactors	Deliberate and effective treatment design Good treatment efficiency Complete breakdown of pollutant	Extended treatment time Disposal space requirement Damage to soil ecosystem Relatively costly	Bioavailability of pollutants Undesired end-products Excavation required Transport and excavation costs
In situ	Biosparging Bioventing Bioaugmentation Biostimulation	Minimal disturbance to contaminated site Cost-effective method Naturally attenuated process, the end product is possible nontoxic	Extended treatment time Distribution of bacteria in subsurface Anaerobic reaction may take place in subsurface, leaving the risk of a residual toxic intermediate	Bioavailability of pollutants Oxygen supply Microbial and pollutants distribution Soil properties

A. Bioaugmentation and Biostimulation

The removal of PAHs is limited by the presence of biodegradable organism (Scullion, 2006; van Hamme, 2004). The natural biota available in a contaminated region may possess the ability to degrade pollutants for sustenance and require minimal concentration of the compounds, well within the limits of concentration beyond which a compound becomes a pollutant. The indigenous bacteria adapt to the availability of these compounds and develop specific degradation pathways, but this process is very slow. Hence, they are unable to treat the large amount of pollutant molecules present in the contaminated regions (Boopathy, 2000). Therefore, culturing bacterial strains that degrade the target chemical compound can significantly enhance bioremediation. Bacterial activity can stimulated by injecting essential nutrients or other required compounds into the contaminated region.

i. Advantages

In-situ bioremediation techniques relying on indigenous bacterial population need longer treatment times as the bacteria needs to adapt its metabolic and enzymatic pathways to degrade the pollutant, either as sole carbon source or as co-metabolite. External supplementation of bacteria capable of degrading the pollutant can reduce or eliminates this delay. Also, specific contaminants can be targeted in this scenario providing a desirable outcome.

ii. Limitations

Degradation requires that contaminant molecules come into contact with bacteria cells. PAHs and long chain alkanes are hydrophobic in nature, and sequester into the hydrophobic voids or non-aqueous phase liquids (NAPLs).

They are also adsorbed onto soil pores and sediments, and thus become inaccessible to bacteria for degradation (Volkering, Breure, & Rulkens, 1997). Introduction of genetically modified bacteria into the natural environment can raise concerns from public, and their use is either prohibited or severely restricted.

B. Surfactants

Surfactants, or amphiphiles, are surface active agents that possess both hydrophobic and hydrophilic moiety. In liquids, addition of surfactant molecules lowers the surface tension. However, when for surfactant concentrations above the critical micelle concentration (CMC) a further decrease in surface tension ceases. At concentrations larger than CMC, the surfactant molecules aggregate to form micelles and vesicles to minimize the free energy of the entire system (Myers 1998). Surfactants can enhance hydrocarbon contaminant bioavailability by reducing interfacial tension between the hydrophobic contaminant and aqueous phase and by trapping the contaminant in hydrophobic micelles.

i. Advantages

A growing body of research has investigated how surface active agents or surfactants can be used to overcome the limits to bioavailability of organic compounds. Some researchers have found that surfactants may increase bioavailability of both sorbed and NAPL bound organics by enhancing their affinity to transfer into aqueous phase and reducing the interfacial tension between the NAPL phase and the aqueous phase (Wan, Veerapaneni, Gadelle, & Tokunaga, 2001). Surfactants have ability to significantly enhance the solubility of organic contaminants by lowering interfacial tension and micelles solution (Bai, Brusseau, & Miller, 1998; Brown, Guha, & Jaffe, 1999; McCray, Bai, Maier, & Brusseau, 2001; Noordman & Janssen, 2002; Ron & Rosenberg, 2002; Tiehm, 1994; Paria, 2008; Tsai, Kao, & Hong, 2008).

ii. Limitations

Surfactant micelles affect different species of bacteria in different ways (Feng, 2010). Some bacteria with hydrophilic surface property become hydrophobic due to attachment of surfactant micelles, and viceversa. This changes the transport behaviour of certain bacterial species and, hence, the interactions between surfactant, bacteria and soil particles need to be understood further. Chemical surfactants pose an issue to the stability of the soil system and hence biosurfactants are preferred. The improved remediation potential is still mitigated by the absence of aerobic conditions for effective bioremediation. Some studies have shown that microbubble generated from surfactants have the potential to supply oxygen for aerobic biodegradation in oxygen deficient environments (Choi, Kim, & Nam, 2009).

C. Foams

Foams have properties of all three forms of matter. Shear is elastically supported as in a solid material, but they can also deform into arbitrary shapes and flow like liquids.

And just like gas, they are highly compressible, thereby making way for innovative applications. Stability, rheology, chemical composition and physical structure all contribute to the uniqueness and complicated behaviour of foam (Kornev, Neimark, & Rozhkov, 1999). Bubble size in typical foams range from 10 – 100 μm (Jauregi, Mitchell, & Varley, 2000). Studies have found that such microbubble suspensions exhibit a larger interfacial area, possess relatively high stability, easily dissociate from the bulk liquid, and have water-like flow properties (Jauregi, Gilmour, & Varley, 1997; Sebba, 1987; Save & Pangarkar, 1994).

i. Advantages

Foam solves the limitation of oxygen deficiency in subsurface regions. The surplus air provided significantly promotes aerobic degradation and formation of non-toxic end products. Since it is produced from surfactants, it possesses all the advantages of surfactants, and their applicability is improved by favourable bacteria-foam interactions. Certain contaminants adsorb onto the microbubbles and are degraded by the bacteria present on the bubble surface (Wan, Veerapaneni, Gadelle, & Tokunaga, 2001). It has also been found that foam is able to penetrate low permeable regions of soil and treat seemingly inaccessible contaminant molecules (Choi, Park, Kim, & Nam, 2008).

ii. Limitations

The transport behaviour of foam is at present poorly understood. Pollutants can possibly be transported out of, and exit, the treatment region via microbubble movement, thereby spreading further into the environment. While microbubble suspension movement is reminiscent of plug flow, complex interactions occur between foam bubbles, bacteria, pollutant and soil. These may or may not enhance biodegradation efficiency. As such, effective bioremediation designs have not made wide use of this promising approach.

This chapter discusses the benefits of surfactants and microbubble suspensions on *in-situ* bioremediation towards treatment of hydrophobic organic compounds, especially polycyclic aromatic hydrocarbons (PAH). Effects of surfactant micelles on bacterial cell surface, influence of surfactant and foam on bacterial distribution, and oxygen supply through foam are also discussed. The transport behaviour is reviewed to understand the fundamental parameters that affect the design of engineered bioremediation approaches.

3. Characteristics of Surfactant

A. Surfactant Structure

Surfactants are amphiphilic chemical compounds that possess a hydrophobic and hydrophilic moiety on either ends of their structure (Figure 1). This special structure allows them to reduce the surface tension between two immiscible liquids (Lang, 2002). The chemical molecules form micelles in dissolved surfactant solutions. These micelles aggregate at the interface between the liquids, or at liquid surface exposed to air (which is a gaseous fluid) in a two-phase system. The hydrophobic end of the micelle associates with the non-aqueous fluid phase and hydrophilic group interacts with the aqueous liquid. This type of bonding reduces the free surface energy of the system at the interface and thereby reduces the surface tension (Brown & Jaffe, 2006). The reduction in tension promotes the non-aqueous phase solutes and solvent molecules to become miscible with the aqueous phase, and increases their apparent solubility in the aqueous medium.

There is a limit to the reduction in surface tension achieved by a particular surfactant. Initially, at low concentrations, the reduction in surface tension is small. As the concentration increases logarithmically, the surface tension reduces linearly.

Beyond the CMC further increases in surfactant concentration do not reduce the surface tension. Physically, the addition of surfactant increases the concentration of micelles in the dissolved surfactant solution.

As micelles run out of free space, they tend to aggregate to form micelles with a spherical structure in the aqueous media, with inward pointing hydrophobic tails and hydrophilic micelle heads protruding outwards (Figure 2). This spherical formation gives rise to a hydrophobic interior region, enhancing the apparent aqueous solubility of several hydrophobic compounds by encapsulating the compound in this hydrophobic region of the micelle sphere. Other conformations are possible at concentrations higher than the CMC value. These usually involve presence of a mixture of singular micelles and compact spherical groups with complex interplay. As a side note, the structure shown is valid for surfactant dissolved in aqueous solution. This structure inverts when mixed with non-aqueous phase solution, like oil. Since the hydrophobic ends have a high affinity to fat (lipophilic), the micelle tails project outside and the hydrophilic heads aggregate towards the centre to form an aqueous core within the spherical structure (Elliot, Singhal, & Swift, 2011).

B. Types of Surfactant

i. Synthetic Surfactants

Synthetic surfactants are classified according the property of the polar head group. They are prepared from long-chain organic compounds by modifying one end with anionic or cationic groups.

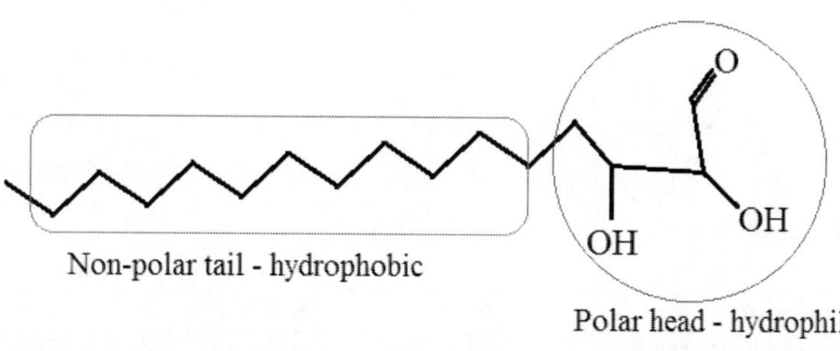

Figure 1. General structure of surfactant. The charge on the polar head gives the surfactant specific behavioural properties. Based on the charge, they can be classified as anionic (-ve charge), cationic (+ve charge), non-ionic or zwitter ionic (both +ve and -ve).

However, these chemicals can be toxic to certain life forms. They are known to interfere with life cycle of aquatic organisms and some chelating agents damage plants. Synthetic surfactants could pose secondary pollution due to their chemical nature (Rothmel, Deflaun, Peters, & St. Martin, 1998; Tiehm, 1994; Hartmann, 1966), which limits their application in bioremediation.

ii. Biosurfactants

Certain bacteria also possess an ability to produce surfactants naturally. These types of surfactants are known as biosurfactants. Due to their biological nature, they are easy to degrade, are non-toxic and do not contribute to pollution (Abalos, Viñas, Sabaté, Solanas, & Manresa, 2004). They are classified according to their major structural features and biological origin. Rhamnolipids are most frequently studied and commercially applicable biosurfactants (Nitschke, Costa, & Contiero, 2005). Biosurfactants have emerged as an alternative to synthetic surfactant owing to their low ecological toxicity and biodegradable nature. The choice of surfactant depends on the desired interaction with bacteria and pollutant, and cost-efficiency. Table 2 lists the various surfactants traditionally used in bioremediation.

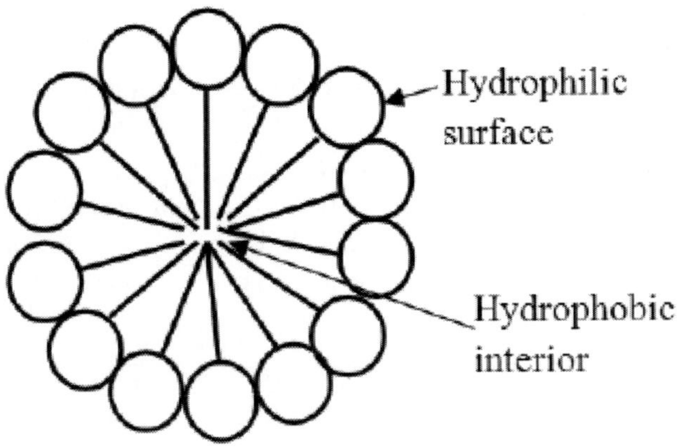

Figure 2. Spherical surfactant micelle structure. The individual surfactant micelles are aligned with the hydrophilic ends pointing into the aqueous solution surrounding it.

Table 2. Characteristics of common surfactants

Surfactant	Chemical structure	CMC value (ppm)	Charge	Reference
Tergitol 5-S-12	$C_{11\text{-}15}H_{23\text{-}31}O(CH_2CH_2O)_{12}H$	104	Nonionic	(Union Carbide, 1996)
SDS	$C_{12}H_{25}OSO_3^-Na^+$	2300	Anionic	(Fendler & Fendler, 1975)
Tween 20	R=monolayrate	60	Nonionic	(Wan & Lee, 1974)
Tween 80	R=monooleate	14	Nonionic	(Wan & Lee, 1974)
Triton-X100	$(4\text{-}C_8H_{17})C_6H_4O(CH_2CH_2O)_{75}H$	130	Nonionic	(Kile & Chiou, 1989)
Rhamnolipid	$C_{26}H_{48}O_9(R_1)C_{32}H_{58}O_{13}(R_2)$	57.7	Anionic	(Clifford, Ioannidis, & Legge, 2007; Zhang & Miller, 1994)

Rhamnolipids have been increasingly used as the surfactant of choice for treating diverse pollutants like phenanthrene, anthracene, hexadecane, pyrene, PAHs, phenol, and crude oil hydrocarbons (Lawniczak, Marecik, & Chrzanowski, 2013).

Besides the advantage presented above, biosurfactants have their peculiar advantages. Firstly, unlike synthetic surfactants resisting degradation and accumulating in the environment, biosurfactants can be degraded naturally by microorganisms. Secondly, biosurfactants have been found to demonstrate specific interface selections. Moreover, biosurfactants can reduce average particle size as well as particle surface properties fundamentally. Furthermore, a wide range of biosurfactants can be produced by microorganism for industrial demand (Cooper, 1986; Oberbremer, Muller-Hurtig, & Wagner, 1990; Cameotra, Makkar, & Mehta, 2010).

C. Surfactant Properties and Effects

i. Interaction with Soil

The presence of dissolved surfactant allows the bulk fluid to overcome resistance to flow through porous media and maintain the same pressure gradient for flow, thereby homogenising the flow rate (Lioumbas, Mouza, & Paras, 2006; Brown & Jaffe, 2006). The surface of porous media provides an adsorption bed for surfactant micelles, which prevents solute particles in bulk fluid from adsorbing onto the porous media. The adsorbed surfactant micelles form a stable coating around the stationary phase (porous media) and enhance reversible mass transfer of solutes between this phase and the mobile phase (bulk fluid). This lowers the loss of solute during transport and visibly reduces its retardation. This property enables the surfactant to sequester into hydrophobic pore spaces of soil and solubilise NAPLs and organic contaminants present there into the bulk aqueous fluid.

ii. Interaction with Pollutant

As discussed before, the presence of dissolved surfactant increases the apparent solubility of hydrophobic organic compounds in the aqueous phase. The formation of spherical micellar structures aid in this process (Shin, Kim, Kim, Lee, & Han, 2008; Guha, Jaffe, & Peters, 1998). Organic molecules prefer sequestration into the hydrophobic void spaces in soil, to avoid the soil aqueous phase. When dissolved surfactant solution is added to the soil, the surfactant micelles are known to adsorb onto soil surfaces. This leads to complex interactions between the surfactant, pollutant molecules and soil. Because the surfactant is able to solubilise the organic pollutant into the micellar structure, the sequestration of pollutants onto the soil surface is reduced. Simultaneously, pollutants would likely partition and solubilise into the surfactant already adsorbed onto the soil particles (Krogh, Halling-Sørensen, Mogensen, & Vejrup, 2003; Pennell, Karagunduz, & Young, 2003). The presence of non-ionic surfactants markedly increased desorption of pyrene and phenanthrene in sand-loamy soil (Cheng & Wong, 2006). Anionic surfactants were shown to increase the solubility of PAHs in the aqueous phase, indicating that surfactants are able to desorb PAH molecules from soil and similar porous media (Grasso, Subramaniam, Pignatello, Yang, & Ratte, 2001). Bacteria delivered to this region tended to be present in the aqueous phase.

The increased concentration of pollutants in the aqueous phase greatly facilitates their contact with bacterial cells, thereby promoting efficient biodegradation.

iii. Interaction with Bacteria

Chen et al. (2004) indicated that bacteria retained on silica sand decrease in the presence of rhamnolipid. More particularly, the interaction between bacteria and sand in porous media decreased with increasing rhamnolipid concentration. When in contact with certain specific species of bacteria, dissolved surfactant tends to change the properties of bacterial cell surface in the culture suspension. Non-ionic surfactants were found to reduce the hydrophilic property of certain *Sphingomonas* species by altering their cell surface, to the extent that some species were altered to have hydrophobic cell surfaces (Brown & Jaffe, 2006). The surfactant micelles associate their hydrophilic moieties towards the hydrophilic cell surface, and the hydrophobic ends on the other side protrude outward to give rise to that characteristic property.

However, this does not fully explain the changes observed to adhesiveness of bacterial cells in the presence of dissolved surfactant. These alterations in surface property increase the efficiency of transport of bacteria in the subsurface environment (Feng, 2010).

4. CHARACTERISTICS OF FOAM

A. Foam Stability and Quality

The stability of foam suspension depends on the type of surfactant used, the concentration of surfactant micelles, and presence of dissolved or suspended solutes in the solution (Feng, 2010). Apaydin and Kovscek (2001) found that decreasing the surfactant concentration results in lowering the foam displacement efficiency and increasing gas mobility, implying that the foam weakens at lower surfactant concentrations. The foam stability is often expressed as the half-drainage time, which is the time required for half of the surfactant liquid to drain from the microbubble suspension.

Foam quality is determined as the gas hold-up in foam suspension. A higher gas fraction in wet foam bubbles is indicative of a more stable foam suspension. Increased gas content is also beneficial from the point of view of aerobic biodegradation. A comparison of foam stability and quality for different traditional surfactants is shown in Table 3

Table 3. Comparing foam stability of different surfactants

Surfactant	Foam stability (seconds)	Bubble size distribution	Air hold up
Tergitol 5-S-12	383-391	Diameter range 30-300um	69%-70%
SDS(a)	341-459	---	69%-72%
Tween 80	413-498	---	---
Rhamnolipid	385-546s	20-140um	67%-72%
Saponins	170-720	30-300	50%-70%

Adapted from Feng, Singhal and Swift (2009).

B. Effects of Foam

i. Overcoming Subsurface Heterogeneity

Mamun et al. (2002) and Wan et al. (2001) observed plug-flow behaviour of foam through porous media. They hypothesized that this should enable foam to penetrate the low permeability zones and flow uniformly through heterogeneous media. Choi et al. (2008) experimentally observed this to be the case and noted the difference from liquid flow, which seemed to avoid low permeability zones by flowing around it. With this behaviour, it seems possible that foam can penetrate liquid-impermeable heterogeneous soil regions and supply bacteria, nutrients and oxygen to trapped hydrophobic pollutants.

ii. Oxygen Supply

A crucial limitation for bioremediation in subsurface is oxygen availability. The release of gas contained in foam can provide oxygen to create aerobic conditions and enhance biodegradation. Michelsen et al. (1984) successfully showed that microbubble suspensions were able to increase the aerobic degradation of phenol. Similar observations were made for p-xylene (Jenkins, Michelsen, & Novak, 1993) and pentachlorophenol (Mulligan & Eftekhari, 2003). By monitoring levels of oxygen and carbon dioxide, it has been experimentally observed that the oxygen supplied by microbubble dispersion was used to degrade phenanthrene (Choi, Kim, & Nam, 2009).

iii. Dispersion

Bubble entrapment during foam flow through porous media alters the dynamics of transport behaviour and this complicates the analysis of solute transport in foam. Gupta et al. (1994) reported difficulty in interpreting tracer results in the presence of air bubbles due to significant retardation observed in the injected helium tracer. Earlier models considered the hydraulic head to be the primary dependent variable, and neglected to account for bubble entrapment and aqueous phase saturation. Attempts to address this issue consider the existence of a continuous and discontinuous (trapped) phase, possessing its respective saturation values (Amos & Mayer, 2006). However, Kaluarachchi and Parker (1992) reported that entrapment of fluid by the non-wetting (gas) phase causes a hysteretic relationship between saturation and capillary pressure. Foam dispersion studies are still incomplete and further investigation is needed to characterise transport of volatile contaminants attached to subsurface in presence of foam. In addition to pollutant transport, bacterial transport is also of interest. Jackson et al. (1998) observed increased transport of bacteria in soil columns under foam flow. It has been known that bacterial cells are partially attached to the bubble surface at the interface between liquid and gas, and foams can contribute to increased transport and dispersion of bacteria (Ripley, Harrison, Betts, & Dart, 2002).

5. Tools for Assessing Performance

The partitioning and transport process (sorption, desorption and dissolution) of both contaminants and surfactants between the soil and water phase affect the degradation of contaminants.

Degradation of pollutant by bacteria is just one of numerous interactions taking place in an *in*-situ bioremediation system. Interactions occur between pollutant and surfactant, surfactant and soil, soil and bacteria, and bacteria and surfactant. All of these interactions influence the efficiency of degradation and the fate of pollutants in the subsurface environment. The transport characteristics of bacteria and pollutants can significantly affect the design and implementation of effective bioremediation approaches.

A. Transport Models

Conducting field experiments to obtain meaningful values of dispersion and retardation of bacteria/pollutant in the delivery fluid, along with effective transport velocities can take decades of observation (Seuntjiens, 2002), especially for compounds with strong sorption tendencies. Due to practical constraints associated with monitoring, transport parameters are often studied experimentally in laboratory using bench-scale soil columns, which can be analysed using one of several one-dimensional transport models that are available to characterise the transport behaviour of solutes in porous medium (van der Zee & van Riemsdijk, 1987).

i. Darcy's Law

Darcy's law is the basic principal governing one dimensional fluid flow through porous media. It is given in equation form as

$$\frac{Q}{A} = -K\frac{dh}{dl} \tag{1}$$

where Q is flow rate of liquid (kg/m^3), A is the area of cross section of flow (m^2), K is the hydraulic conductivity of the solute, and $\frac{dh}{dl}$ is the hydraulic gradient, which is the difference in pressure heads between two points, located axially along the length of flow. The conductivity of fluid is dependent on the characteristics of the porous media like particle size and porosity as well as fluid properties such as viscosity (van Genuchten, 1980).

ii. Convection-Dispersion Equation

The one dimensional convection dispersion equation (CDE) is often used to describe solute transport. Solute is assumed to be transported in fluid by two methods – advection and dispersion. The flow of solute is also retarded due to its interaction with porous media. CDE accounts for these interactions and provides a fairly accurate picture of transport behaviour of conservative solutes, and is given in equation form as

$$\frac{\partial C}{\partial t} = D\frac{\partial^2 C}{\partial x^2} - v\frac{\partial C}{\partial x} \tag{2}$$

where C is the concentration of solute in fluid (kg/m^3), t is time of transport (s), v is pore velocity of the fluid (m/s), x is distance of location of interest from inlet (m), and D is the hydrodynamic dispersion coefficient (m^2/s). The pore velocity is averaged for the cross section and is dependent on the porosity of the porous medium.

The term on the left hand of Equation 2 side describes the changes in concentration of the solute at a particular location with respect to time. The two terms on the right hand side of Equation 2 describe dispersive and advective transport (van Genuchten, 1980). The extended model includes parameters for a reactive solute, where the decay or generation of the solute during transport is also accounted for.

B. Analytical Solutions for Transport Models

Many analytical solutions derived in the past for solute transport can be applied to analyse transport under steady-state conditions. STANMOD (STudio of ANalytical MODels) is a software package containing solutions to the convection-dispersion solute transport equation in porous media for various conditions (Simunek, van Genuchten, Sejna, Toride, & Leij, 2005-2009). The software is powerful and user friendly, allowing input of critical parameters required for estimation of tracer transport. It offers choice of estimating certain parameters as well as keeping parameters within certain constraints during the estimation process. Detailed information about individual models, their application and STANMOD modelling methods can be found in Šimůnek (2007).

CXTFIT has been widely used to model one-dimensional solute transport through homogeneously packed porous media under equilibrium fluid flow (Toride, Leij, & van Genuchten, 1995). Column tracer trials are conducted to observe the progression of solute through the soil column under constant flow. A single-pulse input of tracer solute in injected in the column, and its concentration is noted at different time intervals at different locations along the column. Earlier researchers (Indelman, Touber-Yasur, Yaron, & Dagan, 1998; Zhang, Huang, & Xiang, 1994; Huang, Toride, & Van Genuchten, 1995) preferred to use a step input, where by the tracer solute is continuously injected at a fixed concentration until the end of the experiment. However, pulse input provides a more reliable method for estimating transport parameters as fewer possibilities exist for curve-fitting, thereby narrowing parameter deviations among the small number of eventualities (Wan, Veerapaneni, Gadelle, & Tokunaga, 2001).

C. Observed Success in Bioremediation

Recent studies have shown that surfactants, especially biosurfactants, can enhance the treatment of organics. Kim et al. (2001) reported that the bioavailability of pollutant molecules is the rate limiting step in biodegradation processes, and for a bacterial culture acclimatized to the pollutant they showed that improving phenanthrene desorption enhanced the rate of *in*-situ bioremediation treatment. Laboratory experiments by Hickey et al. (2007) showed significant removal of fluoranthene in presence of surfactants. The degradation efficiency was higher for biosurfactant JBR rhamnolipid (34%) than the synthetic surfactant Tween 80 (17%). They also found that rhamnolipid surfactant at CMC did not inhibit bacterial growth. However, phenanthrene degradation was either delayed or completely inhibited in the presence of synthetic non-ionic surfactants present at concentrations above their CMC value (Yuan, Wei, & Chang, 2000).

Similarly, mixed observations have been made for treatment using biosurfactants. Presence of rhamnolipid was found to increase the degradation of long-chain hydrocarbons from sand and silt loamy soils by 25-70% (Scheibenbogen, Zytner, Lee, & Trevors, 1994). However, using the same surfactant to degrade PAHs showed a decrease in pollutant removal. It has been suggested that in certain experiments, the biosurfactant was used as the preferred carbon source, thereby leaving PAHs untreated (Deschenes, Lafrance, Villeneuve, & Samson, 1996). Recently, a pilot-scale plant was developed and tested for flushing polluted soil region with surfactant solution (Svab, Kubal, Müllerova, & Raschman, 2009). The theoretical models were in agreement with the concentration profiles observed for the pollutant (polychlorinated biphenyls) over 6-12 months. Park et al. (2009) declared that nearly 30% of phenanthrene was removed in 21days with microbubble suspension in sand/clay packed column, while no degradation of phenanthrene occurred in the absence of microbubbles. Biodegradation mainly occurred at the lower part of column, where oxygen delivery by microbubble was most effective.

CONCLUSION

Organic pollutants are highly recalcitrant and hydrophobic compounds. Their presence in the subsurface soil regions poses significant issues for traditional bioremediation approaches. Delivery of nutrients, bacteria and other compounds in water to the subsurface encountered several challenges in so far as the degradation of long chain hydrocarbons and polycyclic aromatic hydrocarbons is concerned. New approaches have been considered to overcome the problem of bioavailability of contaminants to degrader bacteria and improving the environmental conditions for favourable treatment efficiency. Of these, the use of surfactants, especially in the form of foam, seems to be quite promising.

Dissolved surfactant is capable of delivering bacteria and nutrients to the contaminated soil regions. The surfactant micelles are able to solubilise the hydrophobic pollutant molecules in the aqueous phase, thereby making them available for bacteria to act upon. This has a marked increase in the efficiency of bioremediation treatment system. The presence of surfactant also affects the properties of fluid flow and bacterial cells transport.

Foam suspensions (microbubbles) offer an even more advantageous delivery mechanism. Among the improvements is the ability to supply air (oxygen) to oxygen-deprived subsurface regions and the ability to penetrate low permeability regions. The stability and quality of foam generated depends on the specific surfactant used and its concentration in aqueous medium. At present, foam transport behaviour is poorly understood and this needs to be dealt extensively to design an effective bioremediation treatment system.

REFERENCES

Abalos, A., Viñas, M., Sabaté, J., Solanas, A. M., and Manresa, M. A. (2004). Enhanced biodegradation of Casablanca crude oil by a microbial consortium in presence of a rhamnolipid produced by Pseudomonas aeruginosa AT10. *Biodegradation,* 15(4), 249-260.

Amos, R. T. and Mayer, K. U. (2006). Investigating the role of gas bubble formation and entrapment in contaminated aquifers: Reactive transport modelling. *Journal of Contaminant Hydrology,* 87(1-2), 123-154.

Apaydin, O. and Kovscek, A. R. (2001). Surfactant Concentration and End Effects on Foam Flow in Homogeneous Porous Media. *Transport in Porous Media,* 43(3), 511-536.

Baek, K. and Kim, H.-S. (2009). Microbial Community Structure in Hexadecane- and Naphthalene-Enriched Gas Station Soil. *Journal of Microbiology and Biotechnology,* 19 (7), 651-657.

Bai, G., Brusseau, M. L. and Miller, R. M. (1998). Influence of cation type, ionic strength, and pH on solubilization and mobilization of residual hydrocarbon by a biosurfactant. *Journal of Contaminant Hydrology,* 30(3-4), 265-279.

Bassil, K. L., Vakil, C., Sanborn, M., Cole, D. C., Kaur, J. S., and Kerr, K. J. (2007, October). Cancer health effects of pesticides: Systematic review. *Canadian Family Physician,* 53 (10), 1704-1711.

Boopathy, R. (2000). Factors limiting bioremediation technologies. *Bioresource Technology,* 74, 63-67.

Brown, D. G. and Jaffe, P. R. (2006). Effects of Nonionic Surfactants on the Cell Surface Hydrophobicity and Apparent Hamaker Constant of a Sphingomonas sp. *Environmental Science and Technology,* 40, 195-201.

Brown, D. G., Guha, S. G. and Jaffe, P. R. (1999). Surfactant-Enhanced Biodegradation of PAH in Soil Slurry Reactors. *Bioremediation Journal,* 3(3), 269-283.

Cameotra, S. S., Makkar, K. J. and Mehta, S. K. (2010). Synthesis of biosurfactants and their advantages to microorganisms and mankind. *Biosurfactants,* 672, 261-280.

Chen, G., Qiao, M., Zhang, H., and Zhu, H. (2004). Bacterial desorption in water-saturated porous media in the presence of rhamnolipid biosurfactant. *Research in Microbiology,* 155(8), 655-661.

Cheng, K. Y. and Wong, J. W. (2006). Effect of synthetic surfactants on the solubilization and distribution of PAHs in water-soil systems. *Environmental Technology,* 27, 835-844.

Choi, Y. J., Kim, Y.-J. and Nam, K. (2009). Enhancement of aerobic biodegradation in an oxygen-limiting environment using a saponin-based microbubble suspension. *Environmental Pollution,* 157, 2197-2202.

Choi, Y. J., Park, J. Y., Kim, Y.-J., and Nam, K. (2008). Flow Characteristics of Microbubble Suspensions in Porous Media as an Oxygen Carrier. *Clean,* 36(1), 59-65.

Clifford, J. S., Ioannidis, M. A. and Legge, R. L. (2007). Enhanced aqueous solubilization oftetrachloroethylene by a rhamnolipid biosurfactant. *Journal of Colloid and Interface Science,* 305(2), 361-365.

Cooper, D. G. (1986). Biosurfactants. *Microbiological Sciences,* 3(5), 145-149.

Deschenes, L., Lafrance, P., Villeneuve, J. P., and Samson, R. (1996). Adding sodium dodecyl sulfate and Pseudomonas aeruginosa UG2 biosurfactants inhibits polycyclic aromatic hydrocarbon biodegradation in a weathered creosote-contaminated soil. *Applied Microbiology and Biotechnology,* 46, 638-646.

Elliot, R., Singhal, N. and Swift, S. (2011). Surfactants and Bacterial Bioremediation of Polycyclic Aromatic Hydrocarbon Contaminated Soil—Unlocking the Targets. *Critical Reviews in Environmental Science and Technology,* 41, 78-124.

Fendler, J. and Fendler, E. (1975). *Catalysis in Micellar and Macromolecular Systems.* New York: Academic Press.

Feng, W. (2010). *A Study of Rhamnolipid Microbubble Dispersion for Bioremediation Applications: Dispersion Properties and Bacteria/Surfactant/Contaminant Interactions.* PhD Thesis, University of Auckland.

Feng, W., Singhal, N. and Swift, S. (2009). Drainage mechanism of microbubble dispersion and factors influencing its stability. *Journal of Colloid and Interface Science,* 337, 548-554.

Grasso, D., Subramaniam, K., Pignatello, J. J., Yang, Y., and Ratte, D. (2001). Micellar desorption of polynuclear aromatic hydrocarbons from contaminated soil. *Colloids and Surfaces A: Physicochemical and Engineering Aspects,* 194, 65-74.

Grathwohl, P. (2011). Fate and Transport of Organic Compounds in(to) the Subsurface Environment. In: B. Xing, N. Senesi and P. M. Huang (Eds.), *Biophysico-Chemical Processes of Anthropogenic Organic Compounds in Environmental Systems* (pp. 215-231). John Wiley and Sons, Inc.

Guha, S., Jaffe, P. R. and Peters, C. A. (1998). Solubilization of PAH Mixtures by a Nonionic Surfactant. *Environmental Science and Technology,* 32(7), 930-935.

Gupta, S. K., Lau, L. S. and Moravcik, P. S. (1994). Groundwater Tracing with injected Helium. *Ground Water,* 32(1), 96-102.

Hao, R., Wan, H. F., Song, Y. T., Jiang, H., and Peng, S. L. (2007). Polycyclic aromatic hydrocarbons in agricultural soils of the southern Subtropics, China. *Pedosphere,* 17, 673-680.

Hartmann, L. (1966). Effect of surfactants on soil bacteria. *Bulletin of Environmental,* 1(6).

Hickey, A. M., Gordon, L., Dobson, A. D., Kelly, C. T., and Doyle, E. M. (2007). Effect of surfactants on fluoranthene degradation by Pseudomonas alcaligenes PA-10. *Applied Microbiology and Biotechnology,* 74, 851-856.

Huang, K., Toride, N. and Van Genuchten, M. T. (1995). Experimental Investigation of Solute Transport in Large, Homogeneous and Heterogeneous, Saturated Soil Columns. *Transport in Porous Media,* 18, 283-302.

Indelman, P., Touber-Yasur, I., Yaron, B., and Dagan, G. (1998). Stochastic analysis of water flow and pesticides transport in a field experiment. *Journal of Contaminant Hydrology,* 32, 77-97.

Jackson, A., Kommalapati, R., Roy, D., and Pardue, J. (1998). Enhanced transport of bacteria through a soil matrix using colloidal gas aphron suspensions. *Journal of Environmental Science and Health,* 33, 369-384.

Jauregi, P., Gilmour, S. and Varley, J. (1997). Characterisation of Colloidal Gas Aphrons for Subsequent Use for Protein Recovery. *Chemical Engineering Journal,* 65, 1.

Jauregi, P., Mitchell, G. R. and Varley, J. (2000). Colloidal Gas Aphrons (CGA): Dispersion and Structural Features. *AIChE Journal,* 46(1), 24-36.

Jenkins, K. B., Michelsen, D. L. and Novak, J. T. (1993). Application of oxygen microbubbles for in situ biodegradation of p-xylene-contaminated groundwater in a soil column. *Biotechnology Progress,* 9, 394-400.

Kaluarachchi, J. J. and Parker, J. C. (1992). Multiphase Flow with a Simplified Model for Oil Entrapment. *Transport in Porous Media,* 7(1), 1-14.

Katagi, T. (2008). Surfactant Effects on Environmental Behavior of Pesticides. In: D. M. Whitacre (Ed.), *Reviews of Environmental Contamination and Toxicology* (pp. 71-177). Springer.

Kile, D. E. and Chiou, C. T. (1989). Water solubility enhancements of DDT and trichlorobenzene by some surfactants below and above the critical micelle concentration. *Environmental Science and Technology,* 23, 832-838.

Kim, I. S., Park, J. S. and Kim, K. W. (2001). Enhanced biodegradation of polycyclic aromatic hydrocarbons using nonionic surfactants in soil slurry. *Applied Geochemistry,* 16, 1419–1428.

Kornev, K. G., Neimark, A. V. and Rozhkov, A. N. (1999). Foam in porous media: thermodynamic and hydrodynamic peculiarities. *Advances in Colloid and Interface Science,* 82, 127-187.

Krogh, K. A., Halling-Sørensen, B., Mogensen, B. B., and Vejrup, K. V. (2003). Environmental properties and effects of nonionic surfactant adjuvants in pesticides: a review. *Chemosphere,* 50, 871-901.

Lang, S. (2002). Biological amphiphiles (microbial surfactants). *Current Opinion in Colloid and Interdace Science,* 7, 12-20.

Lawniczak, L., Marecik, R. and Chrzanowski, L. (2013). Contributions of biosurfactants to natural or induced bioremediation. *Applied Microbiology and Biotechnology,* 97, 2327-2339.

Lioumbas, J. S., Mouza, A. A. and Paras, S. V. (2006). Effect of surfactant additives on co-current gas-liquid downflow. *Chemical Engineering Science,* 61, 4605-4616.

Luch, A. (2005). *The Carcinogenic Effects of Polycyclic Aromatic Hydrocarbons.* London: Imperial College Press.

Mamun, C. K., Rong, J. G., Kam, S. I., Liljestrand, H. M., and Rossen, W. R. (2002). Extending foam technology from improved oil recovery to environmental remediation. *SPE Annual Technical Conference Exhibition*, (p. SPE 77557). San Antonio, TX.

McCray, J. E., Bai, G., Maier, R. M., and Brusseau, M. L. (2001). Biosurfactant-enhancedsolubilization of NAPL mixtures. *Journal of Contaminant Hydrology,* 48(1-2), 45-68.

Michelsen, D. L., Wallis, D. A. and Sebba, F. (1984). In-situ biological oxidation of hazardous organics. *Environmental Progress,* 3, 103-107.

Miller, G. T. (2004). *Sustaining the Earth* (6th ed.). Pacific Grove, California: Thompson Learning, Inc.

Mulligan, C. N. and Eftekhari, F. (2003). Remediation with surfactant foam of PCP-contaminated soil. *Engineering Geology,* 70, 269-279.

Nitschke, M., Costa, S. G. and Contiero, J. (2005). Rhamnolipid Surfactants: An Update on the General Aspects of These Remarkable Biomolecules. *Biotechnology Progress,* 21(6), 1593-1600.

Noordman, W. H. and Janssen, D. B. (2002). Rhamnolipid stimulates uptake of hydrophobic compounds by Pseudomonas aeruginosa. *Applied and Environmental Microbiology,* 68 (9), 4502-4508.

Oberbremer, A., Muller-Hurtig, R. and Wagner, F. (1990). Effect of the addition of microbial surfactants on hydrocarbon degradation in a soil population in a stirred reactor. *Applied Microbiology and Biotechnology,* 32(4), 485-489.

Paria, S. (2008). Surfactant-Enhanced Remediation of Organic Contaminated Soil and Water. *Advances in Colloid and Interface Science,* 138, 24-58.

Park, J. Y., Choi, Y. J., Moon, S., Shin, D. Y., and Nam, K. (2009). Microbubble suspension as a carrier of oxygen and acclimated bacteria for phenanthrene biodegradation. *Journal of Hazardous Materials,* 163, 761-767.

Pennell, K. D., Karagunduz, A. and Young, M. H. (2003). Impacts of surfactant adjuvants on pesticide availability and transport in soils. In: J. J. Gan, P. C. Zhu, S. D. Aust, and A. T. Lemley (Eds.), *Pesticide Decontamination and Detoxification* (pp. 231-245). Washington, DC: American Chemical Society.

Philp, J. C., Bamforth, S. M., Singleton, I., and Atlas, R. M. (2005). Environmental pollution and restoration: a role for bioremediation. *Environmental Pollution,* 157(8-9), 2197-2202.

Ripley, M. B., Harrison, A. B., Betts, W. B., and Dart, R. K. (2002). Mechanisms for enhanced biodegradation of petroleum hydrocarbons by a microbe-colonized gas-liquid foam. *Journal of Applied Microbiology,* 92, 22-31.

Ron, E. Z. and Rosenberg, E. (2002). Biosurfactants and oil bioremediation. *Current Opinion in Biotechnology,* 13(3), 249-252.

Rothmel, R. K., Deflaun, M. F., Peters, R. W., and St. Martin, E. (1998). Surfactant. *Environmental Science and Technology,* 32(11), 1667-1675.

Sanborn, M., Kerr, K. J., Sanin, L. H., Cole, D. C., and Bassil, K. L. (2007, October). Non-cancer health effects of pesticides: Systematic review and implications for family doctors. *Canadian Family Physician,* 53(10), 1712-1720.

Save, S. V. and Pangarkar, V. G. (1994). Characterisation of Colloidal Gas Aphrons. *Chemical Engineering Communications,* 127, 35.

Scheibenbogen, K., Zytner, R. G., Lee, H., and Trevors, J. T. (1994). Enhanced removal of selected hydrocarbons from soil by Pseudomonas aeruginosa UG2 biosurfactants and some chemical surfactants. *Journal of Chemical Technology and Biotechnology,* 59, 53-59.

Scullion, J. (2006). Remediating polluted soils. *Naturwissenschaften,* 93, 51-65.

Sebba, F. (1987). *Foams and Biliquid Foams: Aphrons.* Hoboken, NJ: John Wiley and Sons.

Seuntjiens, P. (2002). Field-Scale Cadmium Transport in a Heterogenous Layered Soil. *Water, Air, and Soil Pollution,* 140, 401-423.

Shin, K. H., Kim, K. W., Kim, J. Y., Lee, K. E., and Han, S. S. (2008). Rhamnolipid Morphology and Phenanthrene Solubility at Different pH Values. *Journal of Environmental Quality,* 37, 509-514.

Šimůnek, J. (2007). Analytical and Numerical Modeling of Physical and Chemical Processes in the Vadose Zone. In: A. Ebel and T. Davitashvili (Eds.), *Air, Water and Soil Quality Modelling for Risk and Impact Assessment* (pp. 221-233). Springer.

Simunek, J., van Genuchten, M. T., Sejna, M., Toride, N., and Leij, F. J. (2005-2009). STANMOD version 2.08.1130 Studio of Analytical Models for Solving the Convection-Dispersion Equation. *The STANMOD Computer Software for Evaluating Solute Transport in Porous Media Using Analytical Solutions of Convection-Dispersion Equation.* Riverside, California, US. Retrieved from http://www.hydrus3d.com.

Svab, M., Kubal, M., Müllerova, M., and Raschman, R. (2009). Soil flushing by surfactant solution: Pilot-scale demonstration of complete technology. *Journal of Hazardous Materials,* 163, 410-417.

Tiehm, A. (1994). Degradation of polycyclic aromatic hydrocarbons in the presence of synthetic surfactants. *Applied and Environmental Microbiology,* 60(1), 258 -263.

Toride, N., Leij, F. J. and van Genuchten, M. T. (1995). The CXTFIT code for estimating transport parameters from laboratory or field tracer experiments. Version 2.0. *Research Report No. 137.* US Salinity Laboratory, USDA, ARS, Riverside, CA.

Tsai, T. T., Kao, C. M. and Hong, A. (2008). Treatment of tetrachloroethylene-contaminated groundwater rby surfactant-enhanced persulfate/BOF slag oxidation- a laboratory feasibility study. *Journal of Hazardous Materials,* 171(1-3), 571-576.

Union Carbide. (1996). *Tergitol Surfactants.* Danbury.

Van der Zee, S. E. and van Riemsdijk, W. H. (1987). Transport of Reactive Solutes in Spatially Variable Soil Systems. *Water Resources Research,* 23, 2059-2069.

Van Genuchten, M. (1980). A closed form equation for predicting the hydraulic conductivity of unsaturated soils. *Soil Science Society of American Journal,* 44, 892-898.

Van Hamme, J. D. (2004). Bioavailabiilty and biodegradation of organic pollutants – a microbial perspective. In: A. Singh and O. P. Ward (Eds.), *Biodegradation and Bioremediation* (pp. 37-56). Berlin: Springer.

Vidali, M. (2001). An Overview: Bioremediation. *Pure and Applied Chemistry,* 73(7), 163-172.

Volkering, F., Breure, A. and Rulkens, W. (1997). Microbiological aspects of surfactant use for biological soil remediation. *Biodegradation,* 8, 401-417.

Wan, J., Veerapaneni, S., Gadelle, F., and Tokunaga, T. K. (2001). Generation of stable microbubbles and their transport through porous media. *Water Resources Research,* 37 (5), 1173-1182.

Wan, L. S. and Lee, P. F. (1974). CMC of polysorbates. *Journal of Pharmaceutical Sciences,* 63, 136-137.

Yuan, S. Y., Wei, S. H. and Chang, B. V. (2000). Biodegradation of polycyclic aromatic hydrocarbons by a mixed culture. *Chemosphere,* 41, 1463-1468.

Zhang, R., Huang, K. and Xiang, J. (1994). Solute movement through homogeneous and heterogeneous soil columns. *Advances in Water Resources,* 17, 317-324.

Zhang, Y. and Miller, R. M. (1994). Effect of a Pseudomonas rhamnolipid biosurfactant on cell hydrophobicity and biodegradation of octadecane. *Applied and Environmental Microbiology,* 60(6), 2101-2106.

SECTION 2: MICROBIOREMEDIATION

In: Bioremediation: Processes, Challenges and Future Prospects ISBN: 978-1-62948-513-3
Editors: Jesús Bernardino Velázquez-Fernández et al.

Chapter 7

MODELING THE DISCHARGE OF NUTRIENTS FOR BIOREMEDIATION OF OIL-POLLUTED MARINE ENVIRONMENTS: LINEAR AND QUADRATIC PROGRAMMING STRATEGIES

David Parra-Guevara* and Yuri N. Skiba†
Centro de Ciencias de la Atmósfera, Universidad Nacional Autónoma de México, Circuito Exterior, Ciudad Universitaria, México

ABSTRACT

In this work, two variational problems, along with the corresponding linear and quadratic programming problems, are considered with the aim to determine optimal discharge point and optimal discharge rate of a nutrient to be released to a marine environment polluted with oil. The objective is to minimize the total discharge of nutrient into the system provided that their concentrations still reach critical values sufficient to eliminate oil residuals in affected zones through bioremediation. A tridimensional problem for the advection-diffusion equation and its adjoint model are used to simulate, estimate and control the dispersion of nutrient in a limited region. It is shown that all these problems are well posed. In each oil-polluted zone, the mean concentration of nutrient is determined by means of an integral formula in which the adjoint model solution serves as a weight function for discharge rate and initial distribution of the nutrient. Critical values of such mean concentrations are used as the constraints of variational and programming problems (linear and quadratic). In the quadratic variational problem, the analytical expression for determining the optimal discharge rate of nutrient is given as a linear combination of certain values of the adjoint problems solutions calculated for all the zones under consideration. Through this expression, a function has been obtained, whose minimum value is achieved at the optimal point of discharge. In the linear variational problem, some additional constraints are posed to limit not only the local discharge of nutrient, but also the mean concentration of this substance in the whole region. Both constraints serve for environmental protection.

* David Parra-Guevara, E-mail: pdavid@atmosfera.unam.mx.
† Yuri N. Skiba, E-mail: skiba@unam.mx.

The ability of both methods is demonstrated by numerical experiments on the remediation in an oil-polluted channel by using three control zones. In particular, the experiments with the linear programming problem show that the optimal discharge rate can always be obtained with a simple combination of step functions.

Keywords: Bioremediation, oil spill, optimal control, dispersion model, adjoint model

1. INTRODUCTION

Crude oil is one of the most important organic pollutants in marine environments. It has been estimated that worldwide approximately 1.3×10^6 tons of petroleum impact marine waters and estuaries annually (NRC, 2002). Massive releases from pipelines, wells and tankers receive the most public attention, but in fact these account for only a relatively small proportion of the total petroleum entering the environment. Almost 50% comes from natural seeps, and less than 9% emanates from catastrophic releases. Consumption and urban run-off is responsible for almost 40% of the input (NRC, 2002). Independently of the source of pollution, a substantial number of smaller releases of petroleum occur regularly in coastal waters (Head and Swannell, 1999), as a result, oil stranded in shorelines has become a common problem which needs attention.

It is well known that oil is comprised of many different toxic compounds which endanger the marine environment involved in a spill, however there are many natural, native microorganisms which are not only capable, but thrive on the decomposition of these toxic compounds. This process of using microorganisms for such cleanup efforts in shorelines is known as bioremediation, and it has proven to be a successful method for the cleanup of marine areas affected by oil spills (Coulon et al., 2006).

There are two different types of bioremediation used for oil spill cleanup: bioaugmentation and biostimulation. Bioaugmentation is the addition of microorganisms capable of degrading the toxic hydrocarbons, in order to achieve a reduction of the pollutants. Biostimulation is the addition of nutrients needed by indigenous hydrocarbon degrading microorganisms in order to achieve maximum degradation of toxic compounds present in the oil. The degradation of hydrocarbons (biodegradation) begins by the conversion of the alkane chain or polycyclic aromatic hydrocarbon (PAH) into alcohol. Oxidation then converts the compound to an aldehyde and then into an acid and eventually into water, carbon dioxide, and biomass. In the case of the PAH, fission occurs which ultimately leads to mineralization (Venosa, 1998). More than 170 genera of microorganisms have been identified in the environment, which are able to degrade hydrocarbons. Due to such diversity and different conditions at the spill site, the hydrocarbons have different rates of biodegradation, and not all of them are degradable, but estimates for the biodegradability of different crude oils are in the range from 70 to 97%. What remains - is mainly asphaltenes and resin compounds, which are essentially biologically inert (Prince et al., 2003).

Although biodegradation is a particularly important mechanism for removal of non-volatile components of the oil from the environment, it is relatively slow natural process that can take anywhere from several months to several years (ZoBell, 1973; Zhu et al., 2001). The simplest way of stimulating biodegradation, and the only one that has achieved experimental verification in the field, is to carefully add nitrogen and phosphorus nutrients.

It was first used on a large scale in Alaska, following the 1989 spill from the Exxon Valdez (Bragg et al., 1994; Prince and Bragg, 1997; Prince et al., 1994). Two fertilizers were used in the large-scale applications: an oleophilic liquid product designed to adhere to oil, named Inipol EAP22 (Ladousse and Tramier, 1991); and a slow-release granular agricultural product called Customblen (Prince et al., 2003).

The bioremediation was very successful, as shown in a joint monitoring program conducted by Exxon, the USEPA and the Alaska Department of Environmental Conservation (Prince et al., 1994). The fertilizer applications were successful at delivering nutrients throughout the oiled part of the shorelines, microbial activity was enhanced, and oil biodegradation was stimulated 2 to 5 fold. Furthermore, this was achieved with no detectable adverse environmental impact (Bragg et al., 1994; Prince and Bragg, 1997; Prince et al., 1994). Since then, bioremediation has been used on a limited site as part up of the cleanup of the Sea Empress spill (Swannell et al., 1999b), and has been demonstrated on experimental spills in marine or brackish environments on the Delaware Bay (Venosa et al., 1996), a Texas wetland (Mills et al., 1997), a fine-sand beach in England (Swannell et al., 1999a), mangroves in Australia (Ramsay et al., 2000), and an Arctic shoreline in Spitsbergen (Prince et al., 1999).

Due to these successes, it is desirable to include bioremediation in responses to future spills where oil strands on rocky or inaccessible shorelines. Then, the currents can be used to carry the nutrients to polluted zones instead of releasing them directly on the site. An important factor to achieve a successful biostimulation is to obtain the critical concentration of nutrients needed for maximum growth of the organisms and to maintain this concentration as long as possible on the polluted site. This can be a difficult task, since the proper point to release nutrients and some physical influences like those arising from tides, density differences and wave movements are unknown. Tracer studies are often used to examine how the motion of the water and nutrients are influenced under different situations (Boufadel et al., 2006; Boufadel et al., 2007).

In this work, a strategy is proposed for the remediation of oil-polluted marine environments which uses the fluid dynamic in a limited water region D for the distribution of a nutrient (nitrogen or phosphorus) and stimulation of biodegradation in some important internal ecological zones $\Omega_i \subset D$ such as recreation or aquaculture areas. According to the strategy, the nutrient released at point r_0 of domain D with a discharge rate $Q(t)$ spreads by means of currents and turbulent diffusion and reaches all the contaminated zones. Moreover, a critical mean concentration of nutrient c_i (higher than the natural concentration) should be achieved and maintained in each zone Ω_i $(1 \le i \le N)$ within a certain time to properly stimulate the growth of the oil degrading microorganisms (Boufadel et al., 2006). This time interval is denoted below as $[T-\tau, T]$. It should be noted that an adequate release rate $Q(t)$ does not always exist, that is, at times, this strategy fails. In particular, this can happen when the release point r_0 is improperly chosen in respect to the flow and the location of zones, or when the time T is not large enough to let the nutrient to reach all the zones. A strategy is optimal if it solves the problem and, at the same time, minimizes the discharge rate $Q(t)$ to mitigate the impact of the nutrients on the marine environment and to decrease the remediation cost. Thus, by introducing the least amount of nutrients, the optimal control not only cleans the zones, but also protects the whole ecosystem.

Finally, we note that the total mass introduced into the marine enviroment can be assessed by using different norms, which means different control problems. Besides, taking into account some secondary goals for the remediation problem (ecological targets), some alternative constraints could be established.

In this way, the optimal solutions of the corresponding variational problems (control problems) have different features. For example, existence and uniqueness of solutions depend on the strictly convexity of the used norm.

Before we present and analyse the formulations of this control problem defined through the L_1 and L_2 norms, we describe the dispersion model.

2. Dispersion Model

The concentration of nutrient $\phi(r,t)$ in a domain $D \subset \mathbb{R}^3$ and time interval $(0,T)$ is estimated by the following dispersion model:

$$\frac{\partial \phi}{\partial t} + \vec{U} \cdot \nabla \phi - \nabla \cdot \mu \nabla \phi + \sigma \phi + \nabla \cdot \vec{\phi}_s = Q(t)\delta(r - r_0) \tag{1}$$

$$\vec{\phi}_s = -v_s \phi \vec{k} \quad \text{in} \quad D \tag{2}$$

$$\mu \frac{\partial \phi}{\partial n} = \vec{\phi}_s \cdot \vec{n} - \zeta \phi \vec{k} \cdot \vec{n} \quad \text{on} \quad S_T \tag{3}$$

$$\mu \frac{\partial \phi}{\partial n} = 0 \quad \text{on} \quad S^+ \tag{4}$$

$$\mu \frac{\partial \phi}{\partial n} - U_n \phi = 0 \quad \text{on} \quad S^- \tag{5}$$

$$\mu \frac{\partial \phi}{\partial n} = 0 \quad \text{on} \quad S_B \tag{6}$$

$$\phi(r,0) = \phi^0(r) \quad \text{in } D \tag{7}$$

$$\nabla \cdot \vec{U} = 0 \quad \text{in} \quad D \tag{8}$$

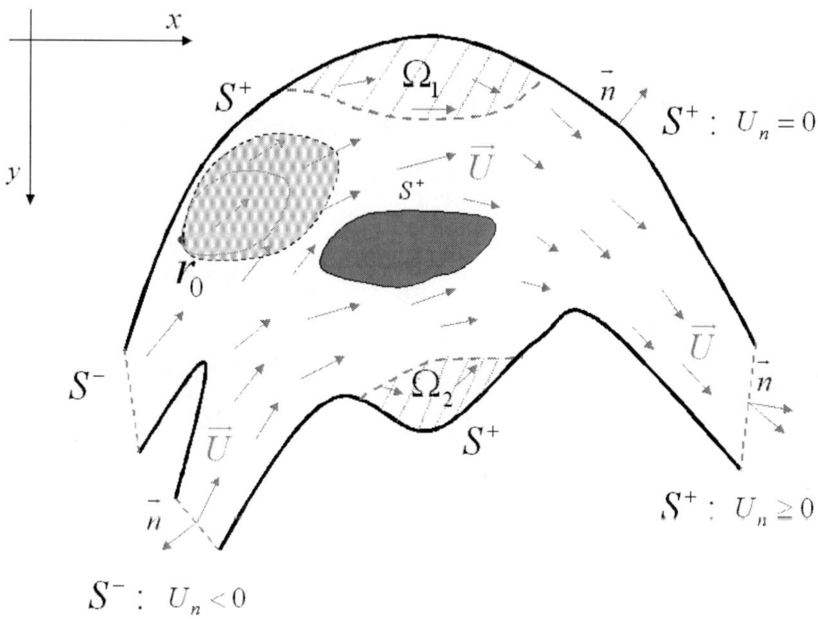

Figure 1. View of domain D from above.

Here (1) is the advection-diffusion equation, $\vec{U}(r,t)$ is the current velocity that satisfies the incompressibility condition (8), $\mu(r,t)$ is the turbulent diffusion coefficient, $\sigma(r,t)$ is the chemical transformation coefficient characterizing the decay rate of nutrient in water. Note that the first-order (linear) kinetics $\sigma\phi$ describing the process of chemical transformation is a reasonable approximation for such nutrients in water as the nitrogen and phosphorus. The term $\nabla \cdot \vec{\phi}_s$ in (1), describes the change of concentration of nutrient per unit time because of sedimentation with constant velocity $v_s > 0$, and $\delta(r - r_0)$ is the Dirac delta centred at the discharge point r_0. Equation (3) is the boundary condition on the free surface S_T of domain D, where $\zeta(r,t)$ is the coefficient characterizing the process of evaporation of nutrient, and (6) represents the boundary condition on the bottom S_B of domain D. Equations (4) and (5) are the corresponding conditions on the lateral boundary of D, besides, S^+ is the rigid or outflow part of boundary where $U_n = \vec{U} \cdot \vec{n} \geq 0$, and S^- is its inflow part where $U_n < 0$ (Figure 1). Finally, the equation (7) represents the initial distribution of the nutrient at $t = 0$. In all equations, $\vec{n}$ is the unit outward normal vector to the boundary $\partial D = S_T \cup S^+ \cup S^- \cup S_B$ of domain D, and $\vec{k} = (0,0,1)'$ is the unit vector directed upward in the Cartesian coordinate system (Figure 2). Note that

$$\vec{k} \cdot \vec{n} = 0 \text{ on } S^+ \cup S^- \text{ and } \vec{U} \cdot \vec{n} = 0 \text{ on } S_T \cup S_B \tag{9}$$

Observe that the boundary conditions (3)-(6) are general (i.e., not only for horizontal free and bottom surfaces S_T and S_B), and hence, the dispersion model can take into account free surface wave motion and marine topography.

First, we show that the solution of dispersion model (1)-(8) satisfies the mass balance equation. Indeed, integrating equation (1) over domain D we get:

$$\frac{\partial}{\partial t}\int_D \phi dr + \int_D \vec{U}\cdot\nabla\phi dr - \int_D \nabla\cdot\mu\nabla\phi dr + \int_D \sigma\phi dr + \int_D \nabla\cdot\vec{\phi}_s dr = \int_D Q(t)\delta(r-r_0)dr$$

Applying the divergence theorem (Kreyszig, 2006), it is possible to rewrite some integrals as follows:

$$\int_D \vec{U}\cdot\nabla\phi dr = \int_D \nabla\cdot\left(\vec{U}\phi\right)dr = \int_{\partial D}\vec{U}\cdot\vec{n}\phi dS$$

$$\int_D \nabla\cdot\mu\nabla\phi dr = \int_{\partial D}\mu\nabla\phi\cdot\vec{n}dS = \int_{\partial D}\mu\frac{\partial\phi}{\partial n}dS$$

$$\int_D \nabla\cdot\vec{\phi}_s dr = \int_{\partial D}\vec{\phi}_s\cdot\vec{n}dS$$

Finally, dividing each integral over boundary ∂D into the four integrals over S_T, S^+, S^- and S_B, and applying equations (3) to (6) and (9), we obtain the mass balance equation:

$$\frac{\partial}{\partial t}\int_D \phi dr = Q(t) - \int_D \sigma\phi dr - \int_{S^+} U_n\phi dS - \int_{S_T}\zeta\phi\vec{k}\cdot\vec{n}dS + \int_{S_B} v_s\phi\vec{k}\cdot\vec{n}dS \tag{10}$$

Since $\vec{k}\cdot\vec{n} > 0$ at S_T and $\vec{k}\cdot\vec{n} < 0$ at S_B, the total mass of the nutrient increases due to the discharge rate $Q(t)$, and decreases because of chemical transformation, advective outflow through S^+, superficial evaporation and sedimentation.

We now show that the dispersion problem (1)-(8) is well posed. Indeed, the model operator is:

$$A\phi = \vec{U}\cdot\nabla\phi - \nabla\cdot\mu\nabla\phi + \sigma\phi + \nabla\cdot\vec{\phi}_s \tag{11}$$

Defining the inner product in $L_2(D)$ as

$$(A\phi,\phi) = \int_D \phi A\phi dr$$

we obtain the expression

$$\left(A\phi,\phi\right)=\int_D \phi\vec{U}\cdot\nabla\phi dr+\int_D \sigma\phi^2 dr-\int_D \phi\nabla\cdot\mu\nabla\phi dr+\int_D \phi\nabla\cdot\vec{\phi}_s dr$$

The divergence theorem allows modifying some integrals in the last equation:

$$\int_D \phi\vec{U}\cdot\nabla\phi dr=\frac{1}{2}\int_{\partial D}\phi^2\vec{U}\cdot\vec{n}dS$$

$$\int_D \phi\nabla\cdot\left(\mu\nabla\phi\right)dr=\int_{\partial D}\phi\mu\frac{\partial\phi}{\partial n}dS-\int_D \mu\left|\nabla\phi\right|^2 dr$$

$$\int_D \phi\nabla\cdot\vec{\phi}_s dr=\frac{1}{2}\int_{\partial D}\phi\vec{\phi}_s\cdot\vec{n}dS$$

Finally, dividing each integral over ∂D into the four integrals over S_T, S^+, S^- and S_B, and applying the conditions (3)-(6) and (9), we get

$$\left(A\phi,\phi\right)=\int_D \sigma\phi^2 dr+\int_D \mu\left|\nabla\phi\right|^2 dr+\int_{S_T}\zeta\phi^2\vec{k}\cdot\vec{n}dS$$

$$+\frac{1}{2}\left\{\int_{S^+}U_n\phi^2 dS-\int_{S^-}U_n\phi^2 dS+\int_{S_T}v_s\phi^2\vec{k}\cdot\vec{n}dS-\int_{S_B}v_s\phi^2\vec{k}\cdot\vec{n}dS\right\} \quad (12)$$

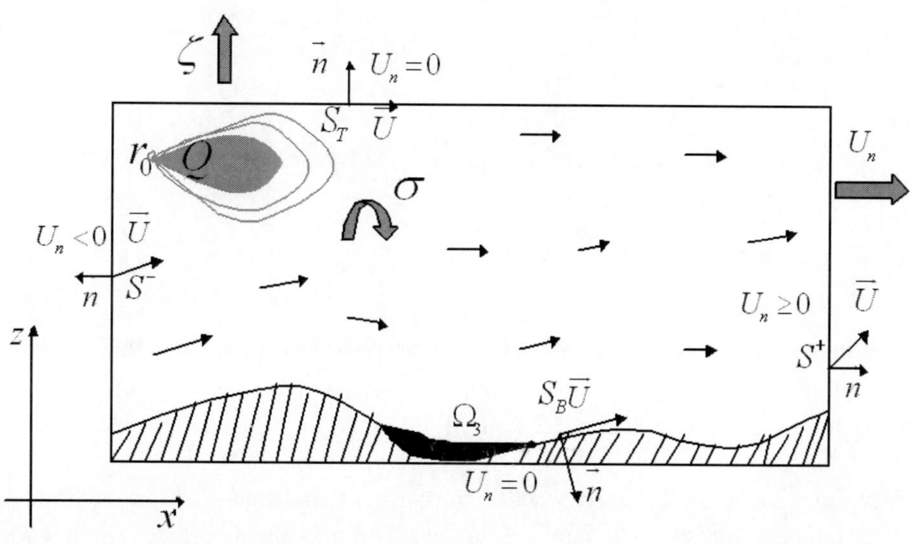

Figure 2. Cross-sectional area of domain D.

Since $U_n < 0$ in S^-, $\vec{k}\cdot\vec{n} > 0$ at S_T and $\vec{k}\cdot\vec{n} < 0$ in S_B, equation (12) can be rewritten as

$$(A\phi,\phi) = \int_D \sigma\phi^2 dr + \int_D \mu|\nabla\phi|^2 dr + \int_{S_T} \zeta\phi^2 \vec{k}\cdot\vec{n} dS + \frac{1}{2}\left\{ \int_{S^+\cup S^-} |U_n|\phi^2 dS + \int_{S_T\cup S_B} v_s\phi^2 \left|\vec{k}\cdot\vec{n}\right| dS \right\}$$

Thus, operator A is positive semidefinite: $(A\phi,\phi) \geq 0$. Taking the inner product of every term of equation (1) with ϕ, we obtain

$$\left(\frac{\partial\phi}{\partial t},\phi\right) = (f,\phi) - (A\phi,\phi), \qquad f(r,t) = Q(t)\delta(r-r_0)$$

Using the condition $(A\phi,\phi) \geq 0$ and Schwarz inequality (Rektorys, 1980), the last equation implies the inequality:

$$\left(\phi,\frac{\partial\phi}{\partial t}\right) \leq \|\phi\|\|f\|, \quad \|\phi\| = \sqrt{(\phi,\phi)}$$

Further,

$$\left(\phi,\frac{\partial\phi}{\partial t}\right) = \frac{1}{2}\frac{\partial}{\partial t}\|\phi\|^2 = \|\phi\|\frac{\partial}{\partial t}\|\phi\|,$$

and hence,

$$\frac{\partial}{\partial t}\|\phi\| \leq \|f\|$$

Finally, the integration over time interval $(0,T)$ leads to

$$\|\phi\| \leq T \max_{0\leq t\leq T} \|Q(t)\delta(r-r_0)\| + \left\|\phi^0(r)\right\| \tag{13}$$

Since the dispersion model (1)-(8) is linear with respect to ϕ, estimation (13) guarantees that the solution of problem (1)-(8) is unique and continuously depends on the initial conditions and forcing. Also, using the method described in Skiba and Parra-Guevara (1999, 2000) it is possible to prove the existence of a generalized solution of problem (1)-(8), that is, the dispersion model (1)-(8) is well posed in the sense of Hadamard (1923). Additionally, the positive semidefiniteness of operator A allows splitting the operator A in coordinate directions.

And finally, we use the numerical schemes by Marchuk (1986) and Crank-Nicolson (1947) to construct unconditionally stable and efficient numerical algorithm of second approximation order in space and time (Skiba, 1993). Such an algorithm for the solution of problem (1)-(8) is described in section 7.

3. ADJOINT MODEL AND THE DUALITY PRINCIPLE

As it is well known in control theory, explicit relationships between the state variables of a system and control variables ease the analysis and solution of a constraint-including variational problem. In the case of bioremediation problem, we are interested in explicit relations between the mean concentration of nutrient ϕ (state variable) and two main variables of control process: the release rate Q and discharge point r_0. In order to establish such explicit dependence, we now introduce one more model which is adjoint to the dispersion model (1)-(8). The operator of new model A^* is adjoint to the operator A of original model (1)-(8) in the sense of Lagrange identity

$$(A\phi, g) = (\phi, A^* g)$$

where $(\cdot,\cdot)$ is the inner product in Hilbert space $L_2(D)$ (Marchuk, 1986; Skiba et al., 2005). The use of the primal model or the dual (adjoint) model (the duality principle) allows getting two equivalent formulae for calculating the mean concentration of the nutrient released in the marine environment.

Let us construct the operator A^* by using the inner product, defined earlier:

$$\left(A\phi, g\right) = \int_D g\vec{U}\cdot\nabla\phi dr + \int_D \sigma g\phi dr - \int_D g\nabla\cdot\mu\nabla\phi dr + \int_D g\nabla\cdot\vec{\phi}_s dr$$

The integrals in the last expression can be rewritten with the divergence theorem as

$$\int_D g\vec{U}\cdot\nabla\phi dr = \int_{\partial D} g\phi\vec{U}\cdot\vec{n}dS - \int_D \phi\vec{U}\cdot\nabla g dr$$

$$\int_D g\nabla\cdot\mu\nabla\phi dr = \int_{\partial D} g\mu\frac{\partial\phi}{\partial n}dS - \int_{\partial D}\phi\mu\frac{\partial g}{\partial n}dS + \int_D \phi\nabla\cdot\mu\nabla g dr$$

$$\int_D g\nabla\cdot\vec{\phi}_s dr = \int_{\partial D} g\vec{\phi}_s\cdot\vec{n}dS - \int_D \phi\nabla\cdot\vec{g}_s dr$$

where $\vec{g}_s = -v_s g\vec{k}$. Then

$$(A\phi, g) = \int_D \phi\left(-\vec{U}\cdot\nabla g - \nabla\cdot\mu\nabla g + \sigma g - \nabla\cdot\vec{g}_s\right)dr$$

$$+\int_{\partial D} g\phi\vec{U}\cdot\vec{n}dS + \int_{\partial D}\phi\mu\frac{\partial g}{\partial n}dS - \int_{\partial D} g\mu\frac{\partial\phi}{\partial n}dS + \int_{\partial D} g\vec{\phi}_s\cdot\vec{n}dS$$

Dividing the integrals over boundary ∂D into four integrals over S_T, S^+, S^- and S_B, and using conditions (3)-(6) and (9), we obtain that

$$(A\phi, g) = \int_D \phi\left(-\vec{U}\cdot\nabla g - \nabla\cdot\mu\nabla g + \sigma g - \nabla\cdot\vec{g}_s\right)dr$$

provided that function g satisfies the boundary conditions (17)-(20) given below. Thus, the Lagrange identity is fulfilled if

$$A^* g = -\vec{U}\cdot\nabla g - \nabla\cdot\mu\nabla g + \sigma g - \nabla\cdot\vec{g}_s$$

On the other hand, multiplying (1) by g and taking the integral over space-time domain $D\times(0,T)$ we get

$$\int_0^T\int_D g\frac{\partial\phi}{\partial t}drdt + \int_0^T\int_D gA\phi drdt = \int_0^T\int_D gQ(t)\delta(r-r_0)drdt$$

Integrating by parts the first integral and using conditions (7) and $g(r,T)=0$ we obtain

$$\int_0^T\int_D g\frac{\partial\phi}{\partial t}drdt = -\int_D g(r,0)\phi^0(r)dr - \int_0^T\int_D \phi\frac{\partial g}{\partial t}drdt$$

Applying now equation (1), Lagrange identity and well-known property of Dirac delta, one can get

$$\int_0^T\int_D \phi\left\{-\frac{\partial g}{\partial t} + A^* g\right\}drdt = \int_0^T Q(t)g(r_0,t)dt + \int_D g(r,0)\phi^0(r)dr \qquad (14)$$

In order to take advantage of equation (14), which (via the function g) explicitly relates the concentration of nutrient $\phi(r,t)$ with the two control variables (the release rate of nutrient $Q(t)$ and discharge point r_0), we consider the following adjoint dispersion model:

$$-\frac{\partial g}{\partial t}-\vec{U}\cdot\nabla g-\nabla\cdot\mu\nabla g+\sigma g-\nabla\cdot\vec{g}_s=p(r,t) \tag{15}$$

$$\vec{g}_s=-v_s g\vec{k} \text{ in } D \tag{16}$$

$$\mu\frac{\partial g}{\partial n}+\zeta g\vec{k}\cdot\vec{n}=0 \quad \text{on } S_T \tag{17}$$

$$\mu\frac{\partial g}{\partial n}+U_n g=0 \quad \text{on } S^+ \tag{18}$$

$$\mu\frac{\partial g}{\partial n}=0 \quad \text{on } S^- \tag{19}$$

$$\mu\frac{\partial g}{\partial n}=-\vec{g}_s\cdot\vec{n} \quad \text{on } S_B \tag{20}$$

$$g(r,T)=0 \text{ in } D \tag{21}$$

Note that the boundary conditions (17)-(20) and "initial" condition (21) imposed on the solution g guarantee the fulfilment of the Lagrange identity. Although the adjoint model is similar to the original model (1)-(8), the main difference between the two models is that the velocities of their dynamic processes (advection, sedimentation, turbulent diffusion) are oppositely directed. That is why the adjoint model (15)-(21) being solved backward in time (from $t=T$ to $t=0$) is also well posed, i.e., it has a unique solution that continuously depends on the forcing $p(r,t)$.

This result can be immediately shown by the transformation of variable $t'=t-T$ (Skiba and Parra-Guevara, 2000).

Moreover, the forcing $p(r,t)$ of equation (15) will be defined so that the mean concentration of nutrient

$$J_i(\phi)=\frac{1}{\tau|\Omega_i|}\int_{T-\tau}^{T}\int_{\Omega_i}\phi(r,t)drdt \tag{22}$$

in an oil-polluted zone Ω_i will be explicitly related with the discharge rate $Q(t)$ and initial concentration of nutrient $\phi^0(r)$ through the adjoint solution g. Indeed, let us take

$$p(r,t) = \begin{cases} \dfrac{1}{\tau|\Omega_i|}, & r \in \Omega_i \text{ and } t \in (T-\tau, T) \\ 0, & \text{otherwise} \end{cases} \quad (23)$$

where $|\Omega_i|$ is the volume of oil-polluted zone Ω_i, and τ is the same as in (22). Then, the use of (23) in (14) leads to

$$J_i(\phi) = \int_0^T g_i(r_0,t)Q(t)dt + \int_D g_i(r,0)\phi^0(r)dr \quad (24)$$

The formulas (22) and (23) are equivalent and also known as the duality principle.

Provided that $\phi^0(r) = 0$ for the first discharge of nutrient, the last formula is reduced to

$$J_i(\phi) = \int_0^T g_i(r_0,t)Q(t)dt \quad (25)$$

The equations (24) and (25) will be applied in sections 5 and 6 in order to stablish more convenient constraints for variational problems. It should be noted that forcing in (15) must be defined according to the functional of the state variables under consideration. Thus, if, instead of the mean concentration of nutrient (22), we study the concentration $\phi(r_j,t_k)$ in a point (r_j,t_k) then, due to (14), the forcing should be chosen as follows

$$p(r,t) = \delta(r-r_j)\delta(t-t_k)$$

where $r_j \in D$, $t_k \in (0,T)$ and δ is the Dirac delta. The use of the last equation in (14) leads to another duality principle

$$\phi(r_j,t_k) = \int_0^T g_i(r_0,t)Q(t)dt + \int_D g_i(r,0)\phi^0(r)dr$$

Nevertheless, for the remediation problem, where extended oil-polluted zones must be cleaned, is more important consider the mean concentration of the nutrient in the whole zone Ω_i. Therefore, hereafter the duality principle (24) will be used in this work.

Finally, it is important to note that all the adjoint solutions $g_i(r_0,t)$ which figure in equations (24) are independent of the discharge rate Q. This non-negative solutions are determined by the dynamic processes in region D and serve in equations (24) as the weight functions which characterize the impact of the discharge of nutrient at point r_0 on each zone Ω_i (see Figure 3).

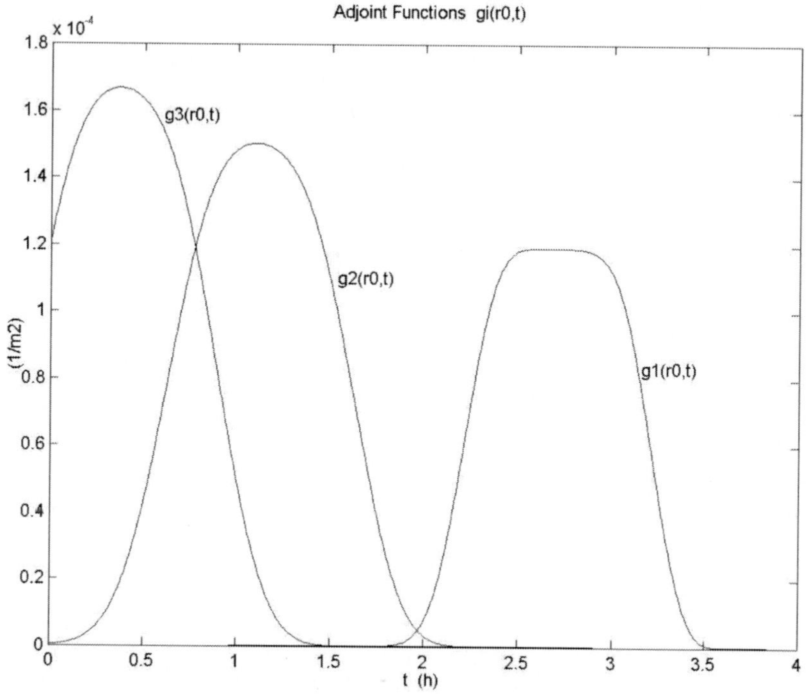

Figure 3. Examples of adjoint functions $g_i(r_0,t),\ \ i=1,2,3$.

In other words, the adjoint solutions are the influence functions (or information functions) in the control theory. Because of this the adjoint problem solutions are widely used in the sensitivity study of various models, and in particular, in the atmosphere and ocean model, weather forecast and climate theory (Marchuk, 1974; Marchuk and Skiba, 1990), data assimilation problems (Marchuk, 1995), problems of identification of unknown pollution sources, like nuclear accidents (Pudykiewicz, 1998; Yee, 2008), simulation of oil pollution (Skiba, 1996; Dang et al., 2012) and optimal control in pollution problems (Marchuk, 1986; Parra-Guevara and Skiba, 2003; Skiba, 2003; Liu et al., 2005; Parra-Guevara and Skiba, 2006; Hinze et al., 2009; Yan and Zhou, 2009; Alvarez-Vazquez et al., 2010; Hongfei, 2010; Parra-Guevara et al., 2010).

4. Peculiarities of Dual Estimates and Sensitivity Formulas

We now discuss the main features of the dual estimates (22) and (24), or its simplification (25), and show the usefulness of adjoint estimates in the study of sensitivity of mean concentration $J_i(\phi)$, to variations in the initial distribution $\phi^0(r)$ of nutrients, discharge rate $Q(t)$ and position r_0 of the operational process (source).

In environmental monitoring, the adjoint estimates (24) are a good complement to the direct estimates (22). One can use either direct or adjoint estimates depending on the specific situation.

Assume, for example, that the mean concentration $J_i(\phi)$ of a nutrient is monitored in N ecologically important zones Ω_i of domain D $(i=1,...,N)$. If the number of zones N is large enough then it is better to solve problem (1)-(8) and use direct estimate (22) in each zone.

On the other hand, if number N is rather small then it is more effective and economical to solve adjoint problem (15)-(21) and use adjoint estimate (24). Unlike (22), the adjoint estimate (24) permits to explicitly evaluate the contribution of each source to value $J_i(\phi)$.

In the case of invariable emission rate ($Q(t)=Q$), evaluation (24) becomes even simpler:

$$J_i(\phi)=w_iQ \tag{26}$$

where

$$w_i=\int_0^T g_i(r_0,t)\,dt \geq 0 \tag{27}$$

Each weight w_i depends only on the adjoint solution and characterizes the contribution of the source with emission rate Q to the mean concentration $J_i(\phi)$ in Ω_i.

What is the basic difference between evaluations (22) and (24)? The direct estimate (22) relating to the solution $\phi(r,t)$ of problem (1)-(8) is independent of a concrete zone Ω, but depends on the discharge rate $Q(t)$ and position r_0 of source, and on the initial distribution of nutrient $\phi^0(r)$ in D.

For this reason the estimate (22) is preferable if one needs to know the concentration of a substance in many zones of D, or in each point of $D\times(0,T)$. However, the direct estimation requires to solve problem (1)-(8) again whenever the parameters $Q(t)$, r_0, or $\phi^0(r)$ vary. In the model sensitivity study, this approach requires much computing time, because $J(\phi)$ must be calculated whenever new values of these parameters are used.

In contrast, the solution $g_i(r,t)$ of adjoint problem depends on Ω_i zone, but is independent of $Q(t)$, r_0, or $\phi^0(r)$. In the adjoint evaluation (24), $g_i(r,t)$ serves as a weight function characterizing the model response to these three parameters. Since the problem is linear, (24) implies that

$$\delta J_i(\phi)=\int_0^T g_i(r_0,t)\delta Q(t)dt+\int_D g_i(r,0)\,\delta\phi^0(r)dr \tag{28}$$

This is the main sensitivity formula relating the mean concentration variation $\delta J_i(\phi)$ in Ω_i with variations $\delta Q(t)$ and $\delta\phi^0(r)$ in emission rates $Q(t)$ and initial distribution of nutrient $\phi^0(r)$. It makes the estimates (24) and (28) rather efficient and computationally economical, because the solution $g_i(r,t)$, once found, can then be used repeatedly in these formulas for different values of $Q(t)$, r_0, or $\phi^0(r)$.

The effect of changing the position of source from r_0 to r_0' is estimated by the formula

$$\delta J_i(\phi) = \int_0^T \{g_i(r_0',t) - g_i(r_0,t)\} Q(t)dt \tag{29}$$

Finally, we give without proof a general sensitivity formula

$$\delta J_i(\phi) = \int_0^T g_i(r_0,t)\delta Q(t)\,dt + \int_D g_i(r,0)\delta\phi^0(r)dr$$

$$-\int_0^T \int_{S_T} g_i(r,t)\phi(r,t)\delta\zeta(r,t)dSdt - \int_0^T \int_D g_i(r,t)\,B(r,t)\,dr\,dt \tag{30}$$

where

$$B(r,t) = \delta \boldsymbol{U}\cdot\nabla\phi + \delta\sigma\,\phi - \nabla\cdot\delta\mu\nabla\phi + \delta v_s \frac{\partial\phi}{\partial z}$$

(Skiba and Parra-Guevara, 2000) taking account of arbitrary variations $\delta Q(t)$ and $\delta\phi^0(r)$ and small variations $\delta\mathbf{U}$, $\delta\sigma$, $\delta\mu$, δv_s and $\delta\varsigma$ in the domain D. Unlike the previous formulas, estimate (30) is more complicated, because it uses solutions of both problems (1)-(8) and (15)-(21) and linearized equations for perturbations.

5. Analytical Results in the Quadratic Control Problem

Taking into account the basic theory developed in sections 2 and 3, we are ready to stablish the first variational formulation of remediation as the following quadratic control problem

$$\text{minimize } m(Q) = \int_0^T Q^2(t)dt \tag{31}$$

subject to: $Q(t) \geq 0, \quad 0 \leq t \leq T$ (32)

$$J_i(\phi) = \int_0^T g_i(r_0, t) Q(t) dt = c_i \quad (i = 1, ..., N) \tag{33}$$

where $m(Q) = \|Q\|^2_{L_2(0,T)}$ is the quadratic functional that represents the total mass of nutrient released into the system within time interval $[0,T]$, $\phi = \phi(r,t)$ is the concentration of nutrient in D (the solution of dispersion model (1)-(8)), and $J_i(\phi)$ is the functional assessing the mean concentration of nutrient in i th zone Ω_i $(i = 1, ..., N)$. For simplicity, all the zones Ω_i considered in this work are nonintersecting. Note that constraints (33) have been introduced in order to achieve critical concentrations c_i of nutrient in the oil-polluted zones. By convenience, such constraints have already been taken according to the duality principle (25). On the other hand, if the initial concentration of nutrient $\phi^0(r)$ is not zero, for example in the case that a repeated discharge of nutrient is needed for degrading oil-residuals, then the variational problem (31)-(33) should be solved only in the zones in which the new restriction parameters

$$c_i' = c_i - \int_D g_i(r, 0) \phi^0(r) dr, \quad i = 1, ..., N'$$

are positive. Indeed, if $c_j' \leq 0$ for some zone Ω_j then it is not necessary to add nutrient to this zone. Thus, the variational problem (31)-(33) is rather general to be considered.

5.1. Construction of the Optimal Discharge Rate

In order to prove the existence and uniqueness of solution of the variational problem (31)-(33), we now analyze some properties of the feasibility space and functional $m(Q)$ when the discharge point r_0 is fixed (Parra-Guevara and Skiba, 2007; Parra-Guevara et al., 2011).

Lemma 1. *The feasibility space*

$$\Theta = \left\{ Q \in L_2(0,T);\ Q(t) \geq 0,\ 0 \leq t \leq T, \text{ and } \int_0^T Q(t) g_i(r_0, t) dt = c_i\ ,\ (i = 1, ..., N) \right\} \tag{34}$$

is a convex set in $L_2(0,T)$.

In fact, let $Q_1, Q_2 \in \Theta$ and $\lambda \in (0,1)$. Then, obviously, $\lambda Q_1 + (1-\lambda) Q_2 \geq 0$, $0 \leq t \leq T$. Besides,

$$\int_0^T \left(\lambda Q_1 + (1-\lambda)Q_2\right) g_i(r_0,t)dt = \lambda c_i + \left(1-\lambda\right)c_i = c_i \ , \ \ (i=1,\ldots,N)$$

and hence, Θ is convex.

Lemma 2. *The feasibility space* Θ *is a closed set in* $L_2(0,T)$.

We must prove that $\Theta = \overline{\Theta}$. Let Q_0 be an element of $\overline{\Theta}$. Then, there is a sequence $\left\{Q_k\right\}_{k=1}^{\infty}$ in Θ such that $\|Q_k - Q_0\| \to 0$ as $k \to \infty$. Assume that $Q_0(t) < 0$ in some interval $I \subset (0,T)$ of positive measure $|I| > 0$. Then

$$\|Q_k - Q_0\|^2 = \int_0^T (Q_k - Q_0)^2 \, dt \geq \int_I (Q_k - Q_0)^2 \, dt \geq \int_I Q_0^2 dt = l > 0$$

The last inequality contradicts the convergence of $\left\{Q_k\right\}_{k=1}^{\infty}$ in $L_2(0,T)$, and hence, Q_0 is a non-negative function in $(0,T)$.

The Schwarz inequality (Rektorys, 1980) leads to

$$\left| c_i - \int_0^T Q_0 g_i(r_0,t)dt \right| = \left| \int_0^T (Q_k - Q_0) \, g_i(r_0,t)dt \right| \leq \|Q_k - Q_0\| \|g_i\|$$

and we get $\int_0^T Q_0 g_i(r_0,t)dt = c_i, \ (i=1,\ldots,N)$ in the limit when $k \to \infty$, that is, $Q_0 \in \Theta$.

Lemma 3. *The functional* $m(Q)$ *of problem (31)-(33) is strictly convex on the feasibility space* Θ.

Indeed, if $Q_1, Q_2 \in \Theta \ \left(Q_1 \neq Q_2\right)$ and $\lambda \in (0,1)$ then

$$\begin{aligned}(\lambda Q_1 + (1-\lambda)Q_2)^2 &= \lambda Q_1^2 + (1-\lambda)Q_2^2 - \lambda(1-\lambda)(Q_1 - Q_2)^2 \\ &< \lambda Q_1^2 + (1-\lambda)Q_2^2\end{aligned}$$

Integrating this inequality over time we get

$$m(\lambda Q_1 + (1-\lambda)Q_2) < \lambda m(Q_1) + (1-\lambda)m(Q_2).$$

Lemma 4. *A non-empty closed convex set in a uniformly convex Banach space possesses a unique point closest to a given point.*

The proof can be seen in Cheney (1966).

It should be noted that Lemmas 1-3 have been proved under the condition that set Θ is non-empty. It is a non-trivial property of the feasibility space (34).

A necessary condition to satisfy this property is that for each i, $g_i(r_0,t)>0$ in an open interval $I_i \subset (0,T)$, $1 \le i \le N$ (see Figure 3). This means that the discharge point r_0 cannot be arbitrary. In fact, it must be chosen so that, for a certain interval $(0,T)$, the nutrient could reach every oil-contaminated zone Ω_i. At the same time, this condition is not sufficient (except for the case $N=1$). Indeed, if

$$0 < g_j(r_0,t) < g_k(r_0,t) \text{ for all } t \in [0,T]$$

and $c_j \ge c_k$, then the feasibility space Θ is empty, and variational problem (31)-(33) has no solution. This set is also empty in the case when two different zones Ω_j and Ω_k possess such special dynamic and geometrical symmetry that

$$0 < g_j(r_0,t) = g_k(r_0,t) \text{ for all } t \in [0,T],$$

but $c_j \ne c_k$. Thus, if the critical concentrations for two zones are different then it is desirable to avoid symmetries in choosing the discharge point r_0. We now prove the existence and uniqueness of solution to problem (31)-(33).

Theorem 1. *If the feasibility space* Θ *is non-empty, then variational problem (31)-(33) has unique solution.*

Since $m(Q)=\|Q\|^2_{L_2(0,T)}$ is a strictly convex functional (Lemma 3) defined on the convex set Θ (Lemma 1), its minimum, if it exists, is global and unique (Cheney, 1966). On the other hand, since $L_2(0,T)$ is the Hilbert space, it is a uniformly convex Banach space (Cheney, 1966).

Moreover, by Lemmas 1 and 2, the feasibility space Θ is a closed convex set in $L_2(0,T)$. Note that $Q \equiv 0 \notin \Theta$, since in the constraints (33) of the control problem, the critical concentrations c_i are positive for any i $(i=1,...,N)$.

Therefore, due to Lemma 4, there is a point $Q^* \in \Theta$ that minimizes the distance between the set Θ and point $Q \equiv 0$. This point is the solution of control problem (31)-(33).

The analytical expression for the optimal discharge rate Q^* can be obtained by applying the Lagrange multipliers method (Smith, 1998) to the variational problem (31)-(33). Let

$$F(Q)=\frac{1}{2}\int_0^T Q^2(t)dt-\sum_{j=1}^N \alpha_j\left\{\int_0^T Q(t)g_j(r_0,t)dt-c_j\right\}$$

be a Lagrange functional and let α_j be the Lagrange multipliers. The first variation of F in the sense of Gateaux (Smith, 1998) is calculated as

$$\delta F(Q;\delta Q) = \frac{\partial}{\partial \varepsilon} F(Q+\varepsilon\delta Q)_{\varepsilon=0} = \int_0^T \left\{ Q(t) - \sum_{j=1}^N \alpha_j g_j(r_0,t) \right\} \delta Q dt$$

where δQ is a variation of Q. A necessary condition for Q^* to be a minimum is $\delta F(Q^*;\delta Q) = 0$ for all δQ (Smith, 1998). Therefore, we have

$$Q^*(t) = \sum_{j=1}^N \alpha_j g_j(r_0,t) \tag{35}$$

where the Lagrange multipliers α_j can be determined from the algebraic linear system

$$\sum_{j=1}^N \alpha_j \left\{ \int_0^T g_i(r_0,t) g_j(r_0,t) dt \right\} = c_i, \quad (i=1,\ldots,N) \tag{36}$$

formed by the integral constraints (33) of variational problem. Now, we show that (35)-(36) are sufficient conditions for the minimum.

Theorem 2. *If $Q^*(t)$, given by (35) and (36), is a non-negative function in $(0,T)$ then it is the optimal discharge rate.*

Let $Q_0 = Q^* + \delta Q$ be a feasible discharge rate $(Q_0 \in \Theta)$, then $\int_0^T \delta Q g_i(r_0,t) dt = 0$, where $\delta Q \neq 0$ is an arbitrary variation of Q^* $(i=1,\ldots,N)$. Then

$$m(Q_0) - m(Q^*) = 2\int_0^T Q^*(t)\delta Q dt + \int_0^T \delta^2 Q dt$$

and (35) leads to

$$m(Q_0) - m(Q^*) = 2\sum_{j=1}^N \alpha_j \int_0^T \delta Q g_j(r_0,t) dt + \int_0^T \delta^2 Q dt = \int_0^T \delta^2 Q dt > 0$$

Thus, $m(Q_0) > m(Q^*)$, and hence, Q^* is the solution of (31)-(33).

On the other hand, we can solve the algebraic linear system (36) with Cramer's rule (Strang, 1980) and use its solution in formula (35) to get the analytical expression for the optimal discharge rate:

$$Q^*(t) = \sum_{j=1}^{N} \frac{\det(\Psi_j)}{\det(\Psi)} g_j(r_0,t), \quad \Psi_j = \begin{pmatrix} \psi_{11} & \dots c_1 & \dots & \psi_{1N} \\ \psi_{21} & \dots c_2 & \dots & \psi_{2N} \\ \vdots & \vdots & \vdots & \vdots \\ \psi_{N1} & \dots c_N & \dots & \psi_{NN} \end{pmatrix} \tag{37}$$

where Ψ is the matrix of system (36) whose entries

$$\psi_{ij} = \int_0^T g_i(r_0,t) g_j(r_0,t) dt, \quad 1 \le i,j \le N \tag{38}$$

are the inner products in $L_2(0,T)$ of adjoint functions (Gram matrix), and matrix Ψ_j is obtained from Ψ by replacing its j th column with the corresponding components of the vector of critical concentrations $\vec{c} = (c_1,\ldots,c_N)'$. It is easily seen from (38) that matrix Ψ is symmetric. Moreover, Ψ is also a positive semidefinite, since

$$\vec{x}'\Psi\vec{x} = \left\| \sum_{i=1}^{N} x_i g_i(r_0,t) \right\|^2 \ge 0$$

for any nonzero vector $\vec{x} \in \mathbb{R}^N$. In the last inequality, $\|\cdot\|$ means the norm in $L_2(0,T)$. In the case when the adjoint functions $\{g_i(r_0,t)\}_{i=1}^N$ are linearly independent, matrix Ψ is positive definite, and hence, non-singular, and the optimal discharge rate Q^* is well defined by (37).

Finally, it should be noted that according to theorem 2, Q^* given by (37) is the optimal discharge rate only if it is a non-negative function in interval $(0,T)$. This condition is satisfied, for example, when the vector of critical concentrations $\vec{c} = (c_1,\ldots,c_N)'$ is in the convex cone (Luenberger, 1984) determined by the columns of matrix Ψ. Then all the coefficients α_j in (35) are non-negative $(j = 1,\ldots,N)$.

In particular, it is the case when the adjoint solutions $\{g_i(r_0,t)\}_{i=1}^N$ are orthogonal functions, that is, $\psi_{ij} = 0$ if $i \ne j$. Then the optimal discharge rate is expressed as

$$Q^*(t) = \sum_{j=1}^{N} \left\{ c_j \Big/ \int_0^T g_j^2(r_0,t)dt \right\} g_j(r_0,t) \tag{39}$$

5.2. Selection of Optimal Discharge Point

We now analyze different cases in order to select the optimal discharge point r_0. The criterion for selecting such a point is to minimize the mass of the nutrient necessary to enter into aquatic system.

If $N=1$ then (35) and (36) lead to

$$c_1 = \int_0^T Q^*(t) g_1(r_0,t)dt = \int_0^T \alpha_1 g^2{}_1(r_0,t)dt$$

and, therefore,

$$Q^*(t) = \frac{c_1}{\|g_1\|^2} g_1(r_0,t)$$

where $\|\cdot\|$ is the norm in $L_2(0,T)$. Applying this norm to both sides of the last equation we get

$$\|Q^*\|^2 = \frac{c_1^2}{\|g_1(r_0,t)\|^2} = \frac{c_1^2}{\int_0^T g^2{}_1(r_0,t)dt} \tag{40}$$

Thus, in order to minimize $\|Q^*\|^2$ we have to maximize the integral $\int_0^T g^2{}_1(r_0,t)dt$. Namely, the optimal discharge point r_0 maximizes the area under the graph of function $g^2{}_1(r_0,t),\ t \in (0,T)$.

If $N>1$ and the adjoint solutions $\{g_i(r_0,t)\}_{i=1}^N$ are orthogonal functions, then due to (40) we have

$$\|Q^*\|^2 = \int_0^T \left(\sum_{j=1}^N \alpha_j g_j(r_0,t) \right) \left(\sum_{j=1}^N \alpha_j g_j(r_0,t) \right) dt = \sum_{j=1}^N \frac{c_j^2}{\|g_j(r_0,t)\|^2} \tag{41}$$

Thus, the optimal discharge point r_0 must be selected so as to minimize the sum defined by (41). Note that the necessary condition

$g_i(r_0,t) > 0$ in an open interval $I_i \subset (0,T)$, $1 \le i \le N$, is implicitly present in equations (40) and (41).

In the general case $(N>1)$, (35) leads to

$$\|Q^*\|^2 = \int_0^T \left(\sum_{j=1}^N \alpha_j g_j(r_0,t)\right)\left(\sum_{j=1}^N \alpha_j g_j(r_0,t)\right) dt = \vec{\alpha}'\Psi\vec{\alpha}$$

where Ψ is given by (38) and $\vec{\alpha} = (\alpha_1,\ldots,\alpha_N)'$ is the vector of Lagrange multipliers. Due to (36), $\vec{\alpha} = \Psi^{-1}\vec{c}$, and since $\Psi = \Psi'$, we obtain

$$\|Q^*\|^2 = \left(\Psi^{-1}\vec{c}\right)' \Psi\left(\Psi^{-1}\vec{c}\right) = \vec{c}'\Psi^{-1}\vec{c} = \vec{c}'\vec{\alpha}$$

Using (37), we can also write

$$\|Q^*\|^2 = \frac{1}{\det(\Psi(r_0))}\sum_{j=1}^N \det(\Psi_j(r_0))c_j \tag{42}$$

Thus, the optimal discharge point r_0 must be selected so as to minimize the sum defined by (42) among all such points $r \in D$ that $\det(\Psi(r)) \neq 0$ (the above mentioned necessary condition is implicitly included here).

Note that in all cases, the objective function $m(Q^*) = \|Q^*\|^2$ is a non-linear function of three real variables (the coordinates x_0, y_0, z_0 of point r_0), and it can be evaluated when all the adjoint functions g_i, $i = 1,\ldots,N$, are already determined. The computation effort of this minimization process can be reduced (see the note in section 6.2).

6. Analytical Results in the Linear Control Problem

In the variational formulation of the remediation problem (31)-(33) presented in section 5, the norm of $L_2(0,T)$ have been used to control the total mass of nutrient introduced into the marine environment. Besides, the necessary critical concentrations for the biostimulation process have been reached by means of the constraints (33). It has also been shown the unique solvability of this problem (Theorem 1), and its solution has been linearly expressed through the adjoint functions. In this section, we establish the second variational formulation of the remediation problem in which the norm of $L_1(0,T)$ will be used to control the total mass of nutrient introduced into the marine environment. At the same time, more general constraints will be considered. As a result, the new problem has some different features to be analyzed.

In particular, the solution of this problem is obtained by using the finite-dimensional subspaces. The new remediation strategy is given by the following linear control problem:

$$\text{minimize} \quad m(Q) = \int_0^T Q(t)dt \tag{43}$$

$$\text{subject to:} \quad c_i \le J_i(\phi) = \int_0^T g_i(r_0,t)Q(t)dt \le c_i + \varepsilon_i, \;\; 1 \le i \le N \tag{44}$$

$$J_0(\phi) = \int_0^T g_0(r_0,t)Q(t)dt \le c_0; \;\; 0 \le Q(t) \le Q_0(t), \;\; 0 \le t \le T \tag{45}$$

where $\phi(r,t)$ is the concentration of nutrient in region D, i.e., the solution of dispersion model (1)-(8), and $m(Q)$ is the functional that represents the total mass of nutrient released into the marine environment within time interval $[0,T]$. The functionals $J_i(\phi)$ and $J_0(\phi)$ are the mean concentrations of nutrient in time interval $[T-\tau,T]$ and in ith zone Ω_i and region D, respectively. For convenience, all the zones Ω_i considered in this paper do not intersect. The constraints in (44) are imposed to maintain $J_i(\phi)$ equal to or slightly above the critical concentration $c_i > 0$ required for successful biodegradation $(1 \le i \le N)$. Also, note that the introduction of small parameters $\varepsilon_i > 0$ increases the number of feasible solutions of problem (43)-(45), and therefore, this model is less restrictive than that described in section 5. The integral condition in (45) represents an ecological constraint that limits the mean concentration of nutrient in the whole region (global constraint), whilst the mass released at the discharge point r_0 is bounded by a function Q_0 (local constraint) according to the inequality in (45). Note that the control problem is simplified if the limiting discharge rate Q_0 and the global restriction c_0 on the mean concentration of nutrient in D satisfy the inequality

$$\int_0^T g_0(r_0,t)Q_0(t)dt \le c_0 \;,$$

since in this case the integral constraint in (45) can be omitted. However, this simplification is not always possible because these parameters can vary with the conditions of the marine environment.

We observe that all the constraints in problem (43)-(45) have been taken according to the duality principle (25). However, if a repeated discharge of nutrient is needed for degrading oil-residuals in the marine environment, then the nonzero initial concentration of the nutrient must be taken into account. It should be noted that, due to the microbial intake of nutrient in the oil-polluted zones and outflow of water from region D, the concentration of nutrient decreases in region D towards its natural level. Therefore, the following conditions for the mean concentration of nutrient must be fulfilled since the moment $t_0 > T$:

$$\frac{1}{|D|}\int_D \phi(r,t)dr < c_0, \quad \frac{1}{|\Omega_i|}\int_{\Omega_i} \phi(r,t)dr < c_i + \varepsilon_i, \quad i = 1,...,N, \quad t \geq t_0 \tag{46}$$

The moment t_0 can be determined by means of monitoring the mean concentration of nutrient in region D, or by using the solution ϕ forecasted by the model (1)-(8) with the forcing $Q(t)$ equal to zero for $t > T$. If conditions (46) are fulfilled then the next time you use this control strategy the initial concentration of nutrient should be chosen as

$$\varphi^0(r) = \phi(r,t_0) \tag{47}$$

and the time interval for such application is $[t_0, t_0 + T]$. Due to conditions (46), the contribution of $\varphi^0(r)$ to the mean concentrations of nutrient during time interval $[t_0 + T - \tau, t_0 + T]$ must be less than the upper bounds c_0 in D and $c_i + \varepsilon_i$ in Ω_i $(i = 1,..., N)$, because otherwise the feasibility space of problem (43)-(45) is empty and there is no solution to the control problem.

If the conditions (46) are met then the second discharge of nutrient can be realized, that is, one can put $t_0 = 0$ and apply the control problem (43)-(45) again with the following positive parameters

$$c'_0 = c_0 - \int_D g_i(r,0)\varphi^0(r)dr \text{ and } c'_i = c_i - \int_D g_i(r,0)\varphi^0(r)dr, \; i = 1,\ldots,N \tag{48}$$

Note that the adjoint functions in (48) must be calculated in time interval $[t_0, t_0 + T]$. Also note that without loss of generality, any negative values that may appear on the left side of constraints (44) are replaced by the zero. With these remarks, the variational problem (43) - (45) is a general strategy for recovery, regardless of whether you use it once or several times.

Finally, it is important to note that all the adjoint solutions $g_i(r_0,t)$ which figure in constraints (44) and (45) are independent of the discharge rate $Q(t)$. Each non-negative adjoint model solution is determined by the flow dynamics in region D and serves in the constraints (44) and (45) as the weight function that characterizes the impact of the discharge of nutrient at point r_0 on the corresponding zone Ω_i (see Figure 1).

6.1. Construction of the Optimal Discharge Rate

In order to determine the features of variational problem (43)-(45) we first analyze this problem for a fixed point $r_0 \in D$, and then consider the problem of choosing the optimal discharge point. The feasibility space Θ for problem (43)-(45) is given as

$$\Theta = \{Q \in L_2(0,T) \text{ such that conditions (44) and (45) are fulfilled}\}$$

where functions Q_0 and $g_i(r_0,t)$, $i=1,...,N$, belong to the Hilbert space $H = L_2(0,T)$. Since each function $g_i(r_0,t)$ is non-negative in $[0,T]$ and $c_i > 0$ $(i=1,...,N)$, it follows from (44) that the feasibility space Θ is non-empty only if

$$g_i(r_0,t) > 0 \text{ in an open interval } I_i \subset (0,T),\ 1 \le i \le N \tag{49}$$

This is a necessary condition for the existence of solution to the control problem, such a condition suggests that the optimal discharge point should be chosen only among such points $r \in D$ that satisfy (49). However, condition (49) is not sufficient for the existence of solution to problem (43)-(45). Indeed, (49) is satisfied if $0 < g_j(r_0,t) < g_k(r_0,t),\ 0 \le t \le T$. However, the feasibility space Θ is empty if $c_j > c_k + \varepsilon_k$.

On the other hand, the feasibility space Θ is a convex, bounded and closed set in H. To show this, we observe that for any Q_1 and Q_2 in H, and $\lambda \in (0,1)$,

$$\int_0^T (\lambda Q_1 + (1-\lambda)Q_2)\, g_i(r_0,t)dt = \lambda \int_0^T Q_1 g_i(r_0,t)dt + (1-\lambda)\int_0^T Q_1 g_i(r_0,t)dt$$

$$\le \lambda(c_i + \varepsilon_i) + (1-\lambda)(c_i + \varepsilon_i) = c_i + \varepsilon_i$$

and

$$0 \le \lambda Q_1 + (1-\lambda)Q_2 \le \lambda Q_0 + (1-\lambda)Q_0 = Q_0, \quad 0 \le t \le T$$

The term $\lambda Q_1 + (1-\lambda)Q_2$ satisfies the constraints (44)-(45) as well, and hence Θ is a convex set.

Also, due to the last constraint in (45), $\|Q\| \le \|Q_0\| = const$ for all Q of Θ, and therefore, the feasibility space Θ is bounded.

Finally, in order to show that Θ is a closed set in H, we now prove that $\overline{\Theta} \subset \Theta$ (Dieudonné, 1969). Let Q be any element of $\overline{\Theta}$. Then there is a sequence $\{Q_k\}_{k=1}^{\infty}$ in Θ such that

$$\|Q_k - Q\| \to 0 \text{ as } k \to \infty \tag{50}$$

Assume that $Q(t) < 0$ in some interval $I \subset (0,T)$ of positive measure $|I| > 0$. Then

$$\|Q_k - Q\|^2 = \int_0^T (Q_k - Q)^2 dt \geq \int_I (Q_k - Q)^2 dt \geq \int_I Q^2 dt = l' > 0$$

The last inequality contradicts the convergence of $\{Q_k\}_{k=1}^{\infty}$ in H, and hence, Q is a non-negative function in $(0,T)$.

Now, we assume that $Q(t) > Q_0(t)$ in some interval $I \subset (0,T)$ of positive measure $|I| > 0$. Then

$$\|Q_k - Q\|^2 = \int_0^T (Q_k - Q)^2 dt \geq \int_I (Q_k - Q)^2 dt \geq \int_I (Q - Q_0)^2 dt = l' > 0$$

The last inequality contradicts the convergence of $\{Q_k\}_{k=1}^{\infty}$ in H, and hence, $Q(t) \leq Q_0(t)$ in $(0,T)$.

On the other hand, the Schwarz inequality (Rektorys, 1980) leads to

$$\left| \int_0^T Q_k(t) g_i(r_0,t) dt - \int_0^T Q(t) g_i(r_0,t) dt \right| \leq \int_0^T |Q_k(t) - Q(t)| g_i(r_0,t) dt \leq \|Q_k - Q\| \|g_i(r_0,t)\|$$

and, due to (50), we get that $\lim_{k\to\infty} \int_0^T Q_k g_i(r_0,t) dt = \int_0^T Q g_i(r_0,t) dt$. By (44), the sequence $\left\{ \int_0^T Q_k g_i(r_0,t) dt \right\}_{k=1}^{\infty}$ belongs to the closed interval $[c_i, c_i + \varepsilon_i]$, and hence, this interval contains its limit point (Dieudonné, 1969). Thus

$$c_i \leq \int_0^T Q(t) g_i(r_0,t) dt \leq c_i + \varepsilon_i, \quad 1 \leq i \leq N$$

The constraint $0 \le \int_0^T Q(t) g_0(r_0, t) dt \le c_0$ is proved in a similar way, and therefore, $Q \in \Theta$.

To prove the existence of solution to problem (43)-(45), we consider a finite-dimensional subspace Γ of H. Since H is a normed space, Γ is a closed set in H (Kreyszig, 1989). So, the intersection $\Pi = \Gamma \cap \Theta$ of closed sets is a closed set in H (Dieudonné, 1969). In fact, Π is a closed set in Γ, as well (Dieudonné, 1969). Also, note that Π is a bounded subset of Γ, because $\Pi \subset \Theta$, and Θ is bounded in the norm of H. Since Γ is a finite-dimensional normed space, and Π is a bounded and closed subset of Γ, Π is a compact subset in Γ (Kreyszig, 1989). On the other hand, Schwarz inequality leads to

$$|m(q) - m(Q)| \le \int_0^T |q(t) - Q(t)| dt \le \sqrt{T} \, \|q - Q\|$$

where $m(Q)$ is the mass functional (43). Using the last inequality, and choosing $\delta = \varepsilon / \sqrt{T}$ for the continuity definition, one concludes that $m(Q)$ is a continuous functional on H, and hence, it is also continuous on Π. Since Π is a compact subset of the metric space Γ, and m is a continuous functional on $\Pi \ne \varnothing$, there exists a point $Q^* \in \Pi$ that minimizes functional m (Kreyszig, 1989). This point $Q^* = Q^*(t; r_0)$ is the optimal discharge rate of the nutrient at the point r_0.

Note that the functional $m(Q)$ is linear and convex, but not strictly convex, and hence, the problem can have more than one optimal solution, that is, Q^* is not necessarily unique. However, the space of optimal solutions is a convex set in Π. To show this, note that Π is a convex set as the intersection of convex sets (Hadley, 1962). Thus, if Q_1 and Q_2 are optimal solutions from Π, and $\lambda \in (0,1)$, then $\lambda Q_1 + (1-\lambda) Q_2$ also belongs to Π, and

$$m(\lambda Q_1 + (1-\lambda) Q_2) = \lambda m(Q_1) + (1-\lambda) m(Q_2) = \lambda m(Q^*) + (1-\lambda) m(Q^*) = m(Q^*),$$

that is, solution $\lambda Q_1 + (1-\lambda) Q_2$ is also optimal.

This property is useful, since it permits choosing, at the same cost $m(Q^*)$, an optimal solution among the functions $\lambda Q_1 + (1-\lambda) Q_2$, $0 < \lambda < 1$, where Q_1 and Q_2 are two given optimal solutions. The criterion for such a choice could be the simplicity of implementation of discharge during the process of bioremediation.

In this work, the finite-dimensional subspace Γ is generated by the 'tent' functions, which are given as

$$\gamma_k(t) = \begin{cases} 1 + (t - t_k)/\Delta t, & t_{k-1} \le t \le t_k \\ 1 - (t - t_k)/\Delta t, & t_k \le t \le t_{k+1} \\ 0, & \text{otherwise} \end{cases} \qquad (51)$$

where the nodes $t_k = k \cdot \Delta t,\ k = 0,1,...,L,\ \Delta t \cdot L = T$, form a regular grid in interval $[0,T]$, and the functions γ_0 and γ_L are equal to zero outside $[0,T]$. The basic functions $\gamma_k,\ k = 0,1,...,L$, have the following useful properties:

$$\gamma_i(t_j) = \begin{cases} 1, & i = j \\ 0, & i \neq j \end{cases},\quad i, j = 0,1,...,L$$

and

$$\int_0^T \gamma_0(t)dt = \int_0^T \gamma_L(t)dt = \frac{\Delta t}{2},\quad \int_0^T \gamma_k(t)dt = \Delta t,\quad k = 1,...,L-1$$

Taking into account these remarks, the discharge rate of nutrient $Q \in \Pi$ can be given as a linear combination of basic functions

$$Q(t) = \sum_{k=0}^{L} x_k \gamma_k(t) \tag{52}$$

and the optimal solution Q^* of control problem can be obtained as the solution of the following linear programming problem

$$\text{minimize}\quad m(x_0,...,x_L) = (\Delta t / 2)\left[x_0 + 2\sum_{k=1}^{L-1} x_k + x_L \right] \tag{53}$$

$$\text{subject to:}\quad c_i \leq \sum_{k=0}^{L} a_{ik} x_k \leq c_i + \varepsilon_i,\quad 1 \leq i \leq N \tag{54}$$

$$\sum_{k=0}^{L} a_{0k} x_k \leq c_0,\quad 0 \leq x_k \leq Q_0(t_k),\quad 0 \leq k \leq L \tag{55}$$

where

$$a_{ik} = \int_{t_{k-1}}^{t_{k+1}} g_i(r_0,t)\gamma_k(t)dt,\quad 0 \leq i \leq N,\quad 0 \leq k \leq L \tag{56}$$

and the feasibility space Π is determined by (52) and (54)-(56).

Thus, the variational problem (43)-(45) reduces to the linear programming problem (53)-(55) with $L+1$ real variables $\{x_k\}_{k=0}^{L}$, and $2(L+N)+3$ constraints. The solution of this problem can be obtained with a large-scale optimization method based on LIPSOL, Linear Interior Point Solver (Zhang, 1995), which is a variant of Mehrotra's predictor-corrector algorithm (Mehrotra, 1992), a primal-dual interior-point method. Also, the medium-scale optimization can be performed with a projection method which is a variation of the well-known simplex method for linear programming (Dantzig et al., 1955; Hadley, 1962). The interior-point method reduces the computing time in the large size problems (Zhang, 1995) and represents a good alternative when the number of grid points in the control problem is too large. In this work, both methods above mentioned have been implemented with the *linprog* subroutine of MATLAB.

Finally, note that in order to simplify the linear programming problem (53)-(55), the coeficients a_{ik} can be calculated by using one of the formulas of second order, for example, by the trapezoidal rule:

$$a_{i0} = (\Delta t/2) g_i(r_0,t_0), \; a_{iL} = (\Delta t/2) g_i(r_0,t_L), \; a_{ik} = \Delta t \cdot g_i(r_0,t_k) \;\; 1 \le i \le N, \;\; 1 \le k \le L.$$

6.2. A Note on the Optimal Discharge Point

Once the optimal discharge rate $Q^* = Q^*(t;r_0)$ is built for any $r_0 \in D$, one can consider the choice of a suitable release site. We remind that the criterion to select the optimal discharge point r_0^* is to minimize the mass of the nutrient that enters into the marine environment. Note that in this case, the objective function is a non-linear function

$$m(Q^*) = \int_0^T Q^*(t;r_0)dt$$

depending on the three real variables (the coordinates x_0, y_0, z_0 of point r_0), and it can be evaluated after determining all the adjoint functions g_i and solving the linear programming model (53)-(55). Besides, while minimizing this function, it is computationally advantageous to reduce the search area only to those points $r_0 \in D$ in which the indicative function

$$\mathrm{P}(r_0) = \prod_{i=1}^{N} \int_0^T g_i^2(r_0,t)dt \tag{57}$$

is positive (see Figure 4), that is, the minimum r_0^* must be searched only within the support of function P (Folland, 1999), where the necessary condition (49) is fulfilled. The case $\mathrm{P}(r_0) = 0$ means that the flow dynamics in D does not allow the nutrient discharged at the point r_0 to reach all the oil-polluted zones during the time interval $[T-\tau, T]$.

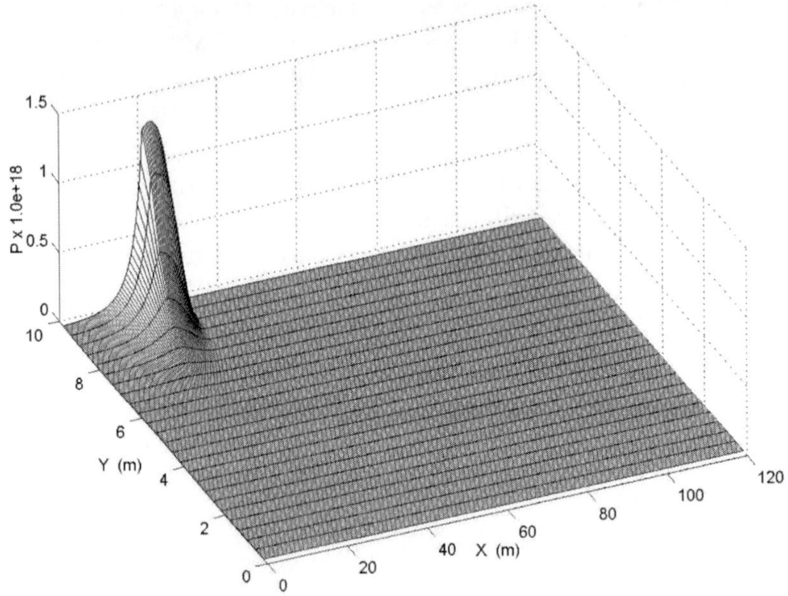

Figure 4. Example of an indicative function.

Finally, note that the adjoint function g_0 is omitted in equation (57) because $\int_0^T g_0^2(r_0,t)dt > 0$ for all $r_0 \in D$, and hence, any discharge point r_0 chosen in D impacts this region. In this way, equation (57) can be applied in both remediation problems: linear and quadratic.

7. Main and Adjoint Numerical Schemes of the Dispersion Problem

In this section, balanced and absolutely stable second-order finite-diference schemes based on the application of the splitting method by Marchuk (1986) and Crank-Nicolson scheme are developed to solve numerically the dispersion model (1)-(8) and its adjoint (15)-(21). Since they were described in detail in Skiba (1993), we give here only basic results.

Using the continuity equation (8), operator A of equation (1) can be written as $A = A_1 + A_2 + A_3$, where

$$A_1\phi = \frac{1}{2}\frac{\partial}{\partial x}(u\phi) + \frac{1}{2}u\frac{\partial \phi}{\partial x} - \frac{\partial}{\partial x}\mu\frac{\partial \phi}{\partial x} + \frac{1}{3}\sigma\phi$$

$$A_2\phi = \frac{1}{2}\frac{\partial}{\partial y}(v\phi) + \frac{1}{2}v\frac{\partial \phi}{\partial y} - \frac{\partial}{\partial y}\mu\frac{\partial \phi}{\partial y} + \frac{1}{3}\sigma\phi \tag{58}$$

$$A_3\phi = \frac{1}{2}\frac{\partial}{\partial z}(\tilde{w}\phi) + \frac{1}{2}\tilde{w}\frac{\partial \phi}{\partial z} - \frac{\partial}{\partial z}\mu\frac{\partial \phi}{\partial z} + \frac{1}{3}\sigma\phi$$

and $\tilde{w} = w - v_s$.

It order to show that each one-dimensional split operator A_i $(i = 1,2,3)$ is positive semidefinite (Skiba, 1993), let us suppose, without loss of generality, that the domain D is a cube $[0,X]\times[0,Y]\times[0,Z]$. Then

$$\int_0^X \phi A_1\phi dx = \frac{1}{3}\int_0^X \sigma\phi^2 dx + \int_0^X \mu\left(\frac{\partial \phi}{\partial x}\right)^2 dx + \left[\frac{1}{2}\phi^2 u - \mu\phi\frac{\partial \phi}{\partial x}\right]_0^X$$

Assume that $u(0) > 0$ and $u(X) > 0$. Then the boundary point $x = 0$ belongs to S^-, while point $x = X$ belongs to S^+. With the boundary conditions (5) and (4) at points $x = 0$ and $x = X$ respectively, the last term can be written as

$$\left[\frac{1}{2}\phi^2 u - \mu\phi\frac{\partial \phi}{\partial x}\right]_0^X = \frac{1}{2}\left[\phi^2(X)u(X) + \phi^2(0)u(0)\right] \geq 0$$

Since $\sigma > 0$ and $\mu > 0$, we conclude that

$$(A_1\phi,\phi)_{L_2(\mathbf{D})} = \int_0^Z\int_0^Y\int_0^X \phi A_1\phi dxdydz \geq 0$$

In the same way one can show that A_2 and A_3 are also positive semidefinite operators. It should be noted that this proof is also true for any region D which represents a union of finite number of cubes.

On the other hand, the operator of the adjoint problem (15)-(21) and (18) is the adjoint of A, and can be presented as the sum $A^* = A_1^* + A_2^* + A_3^*$ where

$$A_1^* g = -\frac{1}{2}\frac{\partial}{\partial x}(ug) - \frac{1}{2}u\frac{\partial g}{\partial x} - \frac{\partial}{\partial x}\mu\frac{\partial g}{\partial x} + \frac{1}{3}\sigma\phi$$

$$A_2^* g = -\frac{1}{2}\frac{\partial}{\partial x}(ug) - \frac{1}{2}u\frac{\partial g}{\partial x} - \frac{\partial}{\partial x}\mu\frac{\partial g}{\partial x} + \frac{1}{3}\sigma\phi \quad (59)$$

$$A_3^* g = -\frac{1}{2}\frac{\partial}{\partial z}(\tilde{w}g) - \frac{1}{2}\tilde{w}\frac{\partial g}{\partial z} - \frac{\partial}{\partial z}\nu\frac{\partial g}{\partial z} + \frac{1}{3}\sigma g$$

The problems (1)-(8) and (15)-(21) are solved in time with the splitting method (Marchuk and Skiba, 1976; Skiba, 1999). Suppose for simplicity that $\mu = \mu(z)$, and define the net functions on different grids:

$$\phi_{ijk} = \phi(x_i, y_j, z_k),\quad u_{ijk} = u(x_{i-1/2}, y_j, z_k),\quad v_{ijk} = v(x_i, y_{j-1/2}, z_k)$$

$$w_{ijk} = w(x_i, y_j, z_{k-1/2}),\quad \mu_k = \mu(z_k),\quad v_{ijk} = v(x_i, y_j, z_{k-1/2}).$$

The second-order discrete approximation of the operators A_i and continuity equation (8) have the following form (invariable indices i, j, k are omitted)

$$(A_1^h\phi)_{ijk} = \frac{1}{2\Delta x}[u_{i+1}\phi_{i+1} - u_i\phi_{i-1}] - \frac{\mu_k}{(\Delta x)^2}[\phi_{i+1} - 2\phi_i + \phi_{i-1}] + \frac{1}{3}\sigma\phi_i \tag{60}$$

$$(A_2^h\phi)_{ijk} = \frac{1}{2\Delta y}[v_{j+1}\phi_{j+1} - v_j\phi_{j-1}] - \frac{\mu_k}{(\Delta y)^2}[\phi_{j+1} - 2\phi_j + \phi_{j-1}] + \frac{1}{3}\sigma\phi_j \tag{61}$$

$$(A_3^h\phi)_{ijk} = \frac{1}{2\Delta z}[\tilde{w}_{k+1}\phi_{k+1} - \tilde{w}_k\phi_{k-1}] - \frac{1}{(\Delta z)^2}[\mu_{k+1}(\phi_{k+1} - \phi_k) - \mu_k(\phi_k - \phi_{k-1})] + \frac{1}{3}\sigma\phi_k \tag{62}$$

$$(u_{i+1} - u_i)/\Delta x + (v_{j+1} - v_i)/\Delta y + (w_{k+1} - w_k)/\Delta z = 0 \tag{63}$$

The adjoint operators $\left(A_i^h\right)^*$ are obtained with the substitution of $-u, -v, -\tilde{w}$ and g for $u, v, \tilde{w}$ and f in (60)-(61), respectively. As to the boundary conditions, we give only one example (see Skiba (1993) for more details). Let u_{ijk} be a positive value of the u-component of the velocity at the left boundary point $M = \left(x_{1/2}, y_j, z_k\right)$ of grid domain. Then, $U_n = -u_{1jk} < 0$, i.e., the point M belongs to S^-, and conditions (5) and (19) are approximated as

$$\mu_k(\phi_{0jk} - \phi_{1jk})/\Delta x + u_{1jk}(\phi_{0jk} + \phi_{1jk})/2 = 0, \qquad g_{0jk} = g_{1jk} \tag{64}$$

For any $i = 1, 2, 3$, the discrete operators A_i^h and $\left(A_i^h\right)^*$ are positive semidefinite, and all of them are skew-symmetric if $\mu = \sigma = 0$ and S is the coast line ($U_n = 0$ everywhere at S). Within each double time step interval $(t_n - \Delta t, t_n + \Delta t)$ the main and adjoint numerical schemes have the form

$$\Phi[n-\frac{3-i}{3}]-\Phi[n-\frac{4-i}{3}]=-\frac{\tau}{2}A_i^h(\Phi[n-\frac{3-i}{3}]+\Phi[n-\frac{4-i}{3}]) \quad (i=1,2)$$

$$\Phi[n+\frac{1}{3}]-\Phi[n-\frac{1}{3}]=-\tau A_3^h(\Phi[n+\frac{1}{3}]+\Phi[n-\frac{1}{3}])+2\tau q[n] \tag{65}$$

$$\Phi[n+\frac{4-i}{3}]-\Phi[n+\frac{3-i}{3}]=-\frac{\tau}{2}A_i^h(\Phi[n+\frac{4-i}{3}]+\Phi[n+\frac{3-i}{3}]) \quad (i=2,1)$$

and

$$G[n+\frac{3-i}{3}]-G[n+\frac{4-i}{3}]=-\frac{\tau}{2}(A_i^h)^*\,(G[n+\frac{3-i}{3}]+G[n+\frac{4-i}{3}]) \quad (i=1,2)$$

$$G[n-\frac{1}{3}]-G[n+\frac{1}{3}]=-\tau(A_3^h)^*\,(G[n-\frac{1}{3}]+G[n+\frac{1}{3}])+2\tau\, p[n] \tag{66}$$

$$G[n-\frac{4-i}{3}]-G[n-\frac{3-i}{3}]=-\frac{\tau}{2}(A_i^h)^*\,(G[n-\frac{4-i}{3}]+G[n-\frac{3-i}{3}]) \quad (i=2,1)$$

where Φ and G are the vectors representing the grid values of solutions ϕ and g at fractional time steps, and q and p are the vectors representing the grid values of functions Q and P at moment t_n, respectively (Skiba, 1993). Each one-dimensional discrete problem is a Crank-Nicolson scheme that can be solved by the efficient direct factorization method for a three-diagonal matrix (see, for example, Marchuk and Skiba, 1976). The stability of schemes (65) and (66) for any time step τ directly follows from the inequalities

$$\|\Phi[n+1]\|\le\|\Phi[n-1]\|+2\tau\|q[n]\| \tag{67}$$

and

$$\|G[n-1]\|\le\|G[n+1]\|+2\tau\|p[n]\| \tag{68}$$

where $\|\cdot\|$ is the Euclidean vector norm (Skiba, 1993). The use of Lagrange identity leads to the equation

$$G^*[n+1]\Phi[n+1]+\tau p^*[n]\left(\Phi[n+\frac{1}{3}]+\Phi[n-\frac{1}{3}]\right)$$

$$= \tau\left(G^*[n+\frac{1}{3}] + G^*[n-\frac{1}{3}] \right) q[n] + G^*[n-1]\Phi[n-1] \tag{69}$$

in each subinterval $[t_n - \Delta t, t_n + \Delta t]$. Summing such relations over all subintervals in $[0, T]$ (i.e., over all n) and using conditions (7) and (21) we derive a discrete version of adjoint estimate (25).

8. Numerical Examples of the Remediation Problem

8.1. Examples Solving the Quadratic Programming Problem

In the case that equation (37) is unable to determine the optimal discharge rate, the solution can be obtained by discretizing the integrals in problem (31)-(**33)** with the middle point formula. It leads to the following quadratic programming problem in $\mathbb{R}^L$:

$$\text{minimize} \quad m(Q_1,\dots,Q_L) = \Delta t \sum_{l=1}^{L} Q_l^2 \tag{70}$$

$$\text{subject to:} \quad Q_l \geq 0, \quad (l = 1,\dots,L) \tag{71}$$

$$\sum_{l=1}^{L} Q_l g_{il} = \frac{c_i}{\Delta t}, \quad (i = 1,\dots,N) \tag{72}$$

where $\Delta t = T/L$, $t_l = l \cdot \Delta t$, $l = 0,1\dots,L$, $Q_l = Q(t_{l-1/2})$ and $g_{il} = g_i(r_0, t_{l-1/2})$, $l = 0,1\dots,L$.

The optimization problem (70)-(72) has unique solution. The existence of solution immediately follows from the fact that the goal function $m(\vec{Q})$ where $\vec{Q} = (Q_1,\dots,Q_L)$ is continuous, and the feasibility set

$$\Phi = \left\{ \vec{Q} \in \mathbb{R}^L;\ Q_l \geq 0,\ l = 1,\dots,L, \text{ and } \sum_{l=1}^{L} Q_l g_{il} = \frac{c_i}{\Delta t},\ i = 1,\dots,N \right\}$$

is a compact set in $\mathbb{R}^L$ (Luenberger, 1984). The uniqueness of solution is the consequence of the facts that $m(\vec{Q})$ is a strictly convex function, and set Φ is convex, too (Luenberger, 1984). We assume here that the feasibility set Φ is non-empty. This optimal solution can efficiently be calculated with a numerical optimization technique, for example, with that provided by the *lsqlin* subroutine of MATLAB.

It should be noted that, when the problem given to *lsqlin* has only upper and lower bounds, i.e., no linear inequalities or equalities are specified, the default algorithm is the large-scale method. Or, if the problem given to *lsqlin* has only linear equalities, i.e., no upper and lower bounds or linear inequalities are specified, the default algorithm is also the large-scale method. This method is a subspace trust-region method based on the interior-reflective Newton method described in Coleman and Li (1996). Each iteration requires the approximate solution of a large linear system using the method of preconditioned conjugate gradients (PCG).

Otherwise, medium-scale optimization is used. In such a case, *lsqlin* uses an active set method, which is also a projection method, similar to that described in Gill, et al. (1981). It finds an initial feasible solution by solving a linear programming problem. Due to the structure of quadratic programming problem (70)-(72), the second method of *lsqlin* routine is applied in the following examples. The coefficients g_{il} were found by solving numerically the adjoint model (15)-(21) for each zone Ω_i, using the splitting method described in section 7.

In order to illustrate the method developed, we now consider some simple examples of remediation in a channel of one hundred and twenty meters long $[0,120]$, ten meters wide $[0,10]$, and four meters deep $[0,4]$, $H=4$. The channel contains three oil-contaminated zones Ω_i ($N=3$). The critical nutrient concentrations c_i $\left(grm^{-3}\right)$ in the zones vary from one experiment to another (Table 1) and generate different optimal discharge rates Q_k^* (Figure5). The zones under consideration are:

$$\Omega_1 = [20,30]\times[9,10]\times[0,4],\ \Omega_2 = [70,80]\times[9,10]\times[0,4]$$

and $\Omega_3 = [95,100]\times[0,2]\times[0,4]$.

The parameters of adjoint model (18)-(24) have been taken as follow: the velocity vector $\vec{U}$ is directed along the channel and is equal to $30\ mh^{-1}\vec{i}$, $\vec{i}=(0,0,1)'$, $\mu = 6\,m^2h^{-1}$, $\sigma = 1\ h^{-1}$ and $\zeta = v_s = 0$. The nutrient is released at the point $r_0 = (3,2.2,2)$ during four hours, the total time interval is 4 hours: $(0,T)\equiv(0,4)$, and the mean concentration is controlled within the last one-hour interval $(3,4)$, i.e., $\tau = 1\ h$. Adjoint functions $g_i(r_0,t)$ for the three zones $(i=1,2,3)$ are given in Figure 3.

The evolution of the mean concentration of nutrient in the zones Ω_i $(i=1,2,3)$ obtained in the 7th experiment is shown in Figure 6, while isolines of the mean concentration of nutrient in domain D at the final moment $T=4h$ are presented in Figure 7. The optimal discharge rate applied is $Q_7(t)$. In every experiment, the optimal discharge rate has successfully been found using equation (37).

Table 1. Critical concentrations of nutrient in zones Ω_i $(i=1,2,3)$

k	$c1$	$c2$	$c3$
1	0.8	0.8	0.8
2	1.0	0.8	0.5
3	0.5	1.0	1.5
4	1.2	0.5	1.2
5	0.6	1.2	0.6
6	0.6	0.6	1.5
7	1.5	0.6	0.6

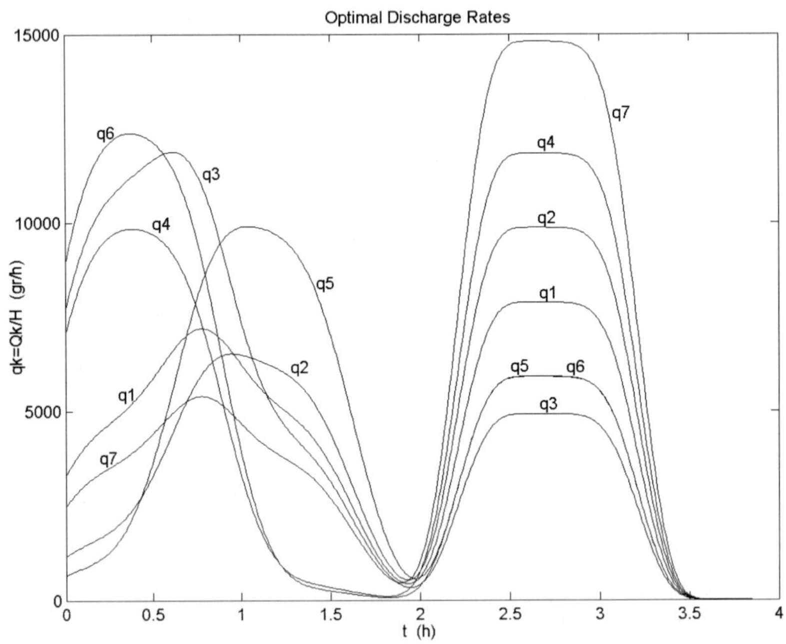

Figure 5. Optimal discharge rates $q_k(t)=Q_k(t)/H$, $k=1,2,...,7$.

The results have been compared with the corresponding numerical solutions of problem (70)-(72) for $L=400$ by means of the *lsqlin* subroutine of MATLAB. It should be noted that the main advantage of equation (37) is that the optimal rate $\vec{Q}_k^*$ in each experiment $(k=1,...,7)$ is calculated by resolving a linear system with a symmetric, positive definite and well-conditioned 3×3 matrix

$$\Psi=10^{-7}\begin{pmatrix}0.1208 & 0.0001 & 0.0000\\ 0.0001 & 0.1788 & 0.0759\\ 0.0000 & 0.0759 & 0.2006\end{pmatrix}$$

(its condition number is 2.35), while numerical formulation (70)-(72) requires to solve a quadratic optimization problem with a 400×400 matrix, since $L=400$.

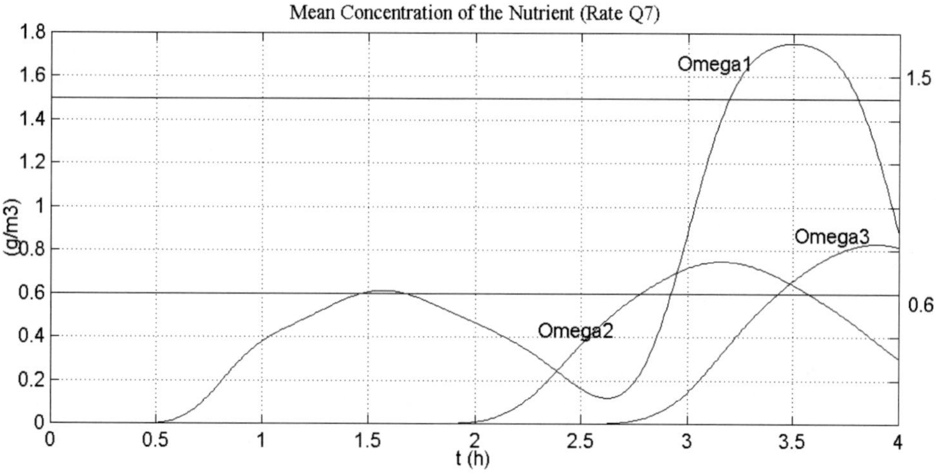

Figure 6. Evolution of mean concentration of nutrient in zones Ω_i, $i=1,2,3$. The optimal discharge rate applied is $Q_7(t)$.

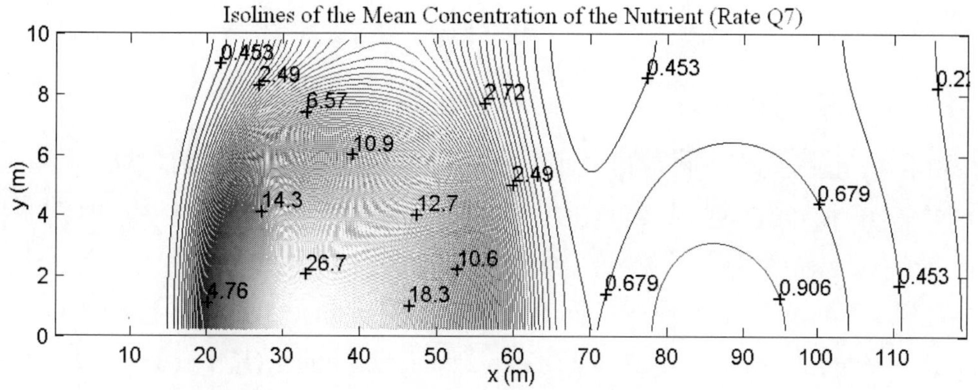

Figure 7. Isolines of mean concentration of nutrient in region D at final moment $T=4\text{h}$. The optimal discharge rate applied is $Q_7(t)$.

However, formulation (70)-(72) is useful when equation (37) leads to a non-positive function Q. This is the case, for example, when the critical nutrient concentrations are $c_1=c_3=0.8\ (grm^{-3})$ and $c_2=0.2\ (grm^{-3})$. Equation (37) then determines $Q=q_e$ (see Figure 8), which cannot be the optimal discharge rate since this function is negative approximately in an interval between the first and the second hour. In contrast, the solution of (70)-(72), $Q=q_{\text{opt}}$ (see Figure 8), is a non-negative function in total time interval $[0,4]$, and according to theorem 2, it must be the optimal discharge rate Q^*.

On the other hand, while minimizing function $\|Q^*\|^2$ on the support of the indicative function P (Figure 4), $\text{supp(P)}\subset[0,40]\times[6,10]$, we got the optimal discharge point $r_0^*=(21.8,6.6,2.0)$ for the case $k=1$ $(c_i=0.8\ grm^{-3},\ i=1,2\text{ and }3)$.

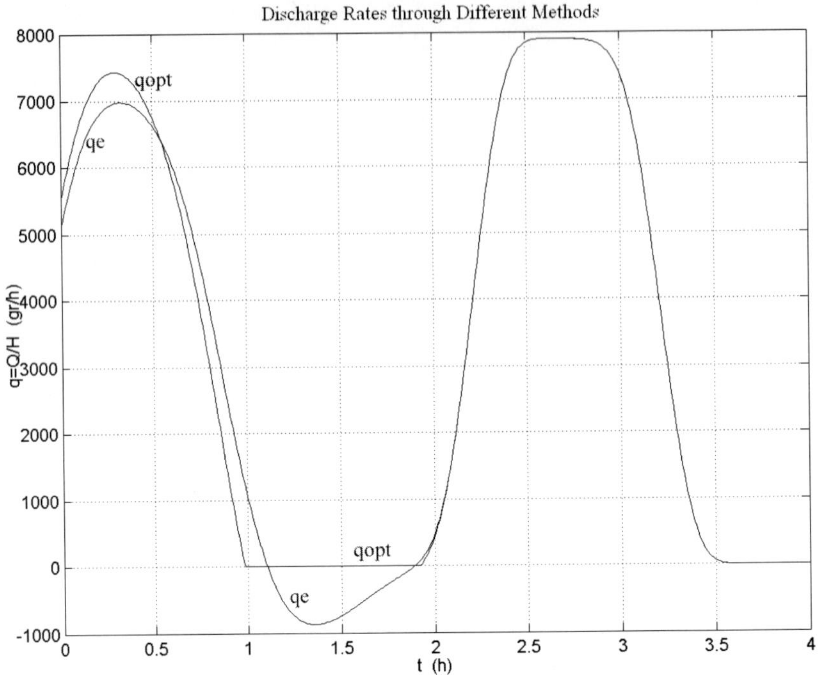

Figure 8. Discharge rates obtained with two different methods.

Also, by means of equation (37), the optimal discharge rate $Q^* = Q\text{opt}$ was determined (Figure 9). As it was expected, $Q\text{opt}$ introduces less mass into the channel than the discharge rate Q_1 ($m(Q_1)/m(Q\text{opt}) = 4.7$), which is the advantage of the method.

The last conclusion can be seen directly from figure 9 where functions $Q\text{opt}$ and Q_1 are shown. Also, it follows from figure 9 that each adjoint function $g_i(r_0^*, t)$ taken at the optimal release point r_0^* has greater area than the corresponding adjoint function $g_i(r_0, t)$ taken at a non-optimal point r_0 (see Figure 3). In other words, the adjoint functions taken at the optimal discharge point show the greatest impact on the time interval $[0, T]$ (see equation 25).

8.2. Examples Solving the Linear Programming Problem

In order to illustrate the method developed in section 6, we now again consider the simple example of bioremediation in a channel of one hundred and twenty meters long $[0,120]$, ten meters wide $[0,10]$, and four meters deep $[0,4]$, $H = 4$. The channel contains three oil-polluted zones: $\Omega_1 = [20,30]\times[9,10]\times[0,4]$, $\Omega_2 = [70,80]\times[9,10]\times[0,4]$ and $\Omega_3 = [95,100]\times[0,2]\times[0,4]$.

The critical nutrient concentrations c_i $\left(grm^{-3}\right)$ in the zones vary from one experiment to another (Table 2) and generate different optimal discharge rates Q_j^* (Figure 10).

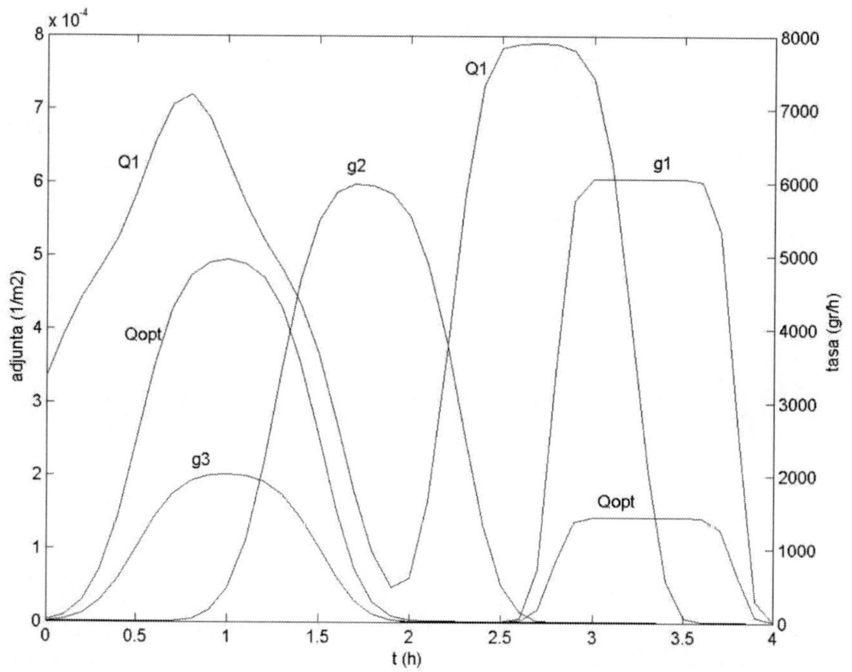

Figure 9. Adjoint functions and optimal discharge rate for the optimal release point r_0^*.

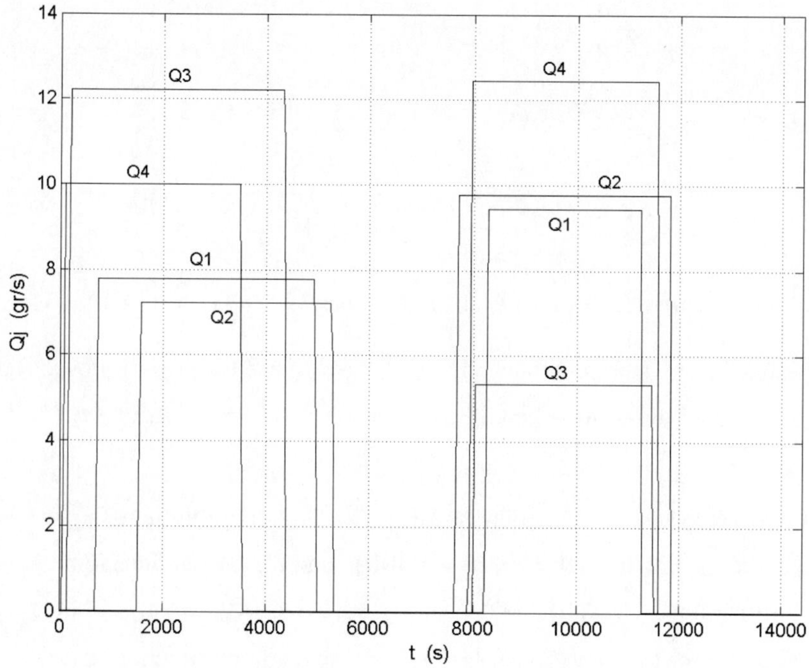

Figure 10. Optimal discharge rates Q_j for data in Table 2, $j = 1, 2, 3$ and 4.

Table 2. Numerical details of linear programming problem solution (53)-(56). For the task a workstation HP-XW8200 was used

j	C_1	C_2	C_3	$m(Q_j^*)$	Simp_meth (s)	Int_pro_meth (s)	Relative error
1	0.8	0.8	0.8	6.1272×10^4	10.6720	0.1090	2.4259×10^{-11}
2	1.0	0.8	0.5	6.7720×10^4	15.0940	0.1869	4.0647×10^{-10}
3	0.5	1.0	1.5	6.9724×10^4	9.5309	0.1250	5.2762×10^{-12}
4	1.2	0.5	1.2	7.9676×10^4	18.1410	0.1570	9.3829×10^{-11}

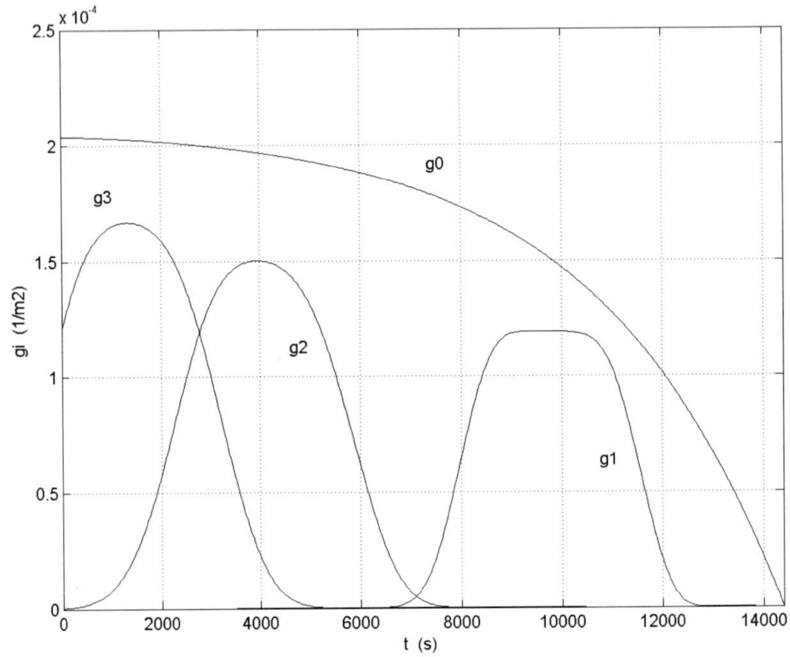

Figure 11. Adjoint functions for region D (g_0) and zones Ω_i $(g_i, i=1,2,3)$.

The parameters of adjoint model (15)-(21) have been taken as follows: The velocity vector $\vec{U}$ is directed along the channel and is equal to $\vec{U}=0.0083\,\vec{i}\ ms^{-1}$, $\vec{i}=(0,0,1)^t$, $\mu=0.0017\ m^2s^{-1}$, $\sigma=0.00027\ s^{-1}$ and $\zeta=v_s=0$.

The nutrient is released at the point $r_0=(3,2.2,2)$ during four hours ($T=14400\ s$), the total time interval is 4 hours: $(0,T)=(0,\ 14400)$, and the mean concentration is controlled within the last one-hour interval $[T-\tau,T]=[10800,\ 14400]$, i.e., $\tau=3600\ s$. The adjoint functions $g_i(r_0,t)$ for the three zones $(i=1,2,3)$, and adjoint function $g_0(r_0,t)$ for region D, are given in figure 11.

In all the examples $(j=1,2,3 \text{ and } 4)$, the linear programming problem (53)-(56) is solved with the following parameters $\varepsilon_i=0.05\ (grm^{-3})$, $c_0=3.5\ (grm^{-3})$,

$$Q_0(t) = \begin{cases} \alpha_j, & 0 \le t < 7200 \\ \beta_j, & 7200 \le t \le 14400 \end{cases}$$

where $\alpha_1 = 7.77$ and $\beta_1 = 9.44$; $\alpha_2 = 7.22$ and $\beta_2 = 9.77$; $\alpha_3 = 12.22$ and $\beta_3 = 5.33$; $\alpha_4 = 10.00$ and $\beta_4 = 12.44$. Finally, the spatial mesh size for numerical solution of adjoint problem (15)-(21) is $\Delta x = \Delta y = 0.4\ m$ with the time step $\Delta t = 36\ s$, and hence, $L = 400$ for the linear programming problem.

The computing time to get each adjoint problem solution using the workstation HP-XW8200 was about $370\ s$. Table 2 shows the time required to solve the linear programming problem (53)-(56) by means of *linprog* subroutine of MATLAB. It is important to note that the interior point method is, approximately, one hundred times faster than the simplex method, and both methods determine the same solution (see relative errors in Table 2). We conclude that, for each example, the computing time required to find the optimal discharge rate was about $1500\ s$. Figure 10 shows the optimal discharge rates Q_j^* for the four examples ($j = 1, 2, 3$ and 4). Each release rate is simply a combination of step functions. The nonzero values of step functions are defined in the intervals where the adjoint functions are positive, besides the height of the steps is determined by the limit function Q_0.

9. Generalization of the Method

We now make a final remark. In this work, we suppose that the nutrient released into the marine environment is a liquid, for example, the product named Inipol EAP22 (Ladousse and Tramier, 1991). Therefore the sedimentation velocity v_s is very small or zero, and the sediment of nutrient mass on the bottom S_B of the marine region D is not significant, namely, the nutrient is totally dissolved in the water. That is why we control the mean concentration of nutrient in the tridimensional marine oil-polluted zones Ω_i, which are located at the shoreline but not at the bottom of the marine region D. On the other hand, when the oil is concentrated on the marine floor, the bioremediation is required in some two-dimensional zones Ω_B^i, located on the sea bottom S_B of region D, the nutrient should be released in granular form, as the Customblen product (Prince et al., 2003). In this case, the sedimentation velocity v_s is not zero, and the mass of the nutrient, deposited on S_B, increases, which favors the remediation process in the polluted zones Ω_B^i. Therefore, the control of the nutrient mass deposited on S_B is more adequate than control of the nutrient concentration per volume. To study this problem, we can also consider the variational problems (31)-(33) or (43)-(45), and dispersion model (1)-(8), but the functional J_i, given by equation (22), is redefined as

$$J_i = -\frac{1}{|\Omega_B^i|}\int_0^T \int_{\Omega_B^i} v_s \phi \vec{k} \cdot \vec{n} dS dt$$

According to the mass balance equation (10), this functional estimates the mass of the deposited nutrient, per superficial unit, on the polluted zone $\Omega_B^i \subset S_B$, during the interval of discharge $[0,T]$. Note that J_i is a non-negative functional because $\vec{k} \cdot \vec{n} < 0$ on S_B.

In order to extend to this case the methodology described in this work, we must change the adjoint model (15)-(21) as follows. The forcing in the transport equation (15) is redefined as

$$p(r,t) = 0, \quad r \in D, \quad 0 < t < T$$

and the boundary condition given by (20) is redefined as

$$\mu \frac{\partial g}{\partial n} + \vec{g} \cdot \vec{n} = \begin{cases} -\dfrac{v_s}{|\Omega_B^i|} \vec{k} \cdot \vec{n}, & \text{if } r \in \Omega_B^i \\ 0, & \text{if } r \in S_B - \Omega_B^i \end{cases} \qquad \text{on } S_B$$

With these modifications, the duality principle (24) also holds for the new functional J_i, and hence, the variational problems (31)-(33) or (43)-(45), and the programming problems (70)-(72) or (53)-(56), can be applied to determine the optimal discharge parameters of the bioremediation problem. We stress that this adjoint model also has unique solution that continuously depends of the parameters determining the flow through the boundary S_B.

CONCLUSION

The main objectives of the mathematical modeling in the environment protection are the prediction of concentrations of different substances (pollutants, cleanears, nutrients, etc.), the development of the methods which help to prevent dangerous pollution levels (control of emissions) and the development of the strategies for the remediation of polluted zones. In this work, we have presented a method of cleaning the oil-polluted marine environment through bioremediation. It is assumed that oil is stranded in some zones at the shoreline and the goal is to release a nutrient into aquatic system in order to increase the amount of indigenous microorganisms which degrade the pollutants in such zones. Thus, the specific objectives are to determine the appropriate parameters of releasing the nutrient, namely, r_0 (discharge site) and Q (discharge rate), in order to reach necessary concentrations of the nutrient in the polluted zones. Both unknown parameters are chosen so that to minimize the total mass of the nutrient, with the aim to minimize the impact on the environment and the cost of remediation.

To this end, for each point r_0, the optimal discharge rate $Q^* = Q^*(t;r_0)$ is obtained through the solution of a linear programming problem, or in case of a quadratic formulation, by means of analytical equations and/or the corresponding quadratic programming problem. We have shown the existence of solutions of all these problems, whilst the uniqueness of solutions depends on the strictly convexity of the norm used to assess the total mass of nutrient introduced into the marine enviroment. Then the optimal discharge site r_0^* is selected as such a point which minimizes the non-linear function $m(Q^*(t;r_0))$. To reduce the computational efforts, the search is limited to the points r_0 in D at which the indicative function P is positive.

To determine the discharge rate of the nutrient $Q^* = Q^*(t;r_0)$ through the linear formulation, it is necessary first to solve $N+1$ adjoint problems (where N is the number of contaminated zones), and then a linear programming problem with $L+1$ positive variables and $2(L+N)+3$ constraints (where $L+1$ is the number of nodes of regular mesh in interval $[0,T]$). This procedure is shown to be computationally efficient. Also, the numerical examples show that in the case when the limit Q_0 is constant, $Q^*(t;r_0)$ has a simple form (combination of step functions). This is important advantage of the method in its practical application in bioremediation.

In case of the quadratic formulation, the discharge rate of nutrient $Q^* = Q^*(t;r_0)$ is determined by means of a linear combination of N adjoint functions, where the coefficients are the solution of a linear system of equations given by the Gram matrix. Alternately, the discharge rate $Q^*(t;r_0)$ can be got as the solution of a quadratic programming problem with L positive variables and $L+N$ constraints. This procedure is shown to be computationally efficient.

This new remediation method is strongly based on the adjoint estimates, but it also uses the direct concentration estimates of the nutrient in polluted zones when multiple discharges of the nutrient are needed. These equivalent estimates complement each other well in the assessment of nutrients and control of pollution. The direct estimates, utilizing the solution of the transport problem, enable making the comprehensive analysis of ecological situation in the whole area. On the other hand, the adjoint estimates use the adjoint problem solutions and explicitly depend on the discharge rate of nutrient and its initial distribution in the region. Besides, the solutions of adjoint problem serve as influence functions, which show the impact of the discharge point location on the concentration of nutrient in each oil-polluted zone. Therefore, the adjoint estimates are effective and economical in the study of the sensitivity of the concentrations of the nutrient to variations in the model parameters.

Owing to special boundary conditions, both the main and adjoint problems are well-posed according to Hadamard, that is, the solution of each problem exists, is unique and stable to initial perturbations. These conditions are reduced to the well-known and natural boundary conditions in the non-diffusion limit (pure advection problem) and in the case of a closed sea basin whose boundary is the coast line.

As a complement, for the solution of the main and adjoint transport problems, we have presented finite difference schemes that are balanced, absolutely stable, of second-order of approximation, and are based on the splitting method and Crank-Nicolson schemes. In the absence of dissipation and sources, each scheme has two conservation laws. The one-dimensional split discrete equations obtained at every fractional step of the numerical algorithm are efficiently solved by the factorization method for a three-diagonal matrix.

Finally, we point out that the adjoint technique described in this work is useful for the solution of the problem of control of industrial emissions, the problem of detecting the enterprises which violate the emission rates prescribed by controls, and it can also be used for estimating the intensity of a pollution source in the case when its position is known. For example, we can consider the case of a nuclear (or chemical) plant accident or nuclear bomb explosion (testing, terrorist attacks, and others). In all these situations, the source position is known or can easily be located (for example, from a satellite). Then, our method gives a lower bound of the source intensity, which can be useful in the assessment of the scale of accident.

ACKNOWLEDGMENTS

This work was supported by the projects: PAPIIT IN103313-2 and PAPIIT IN104811-3 (DGAPA-UNAM, Mexico), and by the grants 25170 and 14539 of The National System of Researchers (SNI, CONACyT, México).

Reviewed by Dr. Denis M. Filatov, Centro de Investigación en Computación (CIC), Instituto Politécnico Nacional (IPN), C. P. 07738, México, D. F., Mexico. E-mail: denisfilatov@gmail.com.

REFERENCES

Alvarez-Vázquez, L. J., García-Chan, N., Martínez, A., Vázquez-Méndez, M. E., and Vilar, M. A. (2010). Optimal control in wastewater management: a multi-objective study, *Communications in Applied and Industrial Mathematics* 1 (2), 62-77.

Boufadel, M. C., Suidan, M. T. and Venosa, A. D. (2006). Tracer studies in laboratory beach simulating tidal influences, *J. Env. Engineering* 132, 616-623.

Boufadel, M. C., Suidan, M. T. and Venosa, A. D. (2007). Tracer studies in a laboratory beach subjected to waves, *J. Env. Engineering* 133, 722-732.

Bragg, J. R., Prince, R. C., Harner, E. J., and Atlas, R. M. (1994). Effectiveness of bioremediation for the Exxon Valdez oil spill, *Nature* 368, 413-418.

Cheney, E. W. (1966). *Introduction to Approximation Theory*, Chelsea Publishing Company, NY.

Coleman, T. F. and Li, Y. (1996). A reflective Newton method for minimizing a quadratic function subject to bounds on some of the variables, *SIAM Journal on Optimization* 6 (4), 1040-1058.

Coulon, F., McKew, B. A., Osborn, A. M., McGenity, T. J., and Timmis, K. N. (2006). Effects of temperature and biostimulation on oil-degrading microbial communities in temperate estuarine waters, *Env. Microbiology* 9(1), 177-186.

Crank, J. and Nicolson, P. (1947). A practical method for numerical evaluation of solutions of partial differential equations of the heat conduction type, *Proc. Cambridge Philos. Soc.* 43, 50-67.

Dang, Q. A., Ehrhardt, M., Tran, G. L., and Le, D. (2012). Mathematical modeling and numerical algorithms for simulation of oil pollution, *Env. Model. Assess.* 17(3), 275-288.

Dantzig, G. B., Orden, A. and Wolfe, P. (1955). Generalized simplex method for minimizing a linear from under linear inequality constraints, *Pacific Journal Math.*, 5, 183-195.

Dieudonné, J. (1969). *Foundations of modern analysis*, Academic Press, NY and London.

Folland, G. B. (1999). *Real analysis: Modern techniques and their applications*, John Wiley and Sons, Inc.

Gill, P. E., Murray, W. and Wright, M. H. (1981). *Practical Optimization*, Academic Press, London, UK.

Hadamard, J. (1923). *Lectures on Cauchy's Problem in Linear Partial Differential Equations*, Yale University Press, New Haven, CT.

Hadley, G. (1962). *Linear programming*, Addison-Wesley, US.

Head, M. and Swannell, R. P. J. (1999). Bioremediation of petroleum hydrocarbon contaminants in marine habitats, *Current Opinion in Biotechnology* 10 (3), 234-239.

Hinze, M., Yan, N. N. and Zhou, Z. J. (2009). Variational discretization for optimal control governed by convection dominated diffusion equations, *Journal of Computational Mathematics* 27 (2-3), 237-253.

Hongfei, F. (2010). A characteristic finite element method for optimal control problems governed by convection-diffusion equations, *Journal of Computational and Applied Mathematics* 235 (3), 825-836.

Kreyszig, E. (2006). *Advanced Engineering Mathematics*, John Wiley and Sons, NJ.

Kreyszig, E. (1989). *Introductory functional analysis with applications*, John Wiley and Sons, Inc.

Liu, F., Zhang, Y. H. and Hu, F. (2005). Adjoint method for assessment and reduction of chemical risk in open spaces, *Env. Model. Assess.* 10 (4), 331-339.

Luenberger, D. G. (1984). *Linear and Non-linear Programming*, Addison Wesley, Reading, MA.

Ladousse, A. and Tramier, B. (1991). Results of 12 years of research in spilled oil bioremediation: Inipol EAP22. In: *Proceedings of the 1991 International Oil Spill Conference* (pp. 577-582), American Petroleum Institute, Washington DC.

Marchuk, G. I. (1995). *Adjoint equations and analysis of complex systems*, Kluwer, Dordrecht.

Marchuk, G. I. (1986). *Mathematical Models in Environmental Problems*, Elsevier, NY.

Marchuk G. I. (1974). *Numerical solution of problems of the dynamics of atmosphere and ocean*, Leningrad, Gigrometeoizdat (in Russian).

Marchuk, G. I. and Skiba, Y. N. (1990). Role of adjoint functions in studying the sensitivity of a model of the thermal interaction of the atmosphere and ocean to variations in input data, *Izvestiya, Atmos. Ocean. Physics* 26, 335-342.

Marchuk, G. I. and Skiba, Yu. N. (1976). Numerical calculation of the conjugate problem for a model of the thermal interaction of the atmosphere with the oceans and continents, *Izvestiya, Atmos. Ocean. Physics* 12, 279-284.

Mehrotra, S. (1992). On the implementation of a primal-dual interior point method, *SIAM Journal on Optimization* 2, 575-601.

Mills, M. A., Bonner, J. S., Simon, M. A., McDonald, T. J., and Autenrieth, R. L. (1997). Bioremediation of a controlled oil release in a wetland. In: *Proceedings of the Twentieth Arctic and Marine Oilspill Program (AMOP) Technical Seminar*, 1, 609-616.

National Research Council, (2002). *Oil in the Sea III: Inputs, Fates and Effects*, National Academy of Sciences, Washington DC.

Parra-Guevara, D. and Skiba, Yu. N. (2003). Elements of the mathematical modelling in the control of pollutants emissions, *Ecological Modelling* 167, 263-275.

Parra-Guevara, D. and Skiba, Yu. N. (2006), On optimal solution of an inverse air pollution problem: Theory and numerical approach, *Math. Comp. Modell.* 43, 766-778.

Parra-Guevara, D. and Skiba, Yu. N. (2007). A variational model for the remediation of aquatic systems polluted by biofilms, *Intern. J. Appl. Math.* 20, 1005-1026, 2007.

Parra-Guevara, D., Skiba, Y. N. and Pérez-Sesma, A. (2010). A linear programming model for controlling air pollution, *Intern. J. Appl. Math.* 23 (3), 549-569.

Parra-Guevara, D., Skiba, Y. N. and Arellano, F. N. (2011). Optimal assessment of discharge parameters for bioremediation of oil-polluted aquatic systems, *Intern. J. Appl. Math.* 24 (5), 731-752.

Prince, R. C., Bare, R. E., Garrett, R. M., Grossman, M. J., Haith, C. E., Keim, L. G., Lee, K., Holtom, G. J., Lambert, P., Sergy, G. A., Owens, E. H., and Guénette, C. C. (1999). Bioremediation of a marine oil spill in the Arctic. In: B. C. Alleman and A. Leeson (Eds.), *In Situ Bioremediation of Petroleum Hydrocarbon and Other Organic Compounds* (pp. 227-232), Battle Press, Columbus, OH.

Prince, R. C. and Bragg, J. R. (1997). Shoreline bioremediation following the Exxon Valdez oil spill in Alaska, *Bioremediation Journal* 1, 97-104.

Prince, R. C., Clark, J. R., Lindstrom, J. E., Butler, E. L., Brown, E. J., Winter, G., Grossman, M. J., Parrish, R. R., Bare, R. E., Braddock, J. F., Steinhauer, W. G., Douglas, G. S., Kennedy, J. M., Barter, P. J., Bragg, J. R., Harner, E. J., and Atlas, R. M. (1994). Bioremediation of the Exxon Valdez oil spill: monitoring safety and efficacy. In: R. E. Hinchee, B. C. Alleman, R. E. Hoeppel and R. N. Miller (Eds.), *Hydrocarbon Remediation* (pp. 107-124), Lewis Publishers, Boca Raton, FL.

Prince, R. C., Lessard, R. R. and Clark, J. R. (2003). Bioremediation of marine oil spills, *Oil and Gas Science and Technology* 58 (4), 463-468.

Pudykiewicz, J. (1998). Application of adjoint tracer transport equations for evaluating source parameters, *Atmospheric Environment* 32, 3039-3050.

Ramsay, M. A., Swannell, R. P. J., Shipton, W. A., Duke, N. C., and Hill, R. T. (2000). Effect of bioremediation on the microbial community in oiled mangrove sediments, *Marine Pollution Bulletin* 41, 413-419.

Rektorys, K. (1980). *Variational Methods in Mathematics, Science and Engineering*, D. Reidel Publishing Company, Dordrecht.

Skiba, Yu. N. (1993). Balanced and absolutely stable implicit schemes for the main and adjoint pollutant transport equations in limited area, *Rev. Intern. Contamin. Ambient.* 9, 39-51.

Skiba, Yu. N. (1996). Dual Oil Concentration Estimates in Ecologically Sensitive Zones, *Env. Monit. Assess.* 43, 139-151.

Skiba, Yu. N. (1999). Direct and adjoint oil spill estimates, *Env. Monit. Assess.* 59, 95-109.

Skiba, Yu. N. and Parra-Guevara, D. (1999). Mathematics of Oil Spills: Existence, Uniqueness, and Stability of Solutions, *Geofísica Internacional* 38 (2), 117-124.

Skiba, Yu. N. and Parra-Guevara, D. (2000). Industrial Pollution Transport. Part I: Formulation of the Problem and Air Pollution Estimates, *Env. Model. Assess.* 5, 169-175.

Skiba, Yu. N. (2003). On a Method of Detecting the Industrial Plants Which Violate Prescribed Emission Rates, *Ecological Modelling* 159, 125-132.

Skiba, Yu. N., Parra-Guevara, D. and Belitskaya V. D. (2005). Air quality assessment and control of emission rates, *Env. Monit. Assess.* 111, 89-112.

Smith, D. R. (1998). *Variational Methods in Optimization*, Dover Publications, NY.

Strang, G. (1980), *Linear Algebra*, Academic Press, NY.

Swannell, R. P. J., Mitchell, D., Jones, D. M., Petch, S., Head, I. M., Wilis, A., Lee, K., and Lepo, J. E. (1999a). Bioremediation of oil-contaminated fine sediment. In: *Proceedings of the 1999 International Oil Spill Conference* (pp. 751-756). American Petroleum Institute, Washington DC.

Swannell, R. P. J., Mitchell, D., Lethbridge, G., Jones, D., Heath, D., Hagley, M., Jones, M., Petch, S., Milne, R., Croxford, R., and Lee, K. (1999b). A field demonstration of the efficacy of bioremediation to treat oiled shorelines following the Sea Empress incident, *Environmental Technology* 20, 863-873.

Venosa, A. D. (1998). Oil spill bioremediation on coastal shorelines: a critique. In: S. K. Sikdar and R. I. Irvine (Eds.), *Bioremediation: Principles and Practice*. Vol. III. Bioremediation Technologies, Technomic, Lancaster, PA, 259-301.

Venosa, A. D., Suidan, M. T., Wrenn, B. A., Strohmeier, K. L., Haines, J. R., Eberhart, B. L., King, D., and Holder, E. (1996). Bioremediation of an experimental oil spill on the shoreline of Delaware Bay, *Environmental Science and Technology* 30, 1764-1775.

Yan, N. N. and Zhou, Z. J. (2009). A priori and a posteriori error analysis of edge Stabilization Galerkin method for the optimal control problem governed by convection-dominated diffusion equation, *Journal of Computational and Applied Mathematics* 223 (1), 198-217.

Yee, E. (2008). Theory for reconstruction of an unknown number of contaminant sources using probabilistic inference, *Boundary-Layer Meteorology* 127 (3), 359-394.

Zhu, X., Venosa, A. D., Suidan, M. T., and Lee, K. (2001). *Guidelines for the Bioremediation of Marine Shorelines and Freshwater Wetlands*, US Environmental Protection Agency, Cincinnati, OH.

Zhang, Y. (1995). *Solving large-scale linear programs by interior-point methods under the MATLAB environment*, Technical Report TR96-01, Department of Mathematics and Statistics, University of Maryland.

ZoBell, C. E. (1973). Microbial degradation of oil: Present status, problems, and perspectives, pp. 3-16, In: Ahearn and Meyers (Eds.), *The Microbial Degradation of Oil Pollutants*, Publication No. LSU-SG-73-01, Proceedings Center for Wetlands Resource, Louisiana State University, Baton Rouge, LA.

In: Bioremediation: Processes, Challenges and Future Prospects ISBN: 978-1-62948-513-3
Editors: Jesús Bernardino Velázquez-Fernández et al.

Chapter 8

EFFECTS OF AGRICULTURAL RESIDUES ON TPH CONCENTRATION AND GENOTOXIC AND MUTAGENIC ACTIVITIES OF A LANDFARMING FACILITY AT AN OIL REFINERY

***T. S. Souza*[1,*]*, F. A. Henckle*in[2]*, D. F. de Angelis*[2]
and C. S. Fontanetti[3]**

[1]Departamento de Biologia, Centro de Ciências Agrárias,
Universidade Federal do Espírito Santo (UFES), Alegre, Espírito Santo, Brazil
[2]Departamento de Bioquímica e Microbiologia, Instituto de Biociências,
UniversidadeEstadual Paulista (UNESP), Rio Claro, São Paulo, Brazil
[3]Departamento de Biologia, Instituto de Biociências,
Universidade Estadual Paulista (UNESP), Rio Claro, São Paulo, Brazil

ABSTRACT

Industrial and agricultural activities produce large quantities of residues. Bioremediation and the recycling of wastes are alternative techniques that are used to minimize the possible environmental effects caused by improper disposal and ineffective treatments. Landfarming is a bioremediation technology designedto reduce the concentration of organic contaminants in the soil. It has been widely used by refineries to treat the sludge produced during oil refinement. Supplementary treatments, such as adding bulking agents and nutrients to the soil, can be used to accelerate this process. Because of their characteristics, the addition of agricultural residues into soils contaminated with petroleum sludge and its by-products may be an environmentally viable recycling alternative. In this chapter, we evaluated the effects of agricultural residues, rice hulls and sugar cane vinasse, on the total petroleum hydrocarbon (TPH) concentration and genotoxicity and mutagenic activity of soil contaminated with petroleum refinery sludge (landfarming soil). The TPH concentration was measured at the beginning and end of theexperiments, as well as the genotoxic and mutagenic activity of the samples using *Allium cepa* bioassay. Our results showed that at the beginning of

* E-mail: tatianas.souza@hotmail.com. Tel: 55 28 3552-8618.

the experiments, a high TPH concentration, 17.28 g/kg, and a high genotoxic and mutagenic activity ($p<0.05$) were detected in soil samples from landfarming site. After 33 days, the TPH concentration in the untreated landfarming soil sample was reduced by 8.77%. In the landfarming soil sample treated with sugar cane vinasse, the TPH concentration decreased by 12.96%, while genotoxic and mutagenic activities were statistically significant ($p<0.05$). In the landfarming soil sample treated with rice hulls alone or rice hulls combined with sugar cane vinasse, there were significant reductions in the TPH concentration, 46.56% and 51.01%, respectively, and in genotoxicity and mutagenicity ($p<0.05$) and may be routinely used in the bioremediation of soils contaminated with petroleum sludge. However, due the deleterious effects obtained for landfarming and rice hulls combined with sugar cane vinasse sample, more studies should be conducted with sugar cane vinasse and soil contaminated with petroleum. It is expected that the study can contribute tothe reuse of wastes.

Keywords: Landfarming, oil refinery, sugar cane vinasse, rice hull, *Allium cepa*

INTRODUCTION

Industrial and agricultural activities produce large quantities of residues that may contain polluting substances. Bioremediation and recycling of wastes are alternative techniques that are used to minimize the possible environmental effects caused by improper disposal and ineffective treatments of these residues.

In petroleum refineries, oily sludge produced during the refining process is spread on the upper layers of soil, in a system termed landfarming. Through aerobic respiration, microorganisms use the organic compounds present in the residues as substrates for growth and energy production. Thus, contaminants are converted into CO_2, H_2O and biomass (Harmsen, 1991). Although very common, landfarming can be inefficient, as organic residues may require long periods of time to be completely biodegraded (Khan et al., 2004). However, several strategies can be used to optimize this process, including the addition of bulking materials and nutrients to the soil (Hamdi et al., 2007).

Rice hulls are organic residues derived from rice milling. The large amount produced annually has raised environmental concerns, since in many countries it has been burned, increasing carbon dioxide emissions (Iranzo et al., 2004; Cai et al., 2007). Rice hulls are rich in organic material (80%) and oxidizable organic carbon (34%), and have a high carbon to nitrogen ratio (ranging from 50 to 150) and high water retention capacity (Iranzo et al., 2004). Addition of rice hulls to the soil can decrease soil compaction, increase porosity, and consequently, can increase the diffusion of oxygen between solid particles, thus accelerating microbial activity (Vasudevan and Rajaram, 2001). Because of these characteristics, the addition of rice hulls into soils contaminated with petroleum sludge and its by-products may be an environmentally viable recycling alternative. Some studies have demonstrated the effectiveness of this procedure (Hamdi et al., 2007).

Vinasse is a by-product of a sugar industry. According to Freire and Cortez (2000), vinasse has a high pollution potential due to its low pH, high corrosivity, high biochemical oxygen demand, and high temperature at discharge. The dark-brownish color of vinasse is rarely considered a form of contamination, despite the damage it causes. This may be associated to the presence of toxic compounds and chromophoric groups, or polymers of high molecular weight formed by the Maillard reaction and phenolic compounds (Ferreira et al., 2011). Nevertheless, in many countries this effluent has been widely recycled as fertilizer in agricultural soils because it is rich organic matter and nutrients, mainly calcium and potassium (España-Gamboa et al., 2001).

Considering that the ethanol production in Brazil is in expansion due to a rising internal and worldwide demand for alternative energy sources and that soils have a limited support capability to receive the vinasse, the search for additional application for the vinasse is very important (Crivelaro et al., 2010). To examine the effects of vinasse on the bioremediation of organic pollutants, Prata et al. (2001) used this effluent to accelerate the degradation of the herbicide ametryn in the soil. Crivelaro et al. (2010) combined sugar cane vinasse with petroleum sludge and observed an increase in the quantity of microorganisms. However, the authors concluded that the effluent not to be adequate to enhance the bioremediation efficiency of the oily sludge at the mineralization level. Mariano et al. (2009) obtained unsatisfactory results when using sugar cane vinasse to bioremediate diesel-contaminated soils and Souza et al. (2013) showed that the sugar cane vinasse potentiated the clastogenic effects of the landfarming soil from oil refinery.

The efficacy of bioremediation of soils contaminated with petroleum sludge has been most commonly evaluated through the rate of disappearance of total contaminants, such as the analysis of total petroleum hydrocarbons (TPH levels) (Marin et al., 2005; Huang et al., 2005). These analyses, however, do not provide information on the toxicity of sites after bioremediation (Plaza et al., 2005). Thus, complementary information, such as that obtained from biological assays, has been increasingly required for the accurate evaluation of the bioremediated soil.

The assessment of mitotic and chromosomal abnormalities, micronuclear and nuclear bud induction in *Allium cepa* cells is considered an efficient method to evaluate environmental quality. This species has few chromosomes (2n = 16), making observation of genetic damage easier (Grant, 1982; Leme et al., 2008). In addition, *A. cepa* is one of the plant species most sensitive to genotoxic and mutagenic agents, due to the size of its chromosomes and the large quantity of metacentric chromosomes (Ma et al., 1995). Also, there are evidences of the correlation between *A. cepa* and others test system. Fiskejo (1985) showed that the *A. cepa* test presents a similar sensitivity to that of the algal and human lymphocyte test system. Rank and Nielsen (1994) showed a correlation of 82% between *A. cepa* test and carcinogenity assays in rodents.

In this chapter, we evaluated the effects of rice hulls and sugar cane vinasse on the TPH concentration and genotoxicity and mutagenicity of soil contaminated with petroleum refinery sludge (landfarming soil).

METHODS

Agricultural Residues

Rice hulls from a processing machine located in Santa Cruz da Estrela, Santa Rita do Passa Quatro, São Paulo State, Brazil were used. Sugar cane vinasse was collected at the Santa Lucia Mill, located in the city of Araras, São Paulo State, Brazil, according to the

Technical Procedure P4.231 of the Company of Environmental Sanitation Technology (CETESB, 2006).

Landfarming Soil and Control Soil

Soil samples from a landfarming facility at an oil refinery were collected in 2007 according to the technical procedure NBR-10.007 of the Brazilian Association of Technical Standards (ABNT, 2004). The total landfarming area of the oil refinery is 40,000 m^2 with a capacity to treat 800 m^3 of petroleum sludge monthly.

Soil samples from the Itirapina Ecological Station, managed by the Forest Institute of São Paulo State, Brazil, were used as control soil (CS). The area of this preserve comprises approximately 2300 hectares of natural vegetation. Soil sampling was carried out according to the Technical Procedure NBR-6.457 of the Brazilian Association of Technical Standards (ABNT, 1986).

Soil samples were dried and passed through a 1.62-mm sieve. Twenty-liter buckets were filled with 10 L of soil (control or landfarming), yielding 30-cm soil columns. Each experiment was conducted in triplicate. Soil moisture was adjusted to 70% of field capacity (USEPA, 1994). Water or sugar cane vinasse was then added to soil samples, depending on the treatment. In the treatment using rice hulls, hulls were added to a concentration of 5% of the total volume of soil in buckets.

The following treatments were set up: control soil (CS); landfarming (L), landfarming and rice hulls (LRH), landfarming and sugar cane vinasse (LV) and landfarming, rice hulls and sugar cane vinasse (LRHV).

Physicochemical Analysis

The sugar cane vinasse was characterized based on pH, total P, N concentration, nutrient concentration (K, Ca, Mg, Al, S, Na, Fe, Mn, Cu, Zn, and B), and heavy metals concentration (Ba, Cd, Cr, Ni, and Pb). The soils collected were examined at the beginning of the experiment (T0) and after 33 days (T33) for texture, pH, organic matter content, nutrient concentration (K, Ca, Mg, Al, S, Na, Fe, Mn, Cu, Zn and B) and heavy metals concentrations (Ba, Cd, Cr, Ni, and Pb). The treatments samples were examined at after days (T33) for pH, organic matter content, nutrient concentration (K, Ca, Mg, Al, S, Na, Fe, Mn, Cu, Zn and B) and heavy metal concentrations (Ba, Cd, Cr, Ni, and Pb).

The metal concentrations were performed using a plasma emission spectrometer (ICP). These analyses were carried out at the Institute of Soil and Fertilizer Analysis of Campinas (ICASA), Campinas, São Paulo State, Brazil.

TPH Concentration

Analyses of TPH concentration in each treatment sample were performed on T0 and after 33 days (T33). Concentrations were expressed using mean values and standard deviation of independent measurements carried out on replicates of control soil and landfarming treated samples. The analyses were conducted by the Analytical Technology Company, São Paulo, Brazil according to the USEPA 8015B method.

The *A. Cepa* Bioassay

The protocol described below was carried out in the beginning of the experiment (T0) and after 33 days (T33).

Aqueous soil extracts were prepared according to Cotelle et al. (1999) for detection of toxic fractions water-soluble (Souza et al. 2009). Approximately 50 g of soil for each replicate (CS, L, LRH, LV and LRHV) were added to 300 mL of ultrapure water. The mixture was stirred for 24 h and then stored for another 24 h at room temperature (25±2 °C). The supernatant was collected and the aqueous extracts without any further filtration process were used for the germination of *A. cepa* seeds.

Seeds of *A. cepa* were allowed to germinate in Petri dishes lined with filter paper that were moistened with aqueous soil extracts or with samples of the soils (direct contact) for detection of water-insoluble contaminants (Souza et al., 2009). Three Petri dishes were used per treatment.

The experiments were carried out in two phases:

1. Continuous experiment: after five days, when the *A. cepa* radicles reached 1 - 2 cm, approximately half of the root tips were collected and fixed in Carnoy I (3:1 methanol + acetic acid).
2. Recovery experiment: the remaining radicles were transferred to another Petri dish containing ultrapure water for the 48-hour recovery experiment.

Experiments consisted of the following combinations:
Continuous experiment, with roots of *A. cepa* germinated in aqueous extract;
Continuous experiment, with roots of *A. cepa* germinated directly on soil samples;
Recovery experiment, with roots of *A. cepa* germinated in aqueous extract;
Recovery experiment, with roots of *A. cepa* germinated directly on soil samples;

After fixation, radicles were hydrolyzed in 1N HCl at 60ºC for 10 min and later stained with Schiff Reagent for 2 h in the dark. The roots tips were sectioned, placed on slides, cover-slipped, and smashed. Cover slips were removed in liquid nitrogen and the slides were mounted with Entellan or Permount. Fifteen slides were prepared for each treatment and a total of 7500 cells were counted (500 cells/slides). The slides were analyzed under a light microscope at a magnification of 400 x.

Several types of abnormalities within different cell division stages (metaphase, anaphase, and telophase) were considered: chromosomal breaks, losses and bridges, adherences, multipolar anaphases, polyploidy, and C-metaphases. All of these abnormalities were classified as mitotic and chromosome abnormalities (MCA). The micronuclei (MN) and nuclear buds (NB) induction in the meristematic cells of *A. cepa* was evaluated separately from mitotic and chromosome abnormalities. Treatments were compared using the non-parametric Kruskal-Wallis test ($p<0.05$). The results were expressed as the mean ± standard deviation of three replicates.

RESULTS

Physicochemical Analysis

The sugar cane vinasse had a pH 3.85, a low organic matter content (3.96 g/kg), 0.22% N, a large amount of nutrients, mainly S, Ca, Mg, Na, Fe, Al and P, and low heavy metal concentrations (Table 1). The control soil was classified as sandy with pH 4, and contained low organic matter content (8 g/Kg), and low metal concentrations (Table 1).

Landfarming soil was classified as clay with pH 5, and contained had 106 g/Kg of organic matter content and high concentrations of micronutrients, particularly S, Na, Ca, Mg, Fe, and Zn. Furthermore, the concentration of heavy metals, particularly Ba, Cr, Pb and Ni were high (Table 1).

At T33 the control soil contained low metal concentrations. The landfarming soil contained had high organic matter content and high concentrations of micronutrients and heavy metals (Table 2).

Table 1. Physical and chemical analyzes at T0

Parameters	CS	L	Sugar cane vinasse
Texture (%)			
Clay (< 0.02 mm)	9.9	47.9	*
Silt (0.002 – 0.06 mm)	4.1	21.9	*
Sand (0.06 – 2 mm)	86	30.2	*
Gravel (2 – 60 mm)	0	0	*
Organic matter (mg/Kg)	8	106	*
pH ($CaCl_2$)	4	5	3.85
P total (ppm)	*	*	65
N (%)	*	*	22
Nutrients (mmolc/dm^3)			
K	0.6	2.8	0.32
Ca	1	66	740
Mg	1	27	210
Al	6	1	72
S	10	852	835
Na	3	320	113
Fe	25	54	97.5
Mn	0.6	7.8	-
Cu	0.9	1.5	-
Zn	0.9	39	7.5
B	0.2	2.7	5
Heavy metals (mg/Kg)			
Ba	14.4	658.1	0.5
Cd	< 0.01	1.4	1.06
Cr	11.9	731.7	0.15
Ni	< 0.01	69.69	0.26
Pb	< 0.01	94.38	< 0.01

CS: control soil, L: landfarming soil

- Result below detection threshold

* Analisis not requested

Table 2. Physical and chemical analyzes at T33

Parameters	CS	L	LRV	LV	LRHV
pH ($CaCl_2$)	4	5.1	5.1	5.1	5.1
Organic matter (mg/Kg)	8	104	80	101	60
Nutrients (mmolc/dm^3)					
K	0.8	2.1	4	9.5	19
Ca	1	63	88	80	61
Mg	1	28	41	36	29
Al	6	1	1	1	-
S	10	878	998	1056	904
Na	3	200	370	260	400
Fe	20	46	58	38	55
Mn	0.7	7.4	9.7	7.7	10.7
Cu	0.9	1.2	1.7	1.2	1.6
Zn	0.7	33	39	56.1	32
B	0.2	2.5	2.6	2.4	3.2
Heavy metals (mg/Kg)					
Ba	14.4	514.9	532	511	563
Cd	< 0.01	1.4	1.6	3.83	1.2
Cr	10	345	390.4	380.1	390.7
Ni	< 0.01	68.6	71.9	62.7	68.8
Pb	< 0.01	70.2	77.4	78.2	86.4

CS: control soil, L: landfarming soil; LRH: landfarming soil and rice hulls; LV: landfarming soil and sugar cane vinasse; LRHV: landfarming soil, rice hulls and sugar cane vinasse

- Result below detection threshold

At T33, landfarming soil and rice hulls (LRH) sample had 80 g/Kg of organic matter content, landfarming soil and sugar cane vinasse (LV) sample had 101 g/Kg of organic matter content and landfarming soil, rice hulls and sugar cane vinasse (LRHV) sample had 60 g/Kg of organic matter content. The treatments showedhigh concentrations of micronutrients and heavy metals (Table 2).

TPH Concentration

The TPH concentration in the control soil sample was below the detection limit, while in the landfarming soil sample, the TPH concentration decreased by 8.77% (15.76 g/kg) compared to the initial concentration (17.28 g/kg). In the LV treatment, the TPH concentration was 15.04 g/kg, a reduction of 12.96%. The LRH and LRHV treatments had the lowest TPH concentrations, 9.23 g/kg and 8.46 g/kg respectively, and consequently, the largest magnitude of hydrocarbon reduction compared to the initial landfarming sample, 46.56% and 51.01% reductions (Figure 1).

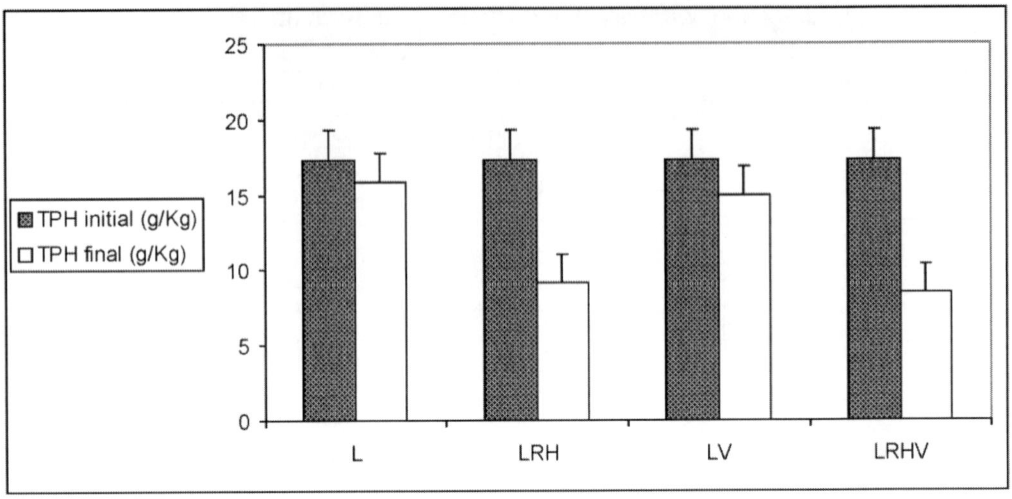

Figure 1. Total petroleum hydrocarbons (TPHs)(g/Kg) in landfarming soil, before and after addition of rice hulls and vinasse. Landfarming soil (L); Landfarmind and rice hulls (LRH); Landfarming and vinasse (LV); Landfarming, rice hulls and vinasse (LRHV).

Allium Cepa Bioassay

The normal mitotic cycle of *A. cepa*is shown in Figure 2.

At T0 all treatments induced significant levels ($p<0.05$) of mitotic and chromosomal abnormalities (Figure 3D-L, Table 3), micronuclei (Figure 3A-B, Table 4), and nuclear bud formation (Figure 3C, Table 5) in *A. cepa* compared to the control soil.

At T33 the genotoxicity (Table 3) and mutagenicity (Table 4 and 5) levels of the LRHV and LRH samples were reduced considerably in all experiments.

At T33 the genotoxicity levels of the L sample were reduced in the continuous experiments/aqueous extract and in the recovery experiments/ aqueous extract and direct contact. By contrast, the mitotic and chromosome abnormalities frequency was statistically significant in continuous experiments/ direct contact (Table 3).

At the micronuclei levels were reduced in the continuous experiments/aqueous extract and direct contact and in the recovery experiments/direct contact. But, the micronucleus frequency was statistically significant in recovery experiment/ aqueous extract (Table 4).

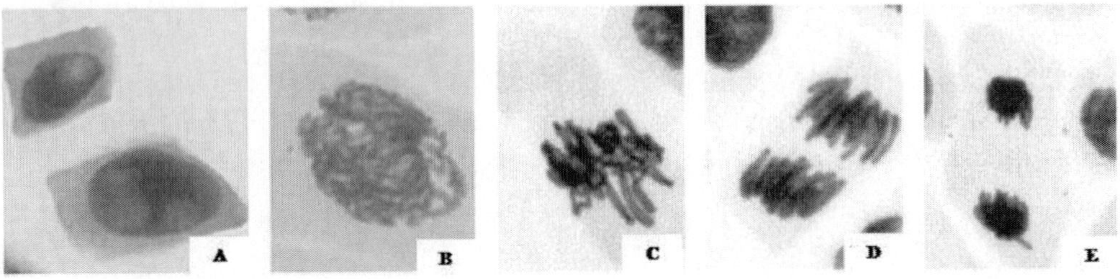

Figura 2. Normal mitotic cycle in *A. cepa.* A. Interphase. B. Prophase. C. Metaphase. D. Anaphase. E. Telophase.

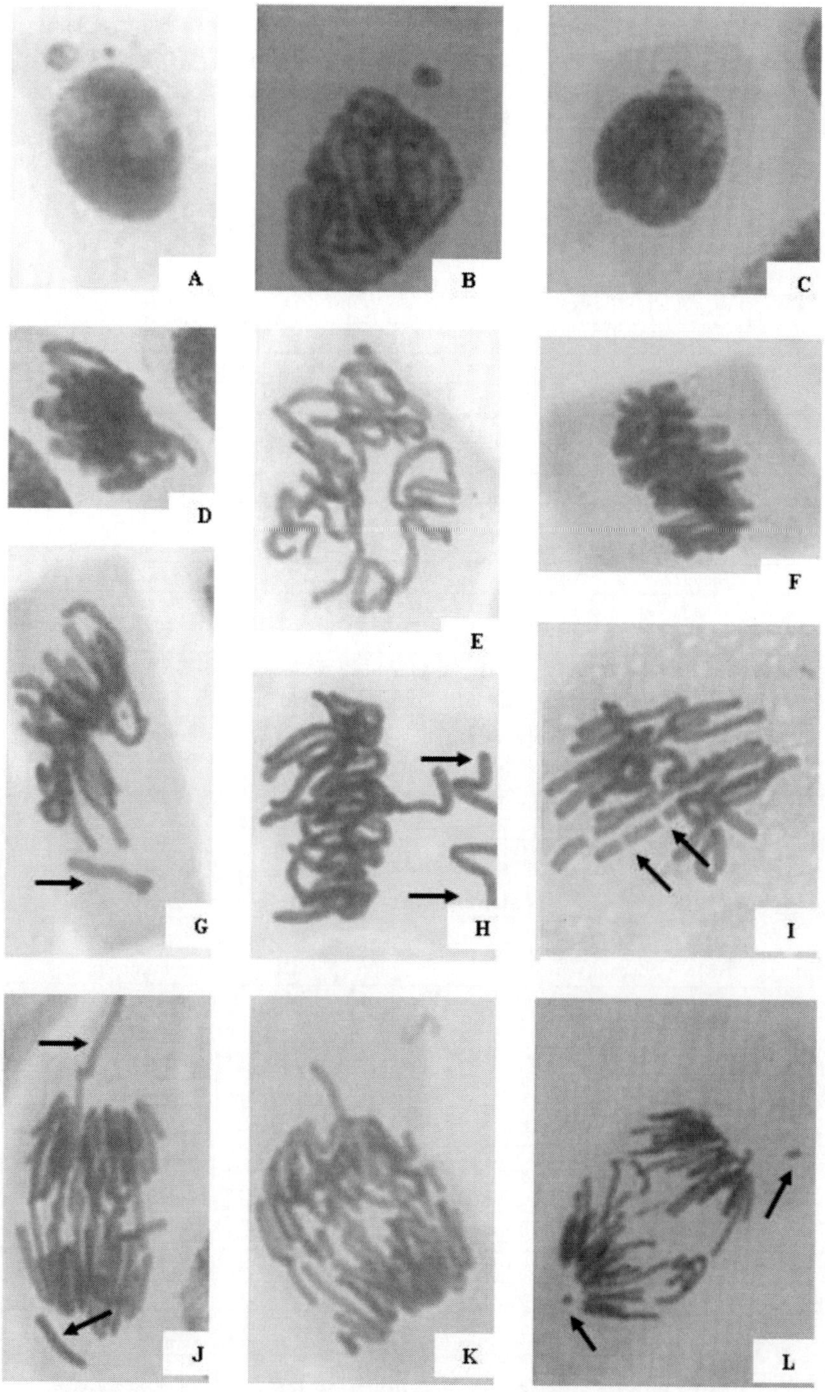

Figura 3. Micronuclei, nuclear buds and mitotic and chromosomal abnormalities in *A. cepa*. A-B. Micronuclei. C. Nuclear bud. D. Metaphase with chromosomal adherence. E. C-metaphase. F. Metaphase polyploid with chromosomal adherence. G-H. Metaphase with chromosomal losses (arrows). I. Metaphase with chromosomal breaks (arrows). J. Anaphase with chromosomal bridges and losses (arrows). K. Anaphase with chromosomal bridges L. Anaphase with chromosomal bridges and breaks (arrows). 400x.

Table 3. Mitotic and chromosome abnormalities (MCA) in root meristem cells of *A. cepa* exposed to landfarming treatments at T0 and T33

Treatments	MCA – continous treatment			MCA – recovery treatmanet		
	T0	T33	Genotoxic reduction (%)	T0	T33	Genotoxic reduction (%)
Aqueous extract						
CS	1.06 ± 1.43	0.86 ± 0.51	-	1.20 ± 1.01	0.89 ± 0.77	-
L	4.04 ± 1.86[a]	2.00 ± 1.13	50.49	4.06 ± 1.75[a]	2.33 ± 1.75	42.61
LRH	4.33 ± 3.26[a]	1.26 ±1.33	70.90	4.89 ± 2.16[a]	2.00 ± 1.88	59.10
LV	4.13 ± 1.88[a]	2.33 ±1.17	43.58	4.13 ± 2.99[a]	2.20 ± 2.11	46.73
LRHV	4.73 ± 1.16[a]	0.46 ± 0.74	90.27	4.06 ± 1.43[a]	1.06 ± 1.53	73.89
Direct contact						
CS	1.00 ± 1.06	0.86 ±0.99	-	1.03 ± 0.98	1.00 ± 1.06	-
L	5.26 ± 2.43[a]	5.26 ± 3.08[a]	-	7.73 ± 2.54[a]	2.80 ± 1.47	63.77
LRH	6.13 ± 3.35[a]	2.00 ± 1.13	67.37	7.13 ± 1.59[a]	2.73 ± 1.90	61.71
LV	6.06 ± 2.93[a]	5.46 ± 3.87[a]	9.90	6.00 ± 2.47[a]	5.00 ± 0.0[a]	16.66
LRHV	7.20 ± 2.75[a]	2.13 ± 1.40	70.41	7.20± 2.24[a]	1.06 ± 1.09	85.27
Trifluraline	6.73 ± 4.36[a]	6.40 ± 3.13[a]	-	10.66 ± 1.98[a]	9.00 ± 2.64[a]	-

CS: control soil; L: landfarming; LRH: landfarming and rice hulls; LV: landfarming and sugar cane vinasse; LRHV: landfarming, rice hulls and sugar cane vinasse

[a] Significant compared to the control soil using the statistical test Kruskal-Wallis (p< 0.05)

Values were expressed as mean ± standard deviation. Data obtained from 7500 cells observed per treatment

Table 4. Micronucleus (MN) in root meristem cells of *A. cepa* exposed to landfarming treatments at T0 and T33

Treatments	MN – continous treatment			MN – recovery treatment		
	T0	T33	Genotoxic reduction (%)	T0	T33	Genotoxic reduction (%)
Aqueous extract						
CS	0.73 ± 1.09	1.00 ± 1.30	-	0.80 ± 0.77	1.00 ± 0.37	-
L	5.40 ± 4.32[a]	3.80 ± 3.36	29.62	3.66 ± 1.04[a]	4.33 ± 4.63[a]	-
LRH	3.33 ± 2.63[a]	1.00 ± 1.36	69.69	3.86 ± 1.40[a]	1.86 ± 1.12	51.81
LV	3.33 ± 1.01[a]	4.93 ± 0.96[a]	-	4.00 ± 1.00[a]	5.20 ± 0.94[a]	-
LRHV	3.13 ± 1.24[a]	0.60 ± 0.98	80.83	3.80 ± 1.74[a]	1.46 ± 0.83	61.57
Direct contact						
CS	0.71 ± 0.95	0.88 ± 0.65	-	0.91 ± 1.00	0.80 ± 0.77	-
L	5.13 ± 2.19[a]	2.46 ± 2.23	52.04	4.73 ± 2.12[a]	2.46 ± 0.63	38.50
LRH	6.86 ± 1.92[a]	1.80 ± 0.94	73.76	4.00 ± 1.06[a]	1.60 ± 1.05	60.00
LV	4.20 ± 1.14[a]	5.00 ± 0.84[a]	-	5.00 ± 1.81[a]	6.00 ± 1.00[a]	-
LRHV	4.26 ± 1.27[a]	0.26 ± 0.45	92.63	4.06 ± 2.89[a]	1.06 ± 0.79	73.89
Trifluraline	6.33 ± 2.09[a]	4.73 ± 5.14[a]	-	6.73 ± 1.98[a]	7.66 ± 2.19	-

CS: control soil; L: landfarming; LRH: landfarming and rice hulls; LV: landfarming and sugar cane vinasse; LRHV: landfarming, rice hulls and sugar cane vinasse

[a] Significant compared to the control soil using the statistical test Kruskal-Wallis (p< 0.05)

Values were expressed as mean ± standard deviation. Data obtained from 7500 cells observed per treatment.

Table 5. Nuclear buds (NB) in root meristem cells of *A. cepa* exposed to landfarming treatments at T0 and T33

Treatments	BN – continous treatment			BN – recovery treatment		
	T0	T33	Genotoxic reduction (%)	T0	T33	Genotoxic reduction (%)
Aqueous extract						
CS	0.53 ± 0.91	0.86 ± 1.06	-	0.86 ± 0.56	1.20 ± 0.67	-
L	3.00 ± 1.69 [a]	2.00 ± 1.36	33.33	2.06 ± 1.38[a]	1.53 ± 0.74	25.72
LRH	3.60 ± 2.06 [a]	1.26 ± 0.96	65.00	3.00 ± 1.41[a]	1.93 ± 0.88	35.66
LV	3.20 ± 2.04 [a]	2.40 ± 2.77	25.00	3.46 ± 1.24[a]	2.60 ± 0.98	24.85
LRHV	3.46 ± 3.50 [a]	0.40 ± 0.90	88.43	2.80 ± 1.01[a]	1.46 ± 0.63	35.00
Direct contact						
CS	0.80 ± 0.77	0.91 ± 1.00	-	0.71 ± 0.95	1.06 ± 0.59	-
L	6.46 ±1.68[a]	1.86 ± 0.74	71.20	7.53 ± 1.92[a]	5.13 ± 1.64[a]	31.87
LRH	3.40 ± 3.60[a]	0.53 ± 0.63	84.41	3.46 ± 1.24[a]	1.00 ± 0.75	71.09
LV	4.00 ± 1.46[a]	1.06 ± 0.88	73.50	4.40 ± 1.29[a]	2.40 ± 1.30	45.45
LRHV	4.00 ± 2.77[a]	0.40 ± 0.50	90.00	3.06 ± 1.79[a]	0.53 ± 0.63	82.67
Trifluraline	4.20 ± 1.08[a]	4.66 ± 1.75[a]	-	10.86 ± 3.33[a]	9.86 ± 1.33[a]	-

CS: control soil; L: landfarming; LRH: landfarming and rice hulls; LV: landfarming and vinasse; LRHV: landfarming, rice hulls and vinasse

[a] Significant compared to the control soil using the statistical test Kruskal-Wallis ($p< 0.05$)

Values were expressed as mean ± standard deviation. Data obtained from 7500 cells observed per treatment

Nuclear buds levels were reduced in the continuous experiments/aqueous extract and direct contact and in the recovery experiments/aqueous extracts. However, the nuclear buds frequency was statistically significant in recovery experiment/ direct contact (Table 5).

At T33 the LV treatment showed the lowest index of reduction of genotoxicity (Table 3)and mutagenicity (Tables 4 and 5).Mitotic and chromosome abnormalities frequency was statistically significant in continuous experiment/ direct contact and recovery experiment/ direct contact (Table 3). Micronucleus frequency was statistically significant in continuous experiment/ aqueous extract and direct contact and recovery experiment/ / aqueous extract and direct contact (Tables 4).

The index reduction of genotoxicity and mutagenitcity was: LRHV > LRH > L > LV (Tables 3, 4 and 5).

DISCUSSION

The results of the physical chemical analyses for the landfarming soil samples examined in this study were associated with a significant induction of genotoxicity and mutagenicity in *A. cepa.*

The physical and chemical characteristics of a soil for bioremediation directly influence microbial activity. In this study, the landfarming soil had pH 5, high organic matter content, and high concentrations of heavy metals and TPH. The ideal pH range for bacterial and fungal growth is between 6 and 8 (USEPA, 1994). Thus, in more acidic soils, biodegradation of hydrocarbons can be negatively affected (Marin et al., 2005). Clay soils have low permeability and are very compact, characteristics that make aeration and even distribution of nutrients difficult (USEPA, 1994). Soils with a high content of organic material tend to adsorb high molecular weight hydrocarbons, decreasing the concentration of these contaminants in the aqueous and gas phases of the soil (Labud et al., 2007), as well as decreasing access to the microorganisms that degrade these hydrocarbons. Soils with TPH concentration between 10 and 50 g/kg can be toxic to most organisms. In this paper, landfarming soil had the lowest reduction of TPH levels, indicating that the fraction of lighter hydrocarbons was rapidly biodegraded and that longer periods of time are required for the biodegradation of more complex hydrocarbons. In addition, landfarming sites, metals accumulate in the upper layers of soil, as they are not biodegraded by microorganisms. High concentrations of these compounds may inhibit bacterial growth and development (USEPA, 1994).

The induction of mitotic and chromosomal abnormalities is a warning signal in environmental monitoring studies, as these abnormalities often lead to the formation of micronucleated cells during the cell cycle. The presence of micronuclei demonstrated the mutagenic potential of the landfarming soil, as micronuclei are believed to form during cell division, as a result of chromosomal breaks and/or disturbances during the mitotic process due to spindle abnormalities (Lindberg et al., 2007). Crott et al. (2001) studied folic acid deficiency in human lymphocytes and confirmed that the presence of nuclear buds may be used as biomarkers of genetic damage. The formation of micronucleated cells, through the elimination of genetic material in nuclear buds has also been suggested (Fenech and Crott, 2002; Lindberg et al., 2007). However, the mechanism is not yet fully understood (Lindberg et al., 2007).

To improve the landfarming techniques of the oil refinery, the effects of two agricultural residues in the TPH concentration and toxicity of the soil were examined.

Landfarming soil without treatment had the lowest reduction of TPH levels, indicating that the fraction of lighter hydrocarbons was rapidly biodegraded and that longer periods of time are required for the biodegradation of more complex hydrocarbons. Marin et al. (2005) showed that 80% of the TPH in landfarming soil were eliminated in 11 months.

The LRH and LRHV treatments had the largest magnitude of hydrocarbon reduction compared to the initial landfarming sample.

The TPH concentration of LV treatment was reduced by only 12.96%. Pinto-Mariano et al. (2009) studied the effects of this effluent in the bioremediation of soils contaminated with petroleum sludge. In their study, sugar cane vinasse was biodegraded primarily during the first 20 days of the experiment, with a striking decrease in the microbial activity after the consumption of this effluent. The increase in the soil microbial population caused by the vinasse was observed by Prata et al. (2001), Mariano et al. (2009) and Crivelaro et al. (2010) in their biodegradation experiments. However, as observed by Hickman and Novak (1989), the total microbial biomass could be a poor predictor for determining the biodegradation potential, mainly because the active biomass could differ in species composition and in metabolic regimes. According to Crivelaro et al. (2010) these additional microorganisms preferentially biodegraded the more labile carbon sources as the glycerol present in the vinasse; otherwise, a marked decrease in the CO_2 production after the consumption of the vinasse would not be observed.Souza et al. (2012) showed the sugar cane vinasse potentiates the clastogenic effects of the landfarming soil. The low pH of sugar cane vinasse may have madeavailable toxic metals.

Rice hulls acts as bulking agent in the soil, promoting aeration and distribution of nutrients and, consequently, microbial activity (Hamdi et al., 2007). Rice hulls are rich in biopolymers that retain soil moisture (Iranzo et al., 2004). According to Nakamura et al. (2003) and Hamdi et al. (2007), the addition of this material to the soil increases active decomposition of cellulose and also increases the degradation hydrocarbons of four or fewer aromatic rings. Hamdi et al. (2007) reported that rice hulls produced higher PAH dissipation rates than those observed in unamended PAH-spiked soils. Souza et al. (2009) showed by Bartha and Pramer method (1965) that in landfarming soil treated with rice hulls, the rate of CO_2 released by microorganisms was 4.21 times higher than in untreated landfarming soil, indicating that rice hulls accelerated the biodegradation of hydrocarbon oil of landfarming facility. Also, the rice hulls decreased TPH concentration of the landfarming soil as well as their genotoxic and mutagenic potential.

In LRHV the rice hulls served as bulking agent and the addition of sugar cane vinasse served as source of nutrients, mainly, K, Na, Mg and B, for microbial activity.

Hencklein (2008) studied the same landfarming treatments that were evaluated in this study. The author showed that the combination of rice hulls and sugar cane vinasse caused a significant increase in the moisture content, increase in the release of CO_2 by optical absorption of infrared light method and increase in the amount heterotrophic bacteria present in soil, followed by landfarming and rice hulls treatment. Certainly, these factors contributed to the striking decrease in the TPH concentration and genotoxicity and mutagenicity in *A. cepa*, observed in the present study. By contrast, the landfarming and sugar cane vinasse treatment obtained had the lowest levels, followed by untreated landfarming (Hencklein, 2008). According this author, biodegradation in landfarming sites can be enhanced if compounds that assist the aeration and the complementation with nutrients are added to the soil.

CONCLUSION

The TPH concentrations and the *A. cepa* assay were adequate to evaluate the quality of landfarming soil. Rice hulls may be recommended for routine use in the bioremediation of soils contaminated with petroleum sludge. Rice hulls combined with sugar cane vinasse promoted depletion of TPH concentration and reduced the genotoxicity and mutagenicity of landfarming soil. However, due the deleterious effects obtained for LV treatment, more studies should be conducted with sugar cane vinasse and soil contaminated with petroleum. It is expected that the study can contribute to the reuse of wastes to minimize it deleterious effects on the environment.

ACKNOWLEDGMENTS

We would like to thank the Foundation of Support the Research of São Paulo State (FAPESP), process number 06/52383-7, and the Foundation for the Development of UNESP

(FUNDUNESP) for financial support, Valdenilson José Alves de Oliveira for collection of soils and Sandra Veloso for technical support.

References

Associação Brasileira de Normas Técnicas. NBR - 6457: Amostras de solo – Preparação para ensaios de compactação e ensaios de caracterização. *Rio de Janeiro*, 1986. 9p.

Associação Brasileira de Normas Técnicas. NBR - 10.007: Amostragem de resíduos sólidos. *Rio de Janeiro*, 2004. 25p.

Bartha, R; Pramer, D, 1965.Features of a flask and method for measuring the persistence and biological effects of pesticides in soil. *Soil Science* 100, 68-70.

Cai, Q.-Y., Mo, C.-H., Wu, Q.-T., Zeng, Q.Y., Katsoyiannis, A., Férard, J.-F., 2007. Bioremediation of polycyclic aromatic hydrocarbon (PAHs)-contaminated sewage sludgeby different composting processes. *JournalofHazardousMaterials* 142, 535-542.

Companhia de Tecnologia e Saneamento Ambiental. Norma Técnica P4.231: Vinhaça – critérios e procedimentos para aplicação em solo agrícola. São Paulo, 2006, 17p.

Cotelle, S., Masfaraud, J.F., Férard, J.F., 1999. Assessment of genotoxicity of contaminated soil with the *Allium/Vicia* – micronucleus and the *Tradescantia* – micronucleus assays. *Mutat. Res.* 426, 67-171.

Crivelaro, S.H.R., Mariano, A.P., Furlan, L.T, Gonçalves, R.A., Seabra, P.N., Angelis, D.F., 2010. Evaluation of the use of vinasse as a biostimulation agent for the biodegradation of oily sludge in soil.*Brazilian Archives of Biology and Technology*53, 1217-1224.

Crott, J.W., Mashiyama, S.T., Ames, B.N. Fenech, M., 2001. Methylenetetrahydrofolate reductase C677T polymorphism does not alter folic acid deficiency-induced uracil incorporation into primary human lymphocyte DNA in vitro. *Carcinogenesis*22, 1019-1025.

España-Gamboa, E., Mijangos-Cortes, J., Barahona-Perez, L., Dominguez-Maldonado, J., Hernández-Zarate, G., Alzate-Gaviria, L., 2011. Vinasses: characterization and treatments. *Waste Manag. Res.* 29, 1235-1250.

Fernandes, T.C.C., Mazzeo, D.E.C., Marin-Morales, M.A., 2007. Mechanism of micronuclei formation in polyploidizated cells of *Allium cepa* exposed to trifluralin herbicid. *Pestic. Biochem. Phys.* 88, 252-259.

Fenech, M., Crott, J.W., 2002. Micronuclei, nucleoplasmic bridges and nuclear buds induced in folic acid deficient human lymphocytes-evidence for breakage-fusion-bridge cycles in the cytokinesis-block micronucleus assay. *Mutat. Res.* 504, 131-136.

Ferreira, L.F.R., Aguiar, M.M., Messias, T.G., Pompeu, G.B., Lopez, A.M.Q., Silva, D. P., Monteiro, R.T., 2011. Evaluation of sugar-cane vinasse treated with *Pleurotus sajor-caju* utilizing aquatic organisms as toxicological indicators. *Ecotox. Environ. Saf.* 74, 132–137.

Fiskejö, G., 1985. The *Allium* test as a standard in environmental monitoring. *Hereditas* 102, 99–112.

Freire, W.J., Cortez, L.A.B., 2000. Vinhaça de cana-deaçúcar. Guaíba: Agropecuária, 203p.

Grant, W.F., 1982. Chromosome aberration assays in *Allium. Mutat. Res.* 99, 273-291.

Hamdi, H., Benzarti, S., Manusadžianas, L., Aoyama, I., Jedidid, N., 2007. Soil bioaugmentation and biostimulation effects on PAH dissipation and soil ecotoxicity under controlled conditions. *Soil Biol. Biochem.* 39, 1926–1935.

Harmsen, J., 1991. Possibilities and limitations of landfarming for cleaning contaminated sites. In: Hinchee R.E. and Olfenbuttel R.F. (Eds.). On-Site bioreclamation: processes for xenobiotic and hydrocarbon treatment. Butterworth-Heinemann, London.

Henckein, F.A., 2008. Estabilização de landfarming de refinaria de petróleo, mediante metabolismo microbiano, e aplicabilidade em solos pré-desertificados. Trabalho de Conclusão de Curso (Ciências Biológicas). Instituto de Biociências, Universidade Estadual Paulista, Rio Claro, São Paulo, Brasil.

Hickman, G.T., Novak, J.T., 1989. Relationship between subsurface biodegradation rates and microbial density. *Environ. Sci. Technol.*23, 525-532.

Huang, X.-D., El-Alawi, Y., Gurska, J., Glick, B.R., Greenber, B.M., 2005. A multi-process phytoremediation system for decontamination of persistent total petroleum hydrocarbons (TPHs) from soils. *Microchem. J.* 81, 139– 147.

Iranzo, M., Cañizares, J.V., Roca-Perez, L., Sainz-Pardo, I., Mormeneo, S., Boluda, R., 2004. Characteristics of rice straw and sewage sludge as composting materials in Valencia (Spain). *Bioresource Technol.* 95, 107-112.

Khan, F.I., Husain, T., Hejazi, R., 2004. An overview and analysis of site remediation technologies. J. Environ. Manage. 71, 95-122.

Labud, V., Garcia, C., Hernandez, T., 2007.Effect of hydrocarbon pollution on the microbial properties of a sandy and a clay soil. *Chemosphere* 66, 1863–1871.

Leme, D.M., Angelis, D.F., Marin-Morales, M.A., 2008. Action mechanisms of petroleum hydrocarbons present in waters impacted by an oil spill on the genetic material of *Allium cepa* root cells. *Aquat. Toxicol.* 88, 214-219.

Lindberg, H.K., Wang, X., Järventaus, H., Falck, G.C.-M., Norppa, H., Fenech, M., 2007. Origin of nuclear buds and micronuclei in normal and folate-deprived human lymphocytes. *Mutat. Res.* 617, 33-45.

Ma, T.-H., Xu, Z., Xu, C., McConnell, H., 1995. The improve *Allium/Vicia* root tip micronucleus assay for clastogenicity of environmental pollutants. *Mutat. Res.* 334, 185-195.

Mariano, A. P., Crivelaro, S. H. R., Angelis, D. F., Bonotto, D. M., 2009. The use of vinasse as an amendment to ex-situ bioremediation of soil and groundwater contaminated with diesel oil. *Braz. Arch. Biol. Technol.*52, 1043-1055.

Marin, J.A., Hernandez, T., Garcia, C., 2005.Bioremediation of oil refinery sludge by landfarming in semiarid conditions: influence on soil microbial activity. *Environ. Res.* 98, 185-195.

Nakamura, A., Tun, C.C., Asakawa, S., Kimura, M., 2003. Microbial community responsible for the decomposition of rice straw in a paddy field: estimation by phospholipid fatty acid analysis. *Biol. Fert. Soils*38, 288–295.

Plaza, G., Nalecz-Jawecki, G., Ulfig, K., Brigmon, R., 2005. The application of bioassays as indicators of petroleum-contaminated soil remediation. *Chemosphere* 59, 289-296.

Prata, F., Lavorenti, A., Reginato, J.B., Tornisielo, V.L., 2001. Degradação e sorção de ametrina em dois solos com aplicação de vinhaça. *Pesqui. Agropecu. Bras.*36, 975-981.

Rank, J., Nielsen, M.H., 1994. Evaluation of the *Allium* anaphase–telophase test in relation to genotoxicity screening of industrial wastewater. *Mutat. Res.* 312, 17–24.

Souza, T.S., Hencklein, F.A., Angelis, D.F., Gonçalves, R.A., Fontanetti, C.S., 2009. The Allium cepa bioassay to evaluate landfarming soil, before and after the addition of rice hulls to accelerate organic pollutants biodegradation. *Ecotox. Environ. Safe.*72, 1363-1368.

Souza, T.S., Hencklein, F.A., Angelis, D.F., Fontanetti, C.S., 2012.Clastogenicity of landfarming soil treated with sugar cane vinasse *Environ. Monit. Assess.*185, 1627-1636.

United States Environmental Protection Agency (USEPA). 1994. Landfarming. In: How to evaluated alternative cleanup technologies for underground storage tank sites: a guide for corrective action plan reviews (EPA 510-B-94-003 and EPA 510-B-95-007)

Vasudevan, N., Rajaram, P., 2001. Bioremediation of oil sludge contaminated soil. *Environ. Int.*26, 409-411.

SECTION 3: PHYTO- AND PHYCOREMEDIATION

In: Bioremediation: Processes, Challenges and Future Prospects ISBN: 978-1-62948-513-3
Editors: Jesús Bernardino Velázquez-Fernández et al. © 2014 Nova Science Publishers, Inc.

Chapter 9

PHYTOREMEDIATION OF HEAVY METAL CONTAMINATED SOIL AND ITS IMPROVEMENT WITH ARBUSCULAR MYCORRHIZA

Tünde Takács*

Hungarian Academy of Sciences, Centre for Agricultural Research,
Institute for Soil Sciences and Agricultural Chemistry,
Budapest, Herman Ottó Str., Hungary

ABSTRACT

Anthropogenic contamination of soils with potential toxic metals has become a global environmental problem. Metal pollution affects crop yield, composition of microbial community and fertility of soils and it is a potential health risk appearing in food production. Phytoremediation is a group of new and promising methods to decrease the harmful effect of soil and water pollution.

Phytoremediation, which is also called green or botanical remediation (phyto (gr) means plant, remedium (lt) means healing, restoration), consists of technologies that applies plants or plant-microbe associations to reduce the concentration or transport of organic or inorganic pollutants to a tolerable level. Research on phytoremediation is diverse because the methods depend on the interaction of the pollutant, the plant, the polluted medium and the climate.

Above all, phytoremedial technologies are presented with their advantages and disadvantages, including the requirements and difficulties of practical application. Both ecological and human health aspects should be considered in planning the effective and efficient phytotechnology of the metal contaminated sites.

Phytoremediation is primarily used for the restoration of metal polluted soils, so this chapter is focused on the treatment of heavy metal polluted media. As the part of the soil-plant interface, the rhizosphere microbial community has a significant role in the practice of soil remediation and field restoration. Soil microorganisms have a potential to improve the efficiency of phytoremedial processes through their effects on heavy metal mobility, availability and transport to plants.

* Hungarian Academy of Sciences, Centre for Agricultural Research, Institute for Soil Sciences and Agricultural Chemistry, H-1022 Budapest, Herman Ottó út 15., Hungary, E-mail: takacs@rissac.hu.

Nowadays, phytoremedial methods assisted by heavy metal-tolerant soil microorganisms are obtaining significant attention. Managed mycorrhization significantly promotes phytoremediation and reuse of damaged fields. The results of our research on enhancing phytoremedial technologies with the application of arbuscular mycorrhizal fungi are also presented.

Keywords: phytoremediation, heavy metals, biotic and abiotic improvers, arbuscular mycorrhiza

INTRODUCTION

Nowadays, more and more attention is paid to environmental friendly and cost effective, plant- and living/dead microorganism-based technologies for bioremediation of contaminated soils. Phytoremediation is a group of environmental friendly, innovative and cost-effective technologies that applies plants for reclamation of soils, sediments or waters contaminated by pollution of primarily anthropogenic origin (Chaney et al., 1997; Cunningham, Berti & Huang, 1995). Compared to the mostly difficult, costly and inefficient traditional physico-chemical clean-up methods, phytoremediation accelerates the process of succession and may improve the soil structure and fertility (Cunningham and Ow, 1996).

Phytoremediation includes many technologies and applications. Currently, scientists have directed their attention toward developing complex phytoremedial processes that involves combined use of physico-chemical and microbe-assisted approaches (Figure 1) (Zaidi, Wani & Khan, 2012). The underlying principles of developing gentle remedial technologies require that the negative impacts of contamination should be alleviated by environmental friendly methods. These innovative approaches are intended to achieve long-term sustainability and to minimize spreading, leaching, uptake and translocation of contaminants to food chain (Marchiol et al., 2011). The expression of phytoremediation has been in use since 1991 as a comprehensive term of plant managed clean-up technologies. Phytoremediation can be used against many kinds of organic and inorganic pollutions *in situ* or *ex situ*. The method can be applied for moderately polluted soils where the aim is to reduce its concentration below a limit value under which the environmental risk can be tolerated (Cunningham and Ow, 1996; McGrath, Zhao & Lombi, 2002).

The major advantage of phytoremediation is its cost which is much lower than that of conventional excavation and reburial. Phytoextraction of metals costs one the tenth compared to conventional technologies. In case of wide area pollutions, phytoremediation is the only economically applicable method. In situ implementation means less soil disturbance and secondary pollution. Phytoremediation is more aesthetical and socially preferable to physical disturbance and chemical treatments than conventional methods. The plant can be easily monitored and it is possible to utilize the biomass even for the recovery and re-use of valuable contaminants. The disadvantage of phytotechnology is that it is time-consuming - twice to seven times longer, depending on the technology (Schnoor, 1998)-, removal of contaminants depends on the species and limited in quantity. Also, remediation is strongly influenced by climatic conditions.

OVERVIEW OF PHYTOREMEDIAL TECHNOLOGIES

Phytoremedial methods are based on the ability of plants to take up, accumulate, convert or degrade pollutants. Depending on the main process, phytoextraction, phytofiltration, hydraulic control, phytovolatilization, phytostabilization and phytodegradation are distinguished within phytoremedial technologies (EPA, 2001) (Table 1).

For phytostabilization, the immobilization on or within the root is the dominant mechanism, for phytoextraction, uptake and translocation are the most important processes. Phytoextraction, -filtration and -volatilization removes metal or organic pollutants from the media.

Applying these methods it is required to promote solution, uptake and transport of pollutants. While implementing phytoremediation, it is important to fit the related processes to strengthen each other and to reduce inhibiting factors. All phytoremedial methods assume a plant that tolerates the pollution.

Plants that are adapted to heavy metal (HM) rich soils are often referred as metallophytes. The phytoremediation potential of plants that are tolerant to extreme concentrations of HMs was characterized by various defensive mechanisms. The excluder or accumulator feature of a plant depends on different regulations of some mechanisms such as HM uptake, binding in cell walls or in exudates, chelation in cytosol, sequestration in vacuoles and HMs root-to-shoot translocation (Hall, 2002; Rascio and Navari-Izzo, 2011). There are plants that can be applied both for immobilization and extraction depending on the quality and quantity of the pollutant.

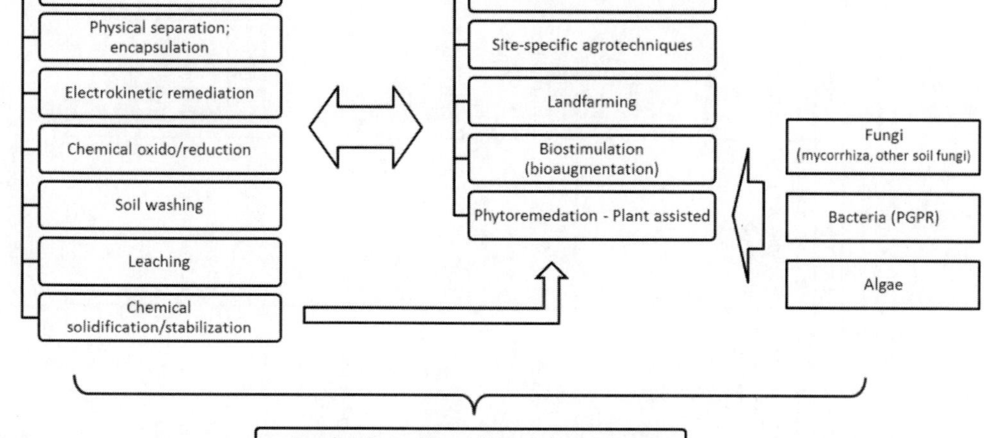

Figure 1. Methods applied in remediation of heavy metal contaminated sites.

Table 1. Summary of phytotechnology mechanisms and applications (modified table of EPA, 2007)

Mechanisms	Process	Media	Contaminants	Application	Plant types	Typical plant species
Phytoextraction (phytoaccumulation)	removal and accumulation in the shoot	soil; sediments; sludge	metals (Ag, Au, As, Cd, Co, Cr, Cu, Hg, Mn, Mo, Ni, Pb, Zn) radionuclides (^{90}Sr; ^{137}Cs, ^{239}Pu; $^{234;238}U$)	vegetative covers; vegetative stands; wetland treatment; hydroponic systems	herbaceous species; grasses; trees; wetland species	*Achillea millefolium, Agrostis castellana, Alyssum* sp., *Brassica juncea, Brassica rapa, Digitalis purpurea, Gleditsia triacanthos, Helianthus annuus, Ilex* spp., Peltaria emarginata *Populus* sp., *Pteris vittata, Salix* sp., *Solidago hispida, Thlaspi* sp., *Viola sp.,*
Phytofiltration (rhizofiltration; hydroponic system)	removal and accumulation in the root	groundwater; wastewater	metals (As, Cd, Cu, Cr, Ni, Pb, Zn), radionuclides (^{90}Sr; ^{137}Cs, ^{239}Pu; $^{234;238}U$)	hydroponic systems	herbaceous species; wetland species	*Azolla pennata, Brassica juncea, Eichorma crassipes, Helianthus annuus, Hydrocotyle umbellata, Lemna minor,*
Phytovolatilization	removal and release to atmosphere	soil; sediments; sludge; groundwater	chlorinated solvents; MTBE, some metals (Se, Hg, As)	vegetative covers; vegetative stands; wetlands treatment; hydroponic systems	herbaceous species; grasses; trees; wetland species	*Brassica juncea Medicago sativa, Robinia pseudoacatia, Populus* sp.,
Phytostabilization (phytorestoration)	immobilization and retention	soil; sediments; sludge	metals (As, Cd, Cr, Cu, Pb, Zn)	vegetative covers; vegetative stands; wetlands treatment; riparian buffers; hydroponic systems	herbaceous species; grasses; trees; wetland species	Poaceae species, *Populus* sp.
Phytodegradation (phytotransformation ; rhizodegradation)	removal and destruction	soil; sediments; sludge; groundwater; surface water	organics compounds, chlorinated solvents; phenols, pesticides; munitions	vegetative covers; vegetative stands; wetland treatment; riparian buffers	herbaceous species; grasses; trees; wetland species	*Agropyron smithii, Betula pendula,* vetiver plants (*Chryzopogon zizanioides*) *Cynodon dactylon, Festuca rubra, Helianthus annuus, Salix* sp. *Sorghum* sp., *Trifolium* sp.,

PHYTOEXTRACTION

During phytoextraction (phytoaccumulation, phytomining) natural or genetically modified plants remove mostly metal, radionuclide and rarely organic pollutants from soil, sediment or groundwater (McGrath, Zhao & Lombi, 2002; Rascio and Navari-Izzo, 2011; Prasad and Freitas, 2003). Heavy metals (HMs) are the most common and perhaps the most dangerous pollutants because they cannot be destructed biologically and therefore persist in the environment.

The content of potentially toxic metals and other trace elements in non-contaminated world soils is showing a wide variety depending on soil types or managements (Table 2). Native HM tolerant plant species can be found on four distinctive type of metal-rich soils that have been characterized by their main HMs: (1) serpentine soils are loaded with Ni, (2) calamine soils have high Zn concentration, (3) seleniferous soils are rich in Se and (4) African copper belt soils contain high amount of Cu, Co, Cr, Ni and Zn (Alford, Pilon-Smits & Paschke, 2010).

Usually a special community of plant species (metallophytes) that are tolerant to metalloids is growing on these metal-rich, nutrient-poor soils (Bothe, 2011). Plants with high bioconcentration factor (BCF is defined by the equilibrium concentrations as C_{plant}/C_{medium}) take the contaminants up and accumulate them mostly in aboveground organs, without exhibiting symptoms of toxicity. Plants with extraordinarily high metal accumulation were first termed hyperaccumulators by Brooks et al. (1977). A phytoremediation method can be based on natural hyperaccumulation or on induced (assisted) hyperaccumulation. The first is called continuous phytoextraction and the latter is promoted with natural or artificial amendments. In the shoot of hyperaccumulators the concentration of the metal is at least 1000 mg kg^{-1} dry weight. In contrast to non-hyperaccumulators, the ratio of the metal concentration of the shoot and the root is always over one (McGrath, Zhao & Lombi, 2001).

Others define hyperaccumulation in relation to the kind of the metal, at least 100 mg kg^{-1} (> 0.01 % dry weight) for Cd and As, 1000 mg kg^{-1} (> 0.1 % dry weight) for Co, Cu, Cr, Ni and Pb, 10000 mg kg^{-1} (> 1 % dry weight) for Mn and Ni (Reeves and Baker, 2000; Wantanabe, 1997). One of the most thoroughly studied metal uptakes is that of Ni, based on it the classification of plants was refined several times. In case of Ni, 1-10 mg kg^{-1} (Reeves and Baker, 2000) or < 100 mg Ni kg^{-1} (Boyd and Jaffré, 2009) are the average HM concentrations, which are found in normal plants (non-accumulators). However, 100-1000 mg Ni can be removed by each kg of dry biomass of Ni accumulator plants (Brooks et al., 1977) or 100-999 mg Ni kg^{-1} were concentrated by plants termed hemi-accumulators (Boyd and Jaffré, 2009). Brooks (1977) has found extreme high Ni rolled up (> 1000 mg kg^{-1}) in some hyperaccumulators and defined the term by this limit. Members of this group includes *Homalium guilllainii* (Saliaceae; 3100 mg kg^{-1}) (Boyd and Jaffré, 2009) and *Alyssum* species (Brassicaceae; 1280-9090 mg kg^{-1}) (Prasad, 2005). Hyper-nickelophore plants were defined by accumulation upwards of 10.000 mg Ni kg^{-1} dry weight (Jaffré and Schmid, 1974). Hyper-nickelophore plants includes *Alyssum argenteum* (Brassicaceae; 29.400 mg kg^{-1}), *Thlaspi cypricum* (Brassicaceae; 52.100 mg kg^{-1}), *Peltaria emarginata* (Brassicaceae; 34.400 mg kg^{-1}) (Prasad, 2005); *Hybanthus austro-caledonicus* (Violaceae) 17.000-20.000 mg kg^{-1} *Sebertia acuminata* (Sapotaceae) 12.000-14.000 mg kg^{-1} *Phyllanthus* sp. (Phyllantaceae); 60.000 mg kg^{-1} (Boyd and Jaffré, 2009). Plants that tolerate extreme amounts of Zn, Ni, Cd, Pb, Mn, Cu and Co, are endemic near bonanzas and grow on metalliferous or serpentine soils (serpentinophytes) (Bothe, 2011). Up to the present time about 400 plant species that hyperaccumulate metals are reported (Prasad and Freitas, 2003). Among these plants the members of Asteraceae, Brassicaceae, Caryopyllaceae, Cyperaceae, Cunouniaceae, Euphorbiaceae, Fabaceae, Flacourtiaceae, Lamiaceae, Poaceae and Violaceae families dominate (Panwar et al, 2010). Brassicaceae has the largest number of hyperaccumulators with 84 species of 11 genera prevailing in moderate climate, in tropical climate the members of the Spurge family (Euphorbiaceae) are the most common. Accumulator species of *Thlaspi*, *Alyssum*, *Sebertia*, *Berkheya* genera can amass more than 0.01% Cd, 0.1% Co, Cu, Pb, Ni, 1 % Mn and Zn in their shoots (Salt, Smith & Raskin, 1998; McGrath, Zhao & Lombi, 2001; Brooks, 1998). The ability and efficiency of species vary metal by metal. Ni hyperaccumulation has been reported in 72 species of seven plant genera, however, Zn accumulation was found in 20 species of three genera. Some plants are not specific to the metal, for example *Thlaspi* species can accumulate different metals simultaneously. *T. caerulescence* is Cd, Ni, Pb and Zn, *T. goesigense* is Ni and Zn, *T. ochroleucum* is Ni and Zn, *T. rotundifolium* is Ni, Pb and Zn accumulator. Cesium (^{137}Cs) and strontium (^{90}Sr) were extracted using sunflowers after the Chernobyl accident. Sunflower and Chinese Brake fern (*Pteris vittata*) can accumulate as Indian mustard (*Brassica juncea*), sunflower (*Helianthus annuus*), ragweed (*Ambrosia artemisiifolia*) and hemp dogbane (*Apocynum cannabinum*) sequester lead in their biomass (Brooks, 1998). Fast growing trees with large biomass that can be used for bioenergy production are preferred for phytoextraction. Willows (*Salix* sp.) are used for Cd, Zn and Cu removal while poplars (*Populus* sp.) can clean lead pollutions up. When used for continuous phytoextraction plants hyper-accumulate the metals gradually during their development. While determining the bioconcentration factor (BCF) and translocation factor (TF) of *Populus alba* and *Morus alba,* it was found that these trees have enormous capability to remove Cd and Ni from contaminated soils (Rafati et al., 2011). In contrast, it was even reported that white poplar is also a suitable plant for long-term monitoring of Cd and Zn movement and their stabilization due to its transfer coefficient from soil to reproductive or vegetative organs (Madejón et al., 2013). Mercury (Hg) is a very toxic HM even in low contamination. Only a few publications report native plants tolerating or accumulating Hg. However, more than ten Hg accumulator native plants were found at a highly contaminated waste area in northern Italy (Massa et al., 2010). In case of Hg polluted soils, phytovolatilization is more common than phytoextraction and the latter requires the detoxification of Hg, too. Nowadays, chloroplast engineering seems to be a promising method to develop Hg accumulation of plastids and its volatilization from cells (Ruiz et al., 2011). Metal uptake by plants is highly influenced by the concentration and the properties of the available ions, the surface of the roots and the amount of the water streaming towards the roots. The uptake depends also on the evaporation and the temperature, in the case of active transport, it is also influenced by respiration (Marschner, 1997). Hyperaccumulators are generally believed to have limited potential in field application because of their small size and slow growth, which limits the rate of metal removal (Cunningham, Berti & Huang, 1995). Without fertilisation, the yield is less than 5 tons in a hectare and the harvest is difficult because of their little size (McGrath, 1998; Robinson et al., 1998). The root system does not penetrate and clean the soil in depth. To eliminate the disadvantages, research aims to transfer the genes responsible for metal tolerance and hyperaccumulation to agricultural plants which have high biomass and rich root system.

Table 2. Averages and ranges of concentrations of potentially toxic elements in soils and in plants

Element	Means and ranges of total concentrations in surface soils [a] (mg kg $^{-1}$)	Threshold limit value [b] (mg kg $^{-1}$)	Average concentrations in crop plants [c d] (in agronomic circules) (mg kg $^{-1}$)			Concentrations in accumulator or hyper -accumulator plants [e] (mg kg $^{-1}$)
			low	normal	toxic	
As	6 (<0.1-66)	15	0.1	1	5	
Cd	0.5 (0.01-2.7)	1		0.2-0.8		100
Co	8 (0.01-70)	30		0.05-0.5		5000
Cr	54 (1-1100)	75		0.2-1.0		
Cu	20 (1-140)	75	<2	4-15	20<	5000
Hg	0.1 (0.01-1.1)	0.5	-	-	-	
Mn	437 (7-9200)	-		15-100		10 000
Mo	2 (0.1-7.4)	7	<0.1	0.5-1.0	100<	
Ni	22 (0.2 450)	40		1	50<	5000
Pb	29 (1.5-176)	100				
Se	0.3 (0.005-1.9)	1		0.02-2.0	50-100	1000
Sr	147 (5-1000)	-		5-3000		
Zn	64 (3.5-770)	200		8-15	200<	10 000

The ranges and the mean concentrations of trace elements are calculated from data of total trace element concentrations measured in podzols (sandy soils), cambisols (silty and loamy soils), redzinas, kastanozems, chernozems and histosols (organic soils) (Kabata-Pendidas and Pendidas; 2001). ([a]Kabata-Pendidas and Pendidas (2001); [b]Hungarian Govermental regulation number 10/2000 (2000); [c]Allaway, (1968); [d]Csathó (1994); [e]Brooks (1998)

Other developments are aimed to influence the metal bioavailability. Metal uptake by plants is highly affected by solubility of the polluting metal, which is specific to the ion and the soil (Sauvé, 2003).

During induced phytoextraction, the metal uptake by the plant - which can be a non-hyperaccumulating one – is enhanced by chelator or other agents which change the bonds and increase the mobility (McGrath, Zhao & Lombi, 2002; Lombi et al., 2001; Salt et al., 1998). Fast growing, high biomass non-hyperaccumulating plants usually concentrate a slight amount of the metals in their roots. Growing in a field polluted moderately with low mobility metals, i.e. Pb, Cr, U, the uptake of these plants is speeded up extremely just before the harvest or the end of the vegetative development. Compared to the continuous phytoextraction, the induced method has a lower risk of metals entering the food chain, but the mobilized metals can contaminate water bodies. Because of that, induced phytoextraction is usually applied *ex situ*, that is, in a separated place, under controlled conditions.

PHYTOFILTRATION

Phytofiltration is a group of methods for cleaning high volumes of slightly polluted water. The absorption and the extraction of the pollutants can be implemented by plants (phytofiltration), plant roots (rhizofiltration) or germs (blastofiltration). During the process plants are raised hydroponically and transplanted to the polluted water and groundwater where roots absorb, remove and concentrate the contaminants (EPA, 2001).

Because of the easy application floating plants, especially aquatic macrophytes - duckweed (*Lemna minor*), water hyacinth (*Eichorma crassipes*), water fern (*Azolla pennata*), pennywort (*Hydrocotyle umbellata*) - are preferred (Brooks, 1998; Rahman and Hasegawa, 2011). Aquatic macrophytes can have a high phytoremedial potential as HM (Ag, As, Cd, Co, Cr, Cu, Mn, Ni, Pb and Zn) accumulators from wastewater due to their direct contact with the polluted surface, fast growth and large biomass production (Rahman and Hasegawa, 2011).

However, some terrestrial plants can be more suitable considering the advanced root system with large surface. Sunflower (*Helianthus annuus*) or Indian mustard (*Brassica juncea*) could significantly decrease the concentration of Cr(VI), Mn, Cd, Ni and Cu (Dushenkov et al., 1997). The duration of the treatment is restricted by the oxygen demand of the roots. Plants that are suitable for rhizofiltration should have massive, high surface root system which immobilizes metals and tolerates the high concentration of metals for a few days at least. Phytochelatins are important in the accumulation of Cd, Cu, Zn and Pb (Salt et al., 1997).

A few publications report that rhizofiltration is applicable against organic pollutants. A new and promising technology is the application of germs and seedlings (blastofiltration) for fast removal of metal pollution. Some cultivated plants can germinate and grow for a short period in aerated water culture without nourishment and light, these plantlets have large specific surface. Besides, some members of the mustard family (Brassicaceae) – Indian mustard (*Brassica juncea*), rapeseed (*B. napus*) and turnip (*B. rapa*) – rice (*Oryza sativa*) and alfalfa (*Medicago sativa*) were applied successfully (Salt et al., 1997). This cheap, fast method can be convenient to clean mining, industrial and waste waters.

PHYTOVOLATILIZATION

By phytovolatilization, contaminants pass from plant leaves into the atmosphere (McGrath, Zhao & Lombi, 2002; EPA, 2001). Se, Hg, As, halides and some other organic compounds are well known to get volatilized by plants. Some plants methylate these metals and make them less harmful. Halides are transformed to monohalomethane or monohalo-thiol compounds. Most experiences are gathered about the phytovolatilization of Se (Chaney, Broadhurst & Centofanti, 2010). The conventional cleanup of Se polluted soil, water or sediment is expensive, difficult and secondary pollution often occurs, so phytoextraction or phytovolatilization should be preferred. Plants with high sulphur content are more efficient volatilizers, i.e., Indian mustard (*B. juncea*), cabbage (*B. oleracea*), cauliflower (*B. oleracea* convar. *botrytis* var. *botrytis)*, rice (*O. sativa*), some wetland plants, sedge (*Carex sp.*) and bulrush (*Thypha* sp.) (Terry et al., 1992). It was reported that the Brassicaceae species can emit up to 40 g Se ha^{-1} day^{-1} in various gaseous compounds. Volatilization of Se can be promoted by selected bacteria, fungi and microalgae (e.g. *Aeromonas* sp., *Flavobacterium* sp., *Pseudomonas fluorescens*; *Acremonium flaciforme*, *Alternaria alternata*, *Aspergillus niger*, *Candida humicola*, *Fusarium* sp., *Penicillum* sp.; *Chlorella* sp.). Contrary to metal bound Se and As, Hg compounds are highly toxic, so the formation of less toxic elemental Hg should be promoted. For higher tolerance, current developments are aimed to insert bacterial mercuric reductase gene in the plant genome. Moderately contaminated soils could be remediated using transgenic plants with the bacterial gene merA-merB (*Nicotiana tabacum, Liriodendron tulipifera, Arabidopsis thaliana*) (Ruiz et al., 2011; Singh et al., 2003). Cabbage can decrease the halide and disulphide concentration of salt affected soils. Poplars and *Pinus taeda* were successfully used for the uptake and the volatilization of trichloroetylene (TCE) or its breakdown products (Newman et al., 1999).

PHYTOSTABILIZATION

In case of moderate, wide area pollution close to the surface, the primary goal is to restrict the spreading of the contamination, which can be caused by deflation, soil erosion, surface water runoff or any other means. Phytostabilization (phytorestoration) immobilizes the metals and decreases the potential solubility by absorption on the root surface, accumulation in the roots and precipitation, resulting in a reduced bioavailability of pollutants (Cunningham, Berti & Huang, 1995; EPA, 2001). Contaminant-tolerant grasses having rich root system – i.e. *Festuca rubra* and *Agrostis tenuis* are preferred for phytostabilization. Poplars (*Populus* sp.) and willows (*Salix* sp.) can be applied against deeper pollution of soils (EPA, 2001). Best stabilizers are drought tolerant, austere perennials. To avoid deflation, erosion and dusting, the area should be covered all the year. When applying this technology, other factors affecting the solubility of the contaminants should be considered too. Absorbents are binding substances inhibiting the uptake and the spreading of the contaminants. Common binding agents include, but are not limited to, cement, lime, limestone, fly ash, slag, clay, activated carbon, mold, ion-exchange resin and gypsum. Beside heavy metals, roots can immobilize some organic pollutants. In small-scale experiments, poplar, maize and grasses could stabilize benzo-α-pyrene, PCB and chlorinated hydrocarbon pollutions (Wenzel et al., 1999). Phytostabilization can generally be applied for the remediation of wide areas where the soil is compact and rich in organic compounds. It is a generally accepted method for the recultivation of spoils, and there are promising investigations on the restoration of small industrial and urban areas.

PHYTODEGRADATION

Urban and industrial sewage is often phytodegraded in constructed wetlands, floating forests. In the course of phytodegradation (phytotransformation) plants and associated microbes transform the pollutants to less harmful compounds or to carbon dioxide and water. Plant organs – primarily roots – are able to partially or wholly degrade oil derivatives, organic microcontaminants (i.e. pesticides, TNT, nitroglycerin, polychlorinated biphenyls, trichloroethylene) in the upper soil. The decomposition can be accomplished either within the root or *ex planta*, by exudates or associated microbes. Rhizodegradation is achieved by rhizosphere microorganisms. This kind of bioremediation is also called plant-assisted degradation, plant assisted bioremediation, plant-aided *in situ* biodegradation and enhanced rhizosphere biodegradation (EPA, 2001). In this technology, microbes play the relevant role, while the plant provides nutrients for the rhizosphere microorganisms.

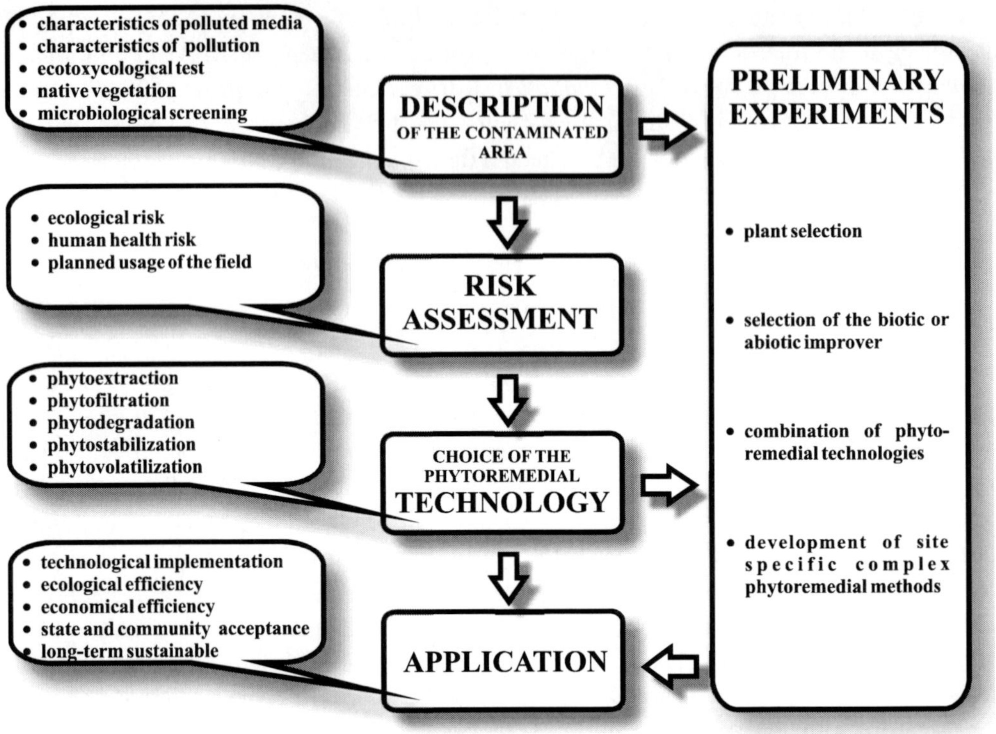

Figure 2. Schema of phytoremedial process.

Remedial Objectives and Phytoremedial System Selection: Considering Site, Plant and Contaminant to Choose Appropriate Phytotechnology

Phytoremediation is always site-specific. The priority of remediation methods are determined mainly by the human health risk (EPA, 2001). The aim of the operations is to prevent harm and to reduce and minimize the risk with the chosen remedial technology. During the planning phase the orientation, the actual and planned usage of the area, the degree and extent of the pollution and the protection of water should be considered. Other important aspects are the origin of the pollution including duration, former conditions and usage, furthermore the properties of pollutant - quality, quantity, volatility, chemical and biological stability and availability. Physical and chemical characteristics of the soil should also be examined. From the possibilities of risk reduction, the proper remedial technologies or their combinations should be chosen according to the environmental, ecological and economic efficiency. The final step of remediation is the recultivation or the restoration of the area.

The main steps to prepare a site specific technology as follows (Figure 2):

1. risk identification through the characterization of the polluted area
2. risk assessment
3. choice of phytoremedial technology according to the risk

4. development of adequate technology
   selection of potential plants
   site-specific optimization of the technology
   selection of biotic and abiotic improvers
5. application and monitoring of the effectivity of the chosen technology

Amendment Assisted Phytoremediation

In practice, the simultaneous application of chemical and biological methods produced the best results both in phytostabilization and phytoextraction. Planting is often preceded by the addition of immobilizing or mobilizer agents. Three main groups of additives are distinguished: inorganic (e.g., lime, ashes, zeolites, silicates), organic (e.g., biosolids, composts, manure, papermill sludge) and mixed (e.g., lime—stabilized biosolids) amendments.

It is generally required that an improver does not make the technology much more expensive. The amendment should easily be purchased or produced and applied. It must not be harmful (toxic, carcinogenic) either to the man or to the environment. Secondary pollution should be avoided. A favourable improver even provides nutrients for the plants, i.e. sewage sludges are rich in N and P. Additives as compost and organic waste can also improve the structure and the water management of the soil. Currently, industrial waste and secondary products are investigated to be used as improvers.

Induced Phytoextraction

Ethylene diamine tetraacetic acid (EDTA), ethylene diamine disuccinate (EDDS), ethylene glycol tetraacetic acid (EGTA), nitrilotriacetic acid (NTA), sulfur and citric acid are widely used compounds in induced phytoextraction (Hegedűsova et al., 2009). In the last ten years more and more attention is paid to the application of natural, "green" chelating agents such as EDDS or NTA that have a shorter degradation and a lesser toxicity compared to synthetic ones. Surinamgrass (*Brachiaria decumbens)*, a fast-growing, productive, aluminum tolerant plant, that has a well-established agronomic system fulfills most of the requirements for chelate-induced phytoextraction. In case of a moderate heavy metal pollution, shoot Cd, Zn and Pb concentrations were increased 2.54, 2.74 and 4.30-fold by EDDS and 1.77, 1.11 and 1.87-fold by EDTA, compared to control plants (Santos et al., 2006). After 5 weeks of low dose citric acid treatment enhanced uptake and translocation of uranium were found in willow (*Salix* spp.) and sunflower (*H. annus* L.) leaves (Mihalík, Tlustoš & Szaková, 2010). Citric acid and ammonium nitrate can raise the availability of radioactive cesium (^{137}Cs). Uptake of organic contaminants can be enforced by natural surfactants (rhamnolipids), artificial tensides (Triton X-100, SDS) and cyclodextrins (HPBCD, RAMEP) (Fenyvesi et al., 2009). The application of cyclodextrins as innovative engineering tools has been suggested for increasing the sensitivity of risk-based bacterial bioassay by enhancing the mobility and availability of soil or water contaminants (Gruiz et al., 2011).

Phytostabilization With Ingredients

In practice, phosphates are proper immobilizers for phytostabilization. Among them, we could find calcium-phosphate, apatite, phosphate rock, fertilizers, industrial secondary products with high phosphate content (Basta and McGowen, 2004; Cao et al., 2003). Additionally, bio-litter, ferro- and manganese oxihydroxides (Lombi et al., 2003), natural and artificial clay minerals (aluminium silicates) and alkalizers (lime, dolomite) (Bolan et al. 2003) are well applicable. Sewage sludges, manures, composts, biosolids, digestates, and some industrial wastes (slags, ashes) are likely amendments (Vangronsveld and Cunningham, 1998). The potential use of red mud as a chemical stabilizer for HMs was published by Feigl et al. (2012). The aims of effective stabilization are: (1) direct reduction of the amounts of soluble and/or exchangeable forms of trace elements (2) limitation of HM transport to native plants or crops (3) to take care of biodiversity and functionality of vegetation and microbial community. With some exception (Pb, Cu), in case of cationic heavy metals (Cd, Hg, Zn, Ni, etc.), the pH of the soil should be kept higher than 6.6, conversely to anionic forms of trace elements (MoO_4^{2-}, Se^{2-}, SeO_3^{2-}, SeO_4^{2-}; AsO_3^{3-}, AsO_4^{3-}) (Fuller, 1977). Solubility of Cu is highly affected by organic complex formers, e.g. humic acids. Pb is immobilized as hardly soluble precipitate. Cationic heavy metals are mainly immobilized by lime, phosphate, clay or aluminium silicates (beringite), Ba and Se are treated with sulfates. Combinations of additives (lime, municipal sewage sludge, natural zeolite) can permanently stabilize the heavy metals (Cd, Cu, Mn, Pb, Zn) in acidic mine spoil (Simon, 2005).

Applied Phytoremediation Over the World and in Europe

The "phytoremediation" term has become used recently, however, the basic methods of the technology have been established a long time ago. In spite of being environment friendly, sustainable and cost effective, use of phytoremediation is less common than traditional cleanup methods. Perhaps, the long term, site specific procedures that require preliminary investigations retard the general usage of phytoremediation. The main application of phytotechnologies –also known as plant promoted practices - has been revitalization, restoration and reuse of large, contaminated sites (e.g., hard rock mining areas, abandoned coal mines, refining, smelting and construction sites). Recently the site specific, multi-stage, complex physico-chemical-biological methods that cover the whole remediation process have been preferred.

The summary of European Environment Agency (EEA) reported that the number of contaminated sites in the EEA member countries were estimated at 250.000. Considering the growth in the number sites that need remediation, it will significantly increase by 2025 (EEA, 2007). In Europe, heavy metals and mineral oils were detected as main soil and groundwater contaminants. 37.3 % of investigated sites were polluted by HMs. Based on the SUMATECS final report (2009) it was found that 25 phytoremedial field trials were realized in 9 European countries between 2000 and 2008 (Marchiol et al., 2011).

Phytoremedial procedures have often multiple purposes. Short-rotation willow coppice was cultivated in Sweden not only to produce biomass for energy but also to treat waste products and take up pollutants from soil and water. About 10.000 ha was covered with different *Salix viminalis*, *S. dasyclados* and *S. schwerinii* clones and hybrids and fertilized with sewage sludge, which contained Cd, other heavy metals and high amounts of phosphorus. Short-rotation tree plantations were often applied in simultaneous waste water management and phytoremediation (Dimitriou and Aronsson, 2004). In 1996, an experimental short rotation culture of 17 poplar and two willow clones was set at an industrial site moderately polluted with heavy metals in Boom (Belgium). During the establishment of phytoextraction potential, a large clonal variance of metal concentrations and biomass production was found. A *Populus trichocarpa* clone removed 1.4 kg ha-1 Al over two years. Other poplar clones showed high accumulation of Cd (47-57 g ha-1) and Zn (2-2.4 kg ha-1) (Laureysens et al., 2005).

The Vetiver System is commonly used for soil and water conservation, pollution control, wastewater treatment and many other environmental protection applications. Vetiver grass (*Chrysopogon* or *Vetiveria zizanioides)* is suitable for phytoremedial technologies because of its tolerance to heavy metals and organic wastes (Danh et al., 2009).

In a cooperation of the Croatian government and the European Union an immobilization project was implemented in Bakar, in an area contaminated with large amounts of total petroleum hydrocarbons (TPHs), polycyclic aromatic hydrocarbons (PAHs) and metals. Besides solidification and stabilization with natural and synthetic zeolites, the mix-in-plant procedure was applied for the remediation of the polluted area.

Phytoremedial technologies are very popular in India and other Asian countries for restoration of mine spoils where the groundwater is contaminated by As and pesticide residues. Both selected native plants and agricultural crops are applied for the cleanup (Prasad, 2004).

Pot and field trials for Ni phytomining have been performed in New Zealand, California and South Africa using Ni hyperaccumulator plants (*Alyssum bertolonii, Berkheya coddii, Streptanthus polygaloides*). 12-22 ton ha^{-1} plant biomass was achieved with a moderate fertilization of the serpentine soil, so 96-122 kg ha^{-1} of Ni could have been removed and provided (Brooks et al., 1999).

Arbuscular Mycorrhizal Fungi Assisted Phytoremedial Processes

Soil microbial community is one of the key components of soil functionality, resilience and also an important indicator of terrestrial ecosystem state (Szili-Kovács et al. 2007). A successful phytoremedial method can also be attributed to plant-associated bacteria, plant growth promoting rhizobacteria (PGPR), soil fungi and mycorrhiza fungi (Di Gregorio, 2006; Kumar, 2012; Kurek and Majewska, 2012; Lebeau, Braud &Jézéquel, 2008; Vassilev et al. 2004; Zhuang et al., 2007).

Arbuscular mycorrhizal fungi (AMF) (Glomeromycota) are important root symbionts which live in strong associations with 80-90 % of higher plants (Harley and Harley, 1987). AMF can significantly increase both water and nutrient uptake of plants and their ability to tolerate stress conditions (drought-, salt-, toxic element etc.) (Brundrett, 2009; Marschner, 1997; Khan, 2005; Takács, Radimszky & Németh, 2005). Arbuscular mycorrhizal (AM) associations are occurring in the roots of most angiosperms, gymnosperms and pteridophytes, and lycopods or gametophytes of some mosses (Harley and Harley, 1987). As obligate symbionts, AM fungi occur in roots of almost all vascular plants, most of tropical trees and several of temperate to boreal trees like Rosaceae and Fabaceae tree species, *Alnus*, *Celtis*, *Fraxinus*, *Populus*, and *Pseudotsuga* sp. In contrast to the diverse array of terrestrial plants that form AM, only ca. 200-250 Glomeromycota species have so far been described, suggesting that mycobiont specificity is low relative to their potential plant hosts. However, some plants from mycorrhiza free plant family (i.e. Brassicaceae, Cactaceae, Cyperaceae, Amaranthaceae, Juncaceae) were found to be infected at habitat under extreme stressed conditions (Füzy et al., 2008; Regvar et al., 2003; Vogel-Mikus, Drobne & Regvar, 2005). It is assumed, that species belonging to these plant families are not or only less dependent on the existence of AM fungi e.g. heavy metal-, drought, salt or root pathogenic tolerant ones, which can be explained by the absence of the plant-fungi symbiosis development (Leyval et al., 1997). The importance of AMF community to plant diversity, productivity and ecosystem stability has also been recognized (Van der Heijden, Scheublin & Brader, 2004).

Metal pollution can even totally block the formation of plant-AMF symbiosis, sporulation, germination of spores and the spreading of hyphae. Several authors have shown that AMF can adapt to long-term metal pollution and heavy metal tolerant AMF strains can be selected by contamination (Orlowska, Godzik & Turnau, 2012; Weissenhorn, Leyval & Berthelin, 1993; Weissenhorn and Leyval, 1995). The mechanism of metal tolerance is complex and not fully understood (Gonzalez-Guerrero et al., 2008). A number of passive and active molecular processes are employed by AMF to maintain metal homeostasis.

Effectiveness of phytoremedial technologies can be enhanced by the application of AMF (Narula et al., 2012). The reports on the role of AMF colonization in host metal uptake and transfer are contradictory (Leyval et al., 2002), both increase and decrease in metal concentration of mycorrhized plants are described (Javaid, 2011; Neagoe, Iordache & Kothe, 2013). While analyzing scientific results, a lot of aspects should be considered to explain the apparently conflicting observations, for example, the chemical and physical properties of the contaminated soil (Killham and Fireston, 1983, Wang and Chao, 1992), the quality, quantity and availability of the polluting metal (El-Kherbawy et al., 1989; Guo, George & Marschner, 1996), the degree and the term of the load (Díaz, Azcón-Aguilar & Honrubia, 1996; Kucey and Janzen, 1987), the plant and fungal species and their ecotypes (Malcova, Vosatka & Gryndler, 2003; Wu et al., 2007) (Figure 3).

The majority of authors consider HM-adapted strains more efficient than those originating from unpolluted soils (Biró and Takács, 2007; Gaur and Adholeya, 2004; Göhre and Paszkowski, 2006; Khan, 2005; Meier et al. 2011; Orlowska, Godzik & Turnau, 2012; Takács et al., 2008; Takács, 2012; Vivas et al. 2003). In the case of soils highly polluted with heavy metals, AMF colonization of roots generally decreases the transport of toxic elements into the root (Leyval, Turnau & Haselwandter, 1997), conversely to moderate contaminations. A 15 year long investigation of zinc wastes in southern Poland revealed that indigenous AMF plays a significant role in HM detoxification and establishment of vegetation in strongly polluted areas during the phytoremedial process (Turnau et al., 2006). The authors consider the state of AMF as a possible toxicity indicator. *Plantago lanceolata* is a highly mycorrhiza dependent plant and its Cd, Zn and Pb uptake can vary with the origin of AMF strains used for inoculation (Orlowska et al., 2005). Heavy metal content of host plants inoculated with HM adapted strains was lower than that of plants treated with non-adapted AMF.

Figure 3. Differences in the effect of *Glomus mosseae* strains on white clover (*Trifolium repens*) host in Cd polluted soils (100 mg kg^{-1}), depending on the origin of strains.

The *in vitro* simulation of outdoor conditions to evolve and sustain phytoremedial effectivity is often referred to as "directed inoculum production process" (DIPP) (Feldmann and Grottkas, 2002; Takács et al., 2008). Comparing to routine propagation methods, by the continuous presence of the stressor DIPP gives a better chance to develop or sustain strains that are successful in remediation. Independently of the adaptation method, to develop HM tolerance in a strain the optimal dose of the selective metal should be assayed. Current developments in gentle bioremediation are intent on establishing specific fungus-plant associations for different purposes, where the selected fungus-plant pairs have a high potential to extract or immobilize the pollutants, and the vitality of the host is also increased (Leyval et al., 2002; Khan, 2005; Göhre and Paszkowski, 2006; Vosatka et al., 2006).

In Institute for Soil Sciences and Agricultural Chemistry (Hungary) methods were developed for site specific phytoremediation of heavy metal polluted areas. In addition to a better biomass production, either the increase of metal uptake or the immobilization was realized by the selection of compatible arbuscular mycorrhizal fungi (AMF) and host plants (Takács, 2012).

A field trial was set up in the industrial area of a metallization factory in Hungary. The total As, Cd, Cu, Cr, Hg, Ni, Pb and Zn concentrations of the soil exceeded the permissible limit for soils several fold. Because of the extreme soil pollution, a pre-cropping maize plantation was applied for phytoextraction first, followed by grass sowing for phytostabilization. The selected grass species (*Lolium perenne*, *Poa pratensis*, *Festuca rupicola*, *Festuca heterophylla*, *Festuca rubra* and *Festuca ovina*) were treated by monoculture and mixed inocula of AMF strains originated from metal contaminated soils.

The organs of maize showed significant differences in accumulation of various metals. Average metal concentrations in plant tissues followed the order: roots > leaves> stem ≥ corn ear for Cd, Cu, Ni, Mn and Pb. The 4-month-old maize took up 0.34-0.52 g plot^{-1} Cd, 2.90-4.45 g plot^{-1} Cr, 0.85-3.44 g plot^{-1} Cu, 8.10-12.20 g plot^{-1} Mn, 0.70-1.06 g plot^{-1} Ni, 0.80-1.20 g plot^{-1} Pb and 8.3-12.5 g plot^{-1} Zn into the biomass. Under the phytostabilization process AMF significantly decreased Cd, Pb, Sr, Hg and Ni uptake of the shoots (Takács et al., 2008). In phytoremedial processes, lignite can also be used as binding agent. Combined application of lignite and AMF can enhance the exclusion of the metals from the food chain.

Conclusion

Because of the environmental and human health problems caused by soil pollution, the development of cost effective and efficient technologies and strategies for remediation is needed in the future. Phytoremediation is a set of cheap, environmental friendly technologies that applies plants to remove or bind contamination. As it appears to be an agricultural activity, it is preferred to the conventional, industrial methods in the society. In addition to reducing risk to human health and environment, it provides a source of bio-energy. For the sake of environment protection phytoremediation should always be preferred to chemical and physical soil cleanup with an exception for the high dose pollutions that must be treated urgently.

Phytoremedial methods are site specific. Because of the complexity of environmental processes careful and detailed investigations are required to choose and adapt a method, which makes phytoremediation a time-consuming technology. The results from ecological, plant physiological, taxonomical and toxicological investigations can promote the development of efficient, site specific phytoremedial methods. The role of arbuscular mycorrhizal fungi (AMF) in HM tolerance and in fitness of host plants is remarkable, so these symbiotic fungi can be important components of phytoremedial systems. Selection of compatible plant and AM fungi oriented for a purpose can be a successful innovation for future phytoremediation technologies.

References

Alford, E.R., Pilon-Smits, E.A.H., Paschke, M.W., Metallophytes- a view from the rhizosphere. *Plant Soil* 337, 33-50.

Allaway, W.H., 1968. Agronomic controls over the environmental cycling of trace elements. *Adv. Agron.* 20, 235-271.

Basta, N.T., McGowen, S.L., 2004. Evaluation of chemical immobilization treatments for reducing heavy metal transport in a smelter contaminated soils. *Environ. Pollut.* 127, 73-82.

Biró, I., Takács, T., 2007. Effects of *Glomus mosseae* strains of different origin on plant macro-micronutrient uptake in Cd-polluted and unpolluted soils. *Acta Agr. Hung.* 55, 183-192.

Bolan, N.S., Adriano, D.C., Mani, A.P., Duraisamy, A. 2003. Immobilization and phytoavailability of cadmium in variable charge soils. II. Effect of lime compost. *Plant Soil* 251, 187-198.

Bothe, H., 2011. Plants in heavy metal soils. In: Sherameti, I., Varma, A., (Eds.), Detoxification of Heavy Metals. Soil Biology Vol. 30. (pp. 35-56.),Berlin Heidelberg, Springer-Verlag.

Boyd, RS., Jaffré, T., 2009 Elemental concentrations of eleven New Caledonian plant species from serpentine soils: Elemental correlations and leaf-age effects. Soil and Biota of Serpentine: A World View Northeastern Naturalist (Special Issue) 16, 5: 93-110.

Brooks, R.R., 1998. In: Brooks, R.R. (Ed.) *Plants that hyperaccumulate heavy metals: their role in phytoremediation, microbiology, archaeology, mineral exploration and phytomining* (2 edn., pp. 1-95.). New York , CAB International.

Brooks, R.R., Anderson, C., Stewart, R., Robinson, B., 1999. Phytomining: growing a crop of a metal. *Biologist* 46(5), 201-205.

Brooks, R.R., Lee, J., Reeves, R.D., Jaffré, T. 1977. Detection of nickeliferous rocks by analysis of herbarium specimens of indicator plants. *J. Geochem. Explor.* 7, 49-57.

Brundrett, M.C., 2009. Mycorrhizal associations and other means of nutrition of vascular plants: understanding the global diversity of host plants by resolving conflicting information and developing reliable means of diagnosis. *Plant Soil* 320, 37-77.

Cao, X.R., Ma, L.Q., Chen, M., Singh, S.P., Harris, W.G., 2003. Phosphate induced metal immobilization in a contaminated site. *Environ. Pollut.* 122,19-28.

Chaney, R.L., Malik, M., Li, Y.M., Brown, S.L., Brewer, E.P., Scott, J.A., Baker, A.J.M., 1997. Phytoremediation of soil metals. *Curr. Opin. Biotechn.* 8, 279-284.

Chaney, R.L., Broadhurst, C.L., Centofanti, T., 2010. Phytoremediation of soil trace elements. In: Hooda, P.S. (1 edn., Ed.), *Trace Elements in Soil* Blackweel, Publishing Ltd.

Cunningham, S.D., Berti, W.R., Huang, J.W., 1995. Phytoremediation of contaminated soils. *Trends. Biotechnol.* 13, 393-397.

Cunningham, S.D., Ow, D.W., 1996. Promises and prospects for phytoremediation. *Plant Physiol.* 110, 715-719.

Csathó, P. 1994. Contamination of the environment with heavy metals and its consequences on agricultural production. RISSAC of Hungarian Academy of Sciences.

Danh, L.T., Trung, P., Mummucari, R., Tran, T, Foster, N., 2009. Vetiver grass, *Vetiveria zizanioides*: a choice plant for phytoremediation of heavy metals and organic wastes. *Int. J. Pytorem.* 11(8), 664-691.

Díaz, G., Azcón-Aguilar, C., Honrubia, M., 1996. Influence of arbuscular mycorrhizae on heavy metal (Zn and Pb) uptake and growth of *Lygeum spartum* and *Anthyllis cytisoides. Plant Soil* 180, 241-249.

Di Gregorio, S., Lampis, S., Malorgio, F., Petruzelli, G., Pezzarossa, B., Vallini, G., 2006. *Brassica juncea* can improve selenite and selenate abatement in selenium contaminated soils through the aid of its rhizospheric bacterial population. *Plant Soil* 285(1-2), 233-244.

Dimitriou, I., Aronsson, P., 2004. Nitrogen leaching from short-rotation willow coppice after intensive irrigation with wastewater *Biomass Bioenergy* 26(5), 433-441.

Dushenkov, S.P.B., Vasudev, D., Kapulnik, Y., Gleba, D., Fleisher, D., Ting, K.C., Ensley, B., 1997. Removal of uranium from water using terrestrial plants. *Environ. Sci. Techn.* 31, 3468-3474.

EEA. 2007. Progress in management of contaminated sites. Report CSI 015, European Environment Agency, Copenhagen, Denmark Online report.

El-Kherbawy, M., Angle, J.S., Heggo, A., Chaney, R.L., 1989. Soil Ph, Rhizobia , And Vesicular Arbuscular Mycorrhizae Inoculation Effects On Growth And Heavy Metal Uptake Of Alfalfa (*Medicago Sativa* L.). *Biol. Fertil. Soil* 8,61-65.

EPA (2001) Brownfields technology primer: Selecting and using phytoremediation for site cleanup. NSCEP Cincinatti, Ohio.

EPA (2007) Treatment technologies for site cleanup: Annual status report. (twelfth edition) EPA, Washington D C.

Feigl, V., Anton, A., Uzigner, N., Gruiz, K., 2012. Red mud as a chemical stabilizer for soil contaminated with toxic metals, *Water, Air Soil Poll.* 223(3), 1237-1247.

Feldmann, F., Grotkass, C., 2002. Direct inoculum production-shall we he able to design populations of arbuscular mycorrhizal fungi to achieve predictable symbiotic effectiveness? In: Gianinazzi, H., Schüepp, H., Barea, J.M., Haselwandter, K, (Eds.), *Micorrhiza Technology in Agriculture from Genes to Bioproducts* (pp. 261-281.), Birkhäuser, Switzerland.

Fenyvesi, E., Molnár, M., Leitgib, L., Gruiz, K., 2009. Cyclodextrin-enhanced soil-remediation technologies. *Land Contam. Reclam.* 17, 585-597.

Fuller, W.H., 1977. Movement of selected metals, asbestos, and cyanide in soil: applications to waste disposal problems, Technical Report No. PB-266905, EPA-68-03-0208, Arizona University, Tucson, USA.

Füzy, A., Biró, B., Tóth, T., Hildebrandt, U., Bothe, H., 2008. Drought, but not salinity determines the apparent effectiveness of halophytes colonized by arbuscular mycorrhizal fungi. *J. Plant Phsyol.* 165, 1181-1192.

Gaur, A., Adholeya, A. 2004. Prospects of arbuscular mycorrhizal fungi in phytoremediation of heavy metal contaminated soils. *Curr. Sci.* 86(4), 528-534.

Gonzalez-Guerrero, M., Melville, L.H., Ferrol, N., Lott, J.N.A., Azcon-Aguilar, C., Peterson, R.L., 2008. Ultrastructural localization of heavy metals in the extraradical mycelium and spores of the arbuscular mycorrhizal fungus *Glomus intraradices*. *Can. J. Microbiol.* 54(2), 103-110.

Governmental regulation. 10/2000(VI.2.) *Magyar Közlöny* 53, 3156-3167.

Göhre, V., Paszkowski, U., 2006. Contribution of the arbuscular mycorrhizal symbiosis to heavy metal phytoremediation. *Planta* 223(6), 1115-1122.

Gruiz, K., Molnár, M., Fenyvesi, E., Hajdu, Cs., Atkári, A., Barkács, K., 2011. Cyclodextrins in innovative engineering tools for risk-based environmental management. *J. Incl. Phenom. Macrocycl. Chem.* 70(3-4): 299-306.

Guo, Y., George, E., Marschner, H., 1996. Contribution of an arbuscular mycorrhizal fungus to the uptake of cadmium and nickel in bean and maize plants. *Plant Soil* 184, 195-205.

Hall, J.L., 2002. Cellular mechanisms for heavy metal detoxification and tolerance. *J. Exp. Bot.* 53(366), 1-11.

Harley, J.L., Harley, E.L. 1987. A check list of mycorrhiza in the British flora. *New Phytol* 105, 1-102.

Hegedűsova, A., Jakabová, S., Hegedűs, O., Vargová, A., 2009. Use of induced phytoextraction for elimination of cadmium content in soil. *Environ. Biotechn.* 5(2), 51-56.

Javaid, A. 2011. Importance of Arbuscular Mycorrhizal Fungi in Phytoremediation of Heavy Metal Contaminated Soil. In: Khan MS et al. (Eds.), *Biomanagement of Metal-Contaminated Soils* (pp. 125-141.), Springer Kabata-Pendidas, A., Pendidas H. 2001. Trace Elements in Soils and Plants (third Edition) CRC Press LLC, Boca Raton.

Khan, A.G., 2005. Role of soil microbes in the rhizospheres of plants growing on trace metal contaminated soils in phytoremediation. *J. Trace Elem. Med. Biol.* 18, 355-364.

Killham, K., Firestone, M.K., 1983. Vesicular–arbuscular mycorrhizal mediation of grass response to acidic and heavy metal depositions. *Plant Soil* 72:39-48.

Kucey, R.M.N., Janzen, H.H., 1987. Effects of VAM and reduced nutrient availability on growth and phosphorus and micronutrient uptake of wheat and field beans under greenhause conditions. *Plant Soil* 104, 71-78.

Kumar, A., 2012. Role of plant-growth-promoting rhizobacteria in the management of cadmium-contaminated soil. In: A. Zaidi, A., et al. (Eds.), *Toxicity of Heavy Metals to Legumes and Bioremediation* (pp.163-178.), Springer Verlag – Wien.

Kurek, E., Majewska, M., 2012. Microbially mediated transformations of heavy metals in rhizosphere. In: A. Zaidi, A., et al. (Eds.), *Toxicity of Heavy Metals to Legumes and Bioremediation.* (pp.129-146.),Springer Verlag –Wien.

Laureysens, I.A., De Temmerman, L.B., Hastir, T.A., Van Gysel, M., Ceulemans, R., 2005. Clonal variation in heavy metal accumulation and biomass production in a poplar coppice culture. II. Vertical distribution and phytoextraction potential. *Environ. Poll.* 133, 541-551.

Lebeau, T., Braud, A., Jézéquel.,K., 2008. Performance of bioaugmentation- assisted phytoextraction applied to metal contaminated soils: A review. *Environ. Poll.* 153, 497-522.

Leyval, C., Joner, E.J., del Va,l C., Haselwandter, K., 2002. Potential of arbuscular mycorrhizal fungi for bioremediation. In: Gianinazzi S et al. (Eds.), *Mycorrhizal technology in Agriculture* (pp. 175-186.)Birkhäuser Verlag, Switzerland.

Leyval, C., Turnau, K., Haselwandter, K., 1997. Effect of heavy metal pollution on mycorrhizal colonization and function: physiological, ecological and applied aspects. *Mycorrhiza* 7, 139-153.

Lombi, E., Zhao, F.J., Dunham, S.J., McGrath, S.P., 2001. Phytoremediation of heavy metal-contaminated soils: natural hyperaccumulation versus chemically enhanced phytoextraction. *J. Environ. Qual.* 30, 1919-1926.

Lombi, E., Hamon, R.E., Mcgrath, S.P., McLaughin, M.J. 2003. Lability of Cd, cu, and zn in polluted soils treated with lime, beringite and red mud and identification of non-labile colloidal fraction of metals using isotope techniques. *Environ. Sci.Techn.* 37, 979-984.

Madejón, P., Ciadamidaro, L., Marañón, T., Murillo, J.M., 2013. Long-term biomonitoring of soil contamination using poplar trees: Accumulation of trace elements in leaves and fruits. *Int. J. Phytoremed.* 15(6), 602- 614.

Malcova, R., Vosatka, M., Gryndler, M., 2003. Effects of inoculation with *Glomus intraradices* on lead uptake by *Zea mays* L. and *Agrostis capillaris* L. *Appl. Soil Ecol.* 23, 55-67.

Marchiol, L., Fellet, G., Pošić, F., Zerbi, G., 2011. A decade of research on phytoremediation in North-East Italy: Lessons learned and future directions. In: Golubev, I.A., (Ed.), *Handbook of Phytoremediation* (pp. 137-184.).Inc. New York, Nova Science Publishers.

Marschner, H., 1997. The soil-root interface (Rhizosphere) in relation to mineral nutrition. In: Marschner, H. (Ed.), *Mineral nutrition of higher plants* (2 edn). London, Academic Press.

McGrath, S.P., 1998. Phytoextraction for soil remediation. In: Brooks, R.R. (Ed.) *Plants that hyperaccumulate heavy metals: their role in phytoremediation, microbiology, archaeology, mineral exploration and phytomining* (2 edn., pp. 261-287.). New York, CAB International.

McGrath, S.P., Zhao, F.J., Lombi, E., 2001. *Plant and rhizosphere processes* involved in phytoremediation of metal-contaminated soils. *Plant Soil* 232(1-2), 207-214.

McGrath, S.P., Zhao, F.J., Lombi, E., 2002. Phytoremediation of metals, metalloids and radionuclides. *Adv. Agron.* 75, 1-56.

Meier, S., Azcón, R., Cartes, P., Borie, F., Cornejo, P. 2011. Alleviation of Cu toxicity in Oenothera picensis by copper-adapted arbuscular mycorrhizal fungi and treated agrowaste residue. *Appl. Soil. Ecol.* 48(2), 117-124.

Mihalík, J., Tlustoš, P., Szaková, J., 2010. Comparison of willow and sunflower for uranium phytoextraction induced by citric acid. *J. Rad. Nucl. Chem.* 285(2), 279-285.

Narula, N., Reinicke, M., Haferburg, G., Kothe, E., Behl, R.K., 2012. Plant- microbe interaction in heavy- metal-contaminated soils. In: Kothe E., Varma, A., (Eds.), *Bio-Geo Interactions in metal- Contaminated Soils* (pp. 143-162.). Springer-Verlag Heidelberg Neagoe, A.M., Iordache, V., Kothe, E., 2013. Upscaling the Biogeochemical Role of Arbuscular Mycorrhizal Fungi on metal Mobility. In: *Fungi as bioremediators.* oltapeh, E.M., Danesh, E.R., Varma, A. (Eds.), *Soil Biology*, Vol. 32. (pp. 285-311.).Springer-Verlag Berlin Heidelberg.

Newman, L.A., Wang, I.A., Muiznieks, G., Ekuan, G., Ruszaj, M., Cortellucci, R., Domroes, D., Karscig, G., Newman, T., Crampton, R.S., Harshmonay, R.A., Yost, M.G., Heilman, P.E., Duffy, J., Gordon, M.P., Strand, S.E., 1999. Remediation of trichloroethylene in an artificial aquifer with trees: a controlled field study. *Environ. Sci. Techn.* 33, 2257-2265.

Orlowska, E., Godzik, B., Turnau, K. 2012. Effect of different arbuscular mycorrhizal fungal isolates on growth and arsenic accumulation in *Plantago lanceolata* L. *Env. Poll.* 168, 121-130.

Orlowska, E., Jurkiewicz, A., Anielska, T., Godzik, B., Turnau, K. 2005. Influence of different arbuscular mycorrhizal fungal (AMF) strains on heavy metal uptake by *Plantago lanceolata* L. *Pol. Bot. Stud.* 19, 65-72.

Panwar, B.S, Márton, L., Kádár, I., Anton, A., Németh, T., 2010. Phytoremediation: A novel green technology to restor soil health. *Acta Agr. Hung.* 58(4), 443-458.

Prasad, M.N.V., 2004. Phytoremediation in India. In: Willey, N. (ed.) *Methods in Biotechnology*, vol 23: Phytoremediation: Methods and Reviews. 1 edn. Humana press Inc, Totowa.

Prasad, M.N.V., 2005. Nickelophilous plants and their significance in phytotechnologies. *Braz. J. Plant Physiol.* 17(1), 113-128.

Prasad, M.N.V., Freitas, H.M.O., 2003. Metal hyperaccumulation in plants- Biodervesity prospecting for phytoremediation technology. *Electr. J. Biotechn.* 6(3), 285-321.

Rafati, M., Khorasani, N., Moattar, F., Shirvany, A., Moraghebi, F., Hosseinzadeh, S., 2011. Phytoremediation potential of *Populus alba* and *Morus alba* for cadmium, chromuim and nickel absorption from polluted soil. *Int. J. Env. Res.* 5(4), 961-970.

Rahman, M.A., Hasegawa, H., 2011. Aquatic arsenic: phytoremediation using floating macrophytes. *Chemosphere* 83(5), 633-646.

Rascio, N., Navari-Izzo, F., 2011. Heavy metal hyperaccumulating plants: how why do they do it? And what makes them so interesting. *Plant Science* 180(2), 169-181.

Reeves, R.D., Baker, A.J.M., 2000. Metal- accumulating plants. In: Raskin, I., Ensley, B.D., (Eds.), *Phytoremediation of toxic metals: using plants to clean-up the environment* (1 edn., pp. 193-229.). New York, John Wiley and Sons.

Regvar, M., Vogel-Mikus, K., Irgel, N., Wraber, T., Hildebrandt, U., Wilde, P., Bothe, H., 2003. Colonization of pennycresses (*Thlaspi* spp.) of the Brassicaceae by arbuscular mycorrhizal fungi. *J. Plant Physiol.* 160(6), 615-626.

Robinson, B.H., Leblanc, M., Petit, D., Brooks, R.R, Kirkman, J.H., Gregg, P.E.H., 1998. The potential of Thlaspi caerulescens for phytoremediated soils. Plant soil 203, 47-56.

Ruiz, O.N., Alvarez, D., Torres, C., Roman, L., Daniell, H., 2011. Metallothionein expression in chloroplasts enhances mercury accumulation and phytoremediation capability. *Plant Biotechnol. J.* 9(5), 609-617.

Salt, D.E., Pickering, I.J., Prince, R.C., Raskin, I., 1997. Metal accumulation by aquacultured seedlings of mustard. *Environ. Sci. Techn.* 31, 468-471.

Salt, D.E., Smith, R.D., Raskin, I., 1998. Phytoremediation. *Annu. Rev. Plant Physiol. Plant Mol. Biol.* 49, 643-668.

Santos, F.S.J., Hernández-Allica, J.M.B., Becerril, J.M., Amaral-Sobrinho, N., Mazur, N., Garbisu, g., 2006. Chelate-induced phytoextraction of metal polluted soils with Brachiaria decumbent. *Chemosphere* 65(1), 43-50.

Sauvé, S., 2003. The role of chemical speciation in bioavailability. In: Bioavaibility, toxicity and risk relationships in ecosystems. In: Naidu, R. et al. (Eds.), Science publhers, Inc. Enfield, NH, USA.

Schnoor, J.L., 1998. Phytoremediation. Technology Evaluation report TE-98-01. Prepared for Ground-Water Remediation Technology Analysis Center, Pittsburgh.

Simon, L., 2005. Stabilization of metals in acidic mine spoil with amendments, additives and red fescue (*Festuca rubra* L.). *Environ. Geochem. Health* 27, 289-300.

Singh, O.V., Labana, S., Pandey, G., Budhiraja, R., Jain, R.K., (2003) Phytoremediation: an overview of metallic ion decontamination from soil. *Appl. Microbiol. Biotech.* 61, 405-412.

SUMATECS 2009. Sustainable management of trace element contaminated soils-Development of a decision tool system and its evaluation for practical application. Final Research Report. http://www.snowman-era.net.

Szili-Kovács, T., Török, K., Tilston, E.L., Hopkins, D.W., 2007. Promoting microbial immobilization of soil nitrogen during restoration of abandoned agricultural fields by organic additions. *Biol. Fert. Soils.* 43(6), 823-828.

Takács, T., 2012. Site-specific optimization of Arbuscular Mycorrhizal Fungi Mediated Phytoremediation. In: Toxicity of Heavy Metals to Legumes and Bioremediation Potential of rhizosphere microbes Zaidi, A., et al. (Eds.), Springer-Verlag Berlin Heidelberg pp. 179-202.

Takács, T., Radimszky, L., Németh, T., 2005. The arbuscular mycorrhizal status of selected poplar clones for phytoremediation of soils with contaminated heavy metals. *Z. Naturforsch. C.* 60, 357-361.

Takács, T., Biró, I., Németh, T., Vörös, I., 2008. Selection and application of infective and effective AMF strains for phytoremediation of metal contaminated soils. In: Feldmann, F., Kapulnik, Y., Baar, J., (Eds.), *Mycorrhiza works* (pp. 267-277.). Germany: Deutsche Phytomedizinishche Gesellschaft, Braunschweig.

Takács, T., 2012. Site-Specific Optimization of Arbuscular Mycorrhizal Fungi Mediated Phytoremediation. In: A. Zaidi, A., et al. (Eds.), *Toxicity of Heavy Metals to Legumes and Bioremediation* (pp.179-202.).Springer Verlag –Wien.

Terry, N., Carlson, C., Raab, T.K., Zayed, A.M., 1992. Rates of Se volatilization among crop species. *J. Environ. Qual.* 21, 341-344.

Turnau, K., Orlowska, E., Ryszka, P., Zubek, Sz., Gawronski, S., Jurikiewicz, A., 2006. Role of mycorrhizal fungi in phytoremediation and toxicity monitoring of heavy metal rich industrial wastes in southern Poland. In: Twardowska, I., et al. (Eds.), *Soil and Water Pollution monitoring, Protection and Remediation* (pp. 3-23.). Springer.

Van der Heijden, M.G.A., Scheublin, T.R., Brader, A., 2004. Taxonomic and functional diversity in arbuscular mycorrhizal fungi - is there any relationship? *New Phytol.* 164, 201-204.

Vangronsveld, J., Cunningham, S.C., 1998. Metal-contaminated soils: in situ inactivation and phytorestoration. Springer Verlag, Berlin-Heidelberg.

Vassilev, A., Schwitzguebél, J.P., Thewys, T., van der Lelie, D., Vangronsveld, J., 2004. The use of plants for remediation of metal-contaminated soils. *Sci. World* J. 4, 9-34.

Vivas, A., Azcón, R., Biró, B., Barea, J.M., Ruiz-Lozano, J.M. 2003. Influence of two bacterial strains from a Pb-polluted soil on plant growth and on arbscular mycorrhizal functioning under Pb toxicity. *Can. J. Microbiol.* 49, 577-588.

Vogel-Mikus, K., Drobne, D., Regvar, M., 2005. Zn, Cd and Pb accumulation and arbuscular mycorrhizal colonisation of pennycress *Thlaspi praecox* Wulf. (Brassicaceae) from the vicinity of a lead mine and smelter in Slovenia. *Environ. Poll.* 133(2), 233-242.

Vosatka, M., Rydlová, J., Sudová, R., Vohník, M., 2006. Mycorrhizal fungi as helping agents in phytoremediation of degraded and contaminated soils. In: Mackova, M., Dowling, D., Macek, T., (Eds.), *Phytoremediation and rhizoremediation.* (pp. 237-257.) Netherlands, Springer.

Wang, Y., Chao, C.C., 1992. Effects of Vesicular-Arbuscular Mycorrhizae and Heavy Metals on the Growth of Soybean and Phosphate and Heavy Metal Uptake by Soybean in Major Soil Groups of Taiwan. *J. Agric. Assoc. China New Se*r. 157, 247-256.

Wantanabe, M.E., 1997. Phytoremediation on the brink of commercialization. *Environ. Sci. Techn.* 31, 182-186.

Weissenhorn, I., Leyval, C. 1995. Root colonization of maize by a Cd-sensitive and a Cd-tolerant *Glomus mosseae* and cadmium uptake in sand culture. *Plant Soil* 175, 233-238.

Weissenhorn, I., Leyval, C., Berthelin, J. 1993. Cd-tolerant arbuscular mycorrhizal (AM) fungi from heavy-metal polluted soils. *Plant Soil* 157(2), 247-256.

Wenzel, W.W., Adriano D.C., Salt, D., Smith, R., 1999. Phytoremediation: a palnt-microbe based remediation system. In: Adriano, D.C. et al. (Eds.), *Bioremediatin of Conatminated Soils. Agronomy Monograph* no 37 ASA, CSSA, SSSA.

Wu, F..Y., Ye, Z.H., Wu, S.C., Wong, M.H., 2007. Metal accumulation and arbuscular mycorrhiza status in metallicolous and nonmetallicolous populations of *Pteris vittata* L. and *Sedum alfredii* Hance. *Planta* 226(6), 1363-1378.

Zaidi, A., Wani, P.G., Khan, M.S, 2012. Bioremediation: a natural method for the management of polluted environment. In: Zaidi, A, et al. (eds.) *Toxicity of Heavy Metals to legumes and Bioremediation* (pp. 101-114.). Springer-Verlag.

Wien Zhuang, X.L., Chen, J., Shim, H., Bai, Z.H., 2007. New advences in plant-promoting rhizobacteria for bioremediation. *Environ. Int.* 33, 406-413.

In: Bioremediation: Processes, Challenges and Future Prospects ISBN: 978-1-62948-513-3
Editors: Jesús Bernardino Velázquez-Fernández et al.

Chapter 10

SELECTED MICROALGAE FOR PETROLEUM BIOREMEDIATION: TOWARDS A BIO-DEPURATION BASED ON VON NEUMANN-LIKE MACHINES

Julia Romero-Lopez, Victoria Lopez-Rodas and Eduardo Costas*
Genetics, Veterinary Faculty, Complutense University, Madrid, Spain

ABSTRACT

Production, transport and use of petroleum hydrocarbons are sources of pollution worldwide. Consequently, numerous chemical, physical and biological methods have been used to remedy this problem. Here we show an efficient bio-remediation procedure based on the principles of the von Neumann self-replicating machine that is capable of manufacturing a copy of itself while doing its work. Long-term contaminations by petroleum spills occur in the Arroyo Minero River, Argentina. After a crude oil spill made in 1915, crude oil was poured continuously into the Arroyo Minero River. Surprisingly, an abundant microalgae biomass (mainly Scenedesmus sp) is living in contact with crude oil. We isolated these microalgae to be used for petroleum bioremediation. Afterwards, we genetically selected the microalgae by increasing the selection pressures to petroleum-resistance. As intended, two efficient petroleum-removing strains were obtained (So3p and Sr4p). To analyze capability for petroleum degradation, 1 x 10^6 microalgae mL^{-1} were cultured in the presence of 5% petroleum over 15 days (final cell density around 13.4 x 10^6 cells). Afterwards, the petroleum amount was measured in the experiments as well as in controls without microalgae according to 418b EPA and ASTM 3921 procedures. Both microalgal strains were able to efficiently biodegrade petroleum (So3p degraded 50% of petroleum whereas Sr4p degraded 63%). These cells are a promising procedure for petroleum bioremediation based on von Newman's self-replicating machine concept.

Keywords: Microalgae, *Scenedesmus*, petroleum, hydrocarbons, biodegradation, bioremediation

* Corresponding Author address: Email: ecostas@ucm.es.

INTRODUCTION

Petroleum hydrocarbon is the main source of energy in the world. Unfortunately, production, transport, use and disposal of petroleum globally have made it a major contaminant in both prevalence and quantity in the environment (Rahman, 2002). Many attempts to remedy this problem have been tried using chemical, physical and biological methods (Kelkar et al., 2007; Chaillan et al., 2006). Among the biological methods, bacterial bioremediation has been widely used (reviewed by Yang et al., 2009).

Nowadays, large quantities of crude oil are extracted, refined and handled each year, and despite improvements in security, transport and containment, it is possible that these activities may cause accidental spills in aquatic ecosystems. Often, crude oil travels great distances, and some discharges inevitably occur from owned transport vehicles, tanks or pipelines.

In addition to occasional oil spills, due to the routine washing of storage tanks, or errors in handling, some accidental major spills occur. For example: i) the crude oil spill from tanker Prestige in 2002 where around 50,000 tons of crude oil were spilled out onto the Spanish Galician coast (Carrera-Martinez et al., 2010); ii) the spill from an off-shore British Petroleum (BP) oil rig in the Gulf of Mexico. The movements of the stains remain difficult to predict for the latest one as the source of the spill is 1600 meters below sea level, and the winds and currents affect the diffusion of crude oil. (Mariano et al., 2011; Dietrich et al., 2012; Venosa & Zhu, 2003).

The consequences of crude oil spills in water are usually catastrophic. In aquatic ecosystems, its main components, hydrocarbons, can stay for more than 5 years causing toxicity (Lindstrom & Braddock 2002; Kobayashi et al., 2009; Pelletier et al., 2004). The magnitude of the catastrophe will depend on the extent, the amount of oil, how it is degraded chemically and physically and the conditions of their displacement, due to currents and tides.

The methods used to reduced spill impacts numerous attempts to reduce the impacts caused by spills can be classified into three large groups:

i) Physical methods: so far the first used in all disasters that occurred, consisting of the removal of crude oil through its mechanical collection with cranes, with networks or even manually.
ii) Chemical methods: consisting of the action of a chemical agent that degrades the components of crude oil in other components easily removable, less toxic, or that allow waste removal from the landfill through physical methods.
iii) Bioremediation methods: degradation of chemicals is done by living organisms, especially microorganisms to obtain energy, cellular metabolite, and biological waste products (Rahman et al., 2002).

The majority of the organisms used in bioremediation have been single species of bacteria such as *Pseudomonas aeruginosa* (Obayori, 2009; Medina-Bellver et al., 2005; Zhang et al., 2005), *Rhodococcus opacus* (Winkelmann et al., 2009), consortia of bacteria (Komukai-Nakamura et al., 1996; Rahman et al., 2002) or genetically-engineered bacteria. In this regard, these genetically modified bacteria have proved to be more efficient than their corresponding wild bacteria. So far, most attempts to elimination of crude oil are assumed to generate hazardous waste (Incardona et al., 2009; Wang et al., 2009).

What if a material capable of degrading petroleum and contributing to a new food chain could be found?

We propose a new approach for the bioremediation of crude oil spills by using primary producers, i.e., initiators of trophic chains, so they do not need to be removed from the aquatic ecosystem once used. In addition, the use of photosynthetic organisms for crude oil degradation reduces the greenhouse effect. The organisms chosen for this purpose were microalgae able to live in environments contaminated with hydrocarbons, specifically *Scenedesmus obtusus* and *Scenedesmus sp*, which are able to efficiently biodegrade petroleum hydrocarbons.

METHODS

We studied an interesting example of extreme contamination by crude oil in the Arroyo Minero river, Río Negro state, Argentina, where an oil exploration has been pouring crude oil since 1915 (Lopez-Rodas et al., 2009). Two microalgal species living in direct contact with the crude oil spill were isolated and grown under laboratory conditions. They were the unicellular chlorophyta *Scenedesmus obtusus* (strain So3P) and *Scenedesmus sp* cf. *S. rapaporti* sp. nov. (strain Sr4P).

The strains obtained from *Scenedesmus obtusus* and *Scenedesmus sp* were cultivated at a temperature of 22°C under continuous light of 80 μmol photons m^{-2} s^{-1} using day-light fluorescent tubes (Phillips TLD 36W/33, France), 400-700 nm wavelength (photosynthetically active light). The cells grew in Pyrex flasks of 250 mL culture (Barloworld Scientific Ltd, Stone, U.K.) with BG-11 medium for inland waters microalgae (Sigma Aldrich Chemie, Taufkirchen, Germany), filtered through a membrane sterile 0.22 μm (Express Plus Membrane, Millipore Iberica S.A.). Cultures were maintained axenically under exponential growth by serial transfers of aliquots to new BG-11 medium once a month. The absence of bacteria in the cultures was confirmed by a fluorescence microscope after acridine orange staining.

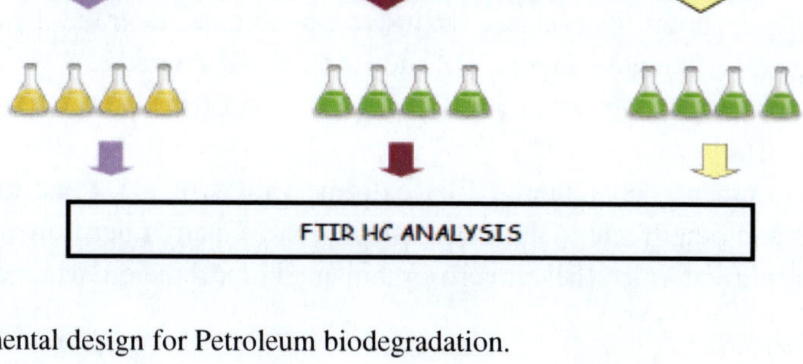

Figure 1. Experimental design for Petroleum biodegradation.

Both strains (So3P and Sr4P) were maintained under laboratory conditions that allow a balance between strong selection intensity, by means of a 1% addition of petroleum to the culture medium and the maintenance of a population size large enough to increase probability of rare spontaneous mutations that confer petroleum adaptation (Romero-Lopez et al., 2012).

The aim of this study was to determine the capability of both strains to degrade hydrocarbons from petroleum. For this purpose, four experiments were established with a 5% Fluka Analytical Petroleum special standard with the following experimental design (Figure 1).

Four controls containing 5% Fluka Analytical Petroleum special standard with BG-11, as well as 2 series (one for each microalgae strain) of four replicates each, containing 50 mL of microalgae biomass (10^6 cells /mL), 45 mL of BG-11 (Sigma Aldrich Chemie, Taufkirchen, Germany) and 5 mL of Fluka Analytical Petroleum Special were prepared. Preparations were sonicated (4 pulses every 20 seconds, with a power of 40 watts and at a frequency of 16 kHz) with vibrating Cell (Sonics & Materials Inc., Danbury CT, USA) prior to microalgal addition to assure fuel homogenization. After the microalgae addition, controls and samples were incubated in a culture chamber for 15 days at 22°C and 80 μmol photons m^{-2} s^{-1} of continuous light. Afterwards, the samples were filtered using a 0.22 μm filter to remove algal biomass and any other solid matter.

Obtained aqueous extracts were analyzed according to EPA 418b (Petroleum Hydrocarbons (Spectrophotometric, Infrared) and ASTM 3921 (Standard Test Method for Oil and Grease and Petroleum Hydrocarbons in Water) methods, for measuring the hydrocarbon degradation rate.

Results and Discussion

While anthropogenic crude oil spills usually have dramatic effects on wildlife, microalgae are able to proliferate in the mix of crude oil and fresh water (9.2ºC, pH=8.1, 11.71% v/v hydrocarbons/water) that emerged from the underground in the Arroyo Minero (Lopez-Rodas et al., 2009, Carrera-Martinez et al., 2011).

After isolation of two crude oil-resistant microalgal strains from Arroyo Minero, they were maintained for four years under petroleum exposure in laboratory conditions for two reasons: i) to assure a strong selection for crude oil resistance (by the addition of petroleum to culture medium) and, ii) to increase the probability of petroleum-resistant mutants occurrence (by maintaining a very large population). The result was that both strains (So3P and Sr4P) proliferated successfully under high concentrations (5%) of petroleum.

As controls show, crude oil does not degrade spontaneously under experimental conditions. However, both microalgae strains were able to degrade hydrocarbons from petroleum with great efficiency (Figure 2). The strain So3P degraded 49% of the petroleum standard in 15 days, whilst the strain Sr4P degraded 74% of the petroleum standard in that same time (Figure 2).

Extreme environments, characterized by extreme values of pH, toxic materials, high or low temperatures and other factors, allow the adaptation of many micro-organisms to unusual conditions. The proliferation of these micro-organisms in extreme environments is motivated by the natural selection of spontaneous mutants which occurs randomly and without influence of the selective agent.

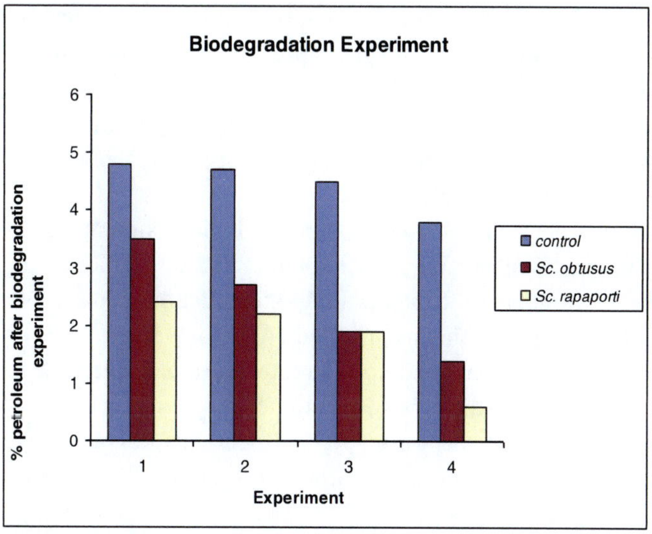

Figure 2. Data of biodegradation of crude oil by microalgae.

Consequently, most of the crude oil-degrading species have been found in the vicinity of petroleum discharges (Yang et al., 2009; Dasa et al., 1999; Mackew et al., 2007; Zinjarde & Pant, 2002). Usually, they are bacterial species but a significant microalgal biomass occurs in the most polluted areas of Arroyo Minero.

Our results indicate that the microalgae *Scenedesmus* strains So3P and Sr4P could be an interesting alternative for petroleum spill remediation.

In general, physical crude oil elimination methods applied to the crude oil and other fuel spills are complex to use, due to expansions of spills and remediation of such responsiveness (Trejo-Hernández et al., 2007; Hughey et al., 2007; Nikolopouou & Kalogerakis 2008; Chaîneau et al., 2005). In addition, these methods are only effective in the early stages after the spill, and they are generally problematic because the collected waste is toxic and its storage is difficult. Chemical methods, in the same way, have serious problems. Used dispersants can be an effective alternative, but they are very selective agents, in which the ability to act, i.e., the chemical reaction produced, require specific conditions of temperature and kinetics to provide adequate yields (Lindstrom et al., 2002; Li et al., 2007). Even in the best cases, hydrocarbons, and the crude oil itself, are not completely degraded or are transformed into toxic products with catastrophic results. Many experts argue that the combination of dispersants and their own aggregates can be more toxic for the ecosystem than the crude itself only. After the last spill in the Gulf of Mexico, the EPA did studies that showed that dispersants are not a good alternative to alleviate the effects of discharges (Li et al., 2007; Chaineau et al., 2005; Kelkar et al., 2007; Saeki et al., 2009; Stockstad, 2010).

Biodegradation seems to be an effective procedure to remove the oil from oil spills in aquatic environments. In cases of greater discharges, as in the case of the ship Exxon Valdez spill in 1989, or the Prestige in 2002, around 50% of the crude oil present in the water was biodegraded during the first three years after the spill (Vega et al., 2009). Several works have been performed in biodegradation of crude oil (Kelkar et al., 2007; Obuekwe et al., 2009; Reddy & Quinn, 1999). This biodegradation can be accelerated by the addition of nutrients to the environment to stimulate the growth of populations of degrader microorganisms (Chaineau et al., 2005; Atlas, 1995). Bioremediation can be a useful technique for discharges of light crude oils, while heavy crude oil spills require physical methods of absorption before bioremediation (Mohajeri et al., 2010). The decision to employ physical or chemical methods with bioremediation in a discharge of crude oil in aquatic environments depends on the spill area, physical conditions, the protection habitat objectives and any other factor that could have an impact on the selected process.

The election of microorganisms used in bioremediation is complex. Most of the studies were carried out with bacterial organisms (Chaillan et al., 2006; Ijah et al., 2003; Rahman et al., 2002) but use of non-native bacteria in biodegradation could start problems in the ecosystems where these microorganisms are used (Anderson et al., 2009; Wang et al., 2009). Sometimes bacteria used in bioremediation treatment are hazardous for native species, making it necessary to remove them once the toxic agent is degraded (Saeki et al., 2009). Several authors have stated that the wastes of hydrocarbons degraded by bacteria are more toxic to aquatic organisms (fish, shrimps and mussels embryo) than expected (Pelletier et al., 2004; Dasa et al., 2007). In addition to this, the reserves of oxygen and nutrients available in the ecosystem are altered by the presence of large amounts of bacteria stimulated by the use of hydrocarbons as energy sources (Chaineau et al., 2005).

In this regard, if we consider microalgae for crude oil degradation, we find important advantages over the use of bacteria. The bacterial degradation products from crude oil spills are less complex than those of microalgae and in some cases are even toxic (Pelletier et al., 2004). Microalgae can degrade hydrocarbons from crude oil very efficiently. A massive growth of the microalgae population has fewer problems than those of bacterial growth because microalgae constitute the most important group of primary producers in the food chain from aquatic ecosystems.

CONCLUSION

1. Both microalgal *Scenedesmus* strains (So3P and Sr4P) isolated from the crude oil spill from Arroyo Minero resist very high petroleum concentrations under laboratory conditions.
2. Both strains degrade crude oil very efficiently (Sr4P up to 74% and So3P until 49% of the total petroleum after 15 days).
3. Using microalgae for crude oil bioremediation has significant advantages in efficiency and impact on the environment.

REFERENCES

Anderson, B.S., Arenella-Parkenson, D., Phillips, B., Tjeerdema, R., Crane, D., 2009 Preliminary investigation of the effects of dispersed Prudhoe Bay Crude Oil on developing topsmelt embryos, Atherinops affinis. *Environmental Pollution,* 157, 1058-1061.

Atlas, R.M., 1995. Petroleum biodegradation and oil spill bioremediation. *Mar. Pollut. Bull.,* 31, 178-182.

Carrera-Martinez, D., Mateos, A., Lopez-Rodas, V. & Costas, E., 2010. Microalgae response to petroleum spill: An experimental model analysing physiological and genetic response of Dunaliella tertiolecta (Clorophyceae) to oil samples from the tanker Prestige. *Aquat. Toxicol.,* 97, 151-159.

Carrera-Martinez, D., Mateos, A., Lopez-Rodas, V., Costas, E., 2011. Adaptation of microalgae to a gradient of continuous petroleum contamination *Aquat. Toxicol.,* 101, 342-350.

Chaillan, F., Gugger, M., Saliot, A., Coute, A. & Oudot, J., 2006. Role of cyanobacteria in the biodegradation of crude oil by a tropical cyanobacterial mat. *Chemosphere,* 62, 1574-1582.

Chaineau, C.H., Rougeux, G., Yepremian, C. & Oudot, J., 2005. Effects of nutrient concentration on the biodegradation of crude oil and associated microbial populations in the soil. *Soil. Biol. Biochem.,* 37, 1490-1497.

Dasa, K. & Mukherjee, A. .1999. Crude petroleum-oil biodegradation by naturally Inhabiting Mixed Bacterial Culture under Different Environmental Factors. *Pak. J. Med. Sci.,* 4, 1211-1218.

Dasa, K. & Mukherjee, A. 2007. Crude petroleum-oil biodegradation efficiency of Bacillus subtilis and Pseudomonas aeruginosa strains isolated from a petroleum-oil contaminated soil from Northeast India. *Bioresource Technol.,* 98, 1339-1345.

Dietrich, J.C., Trahan, C.J., Howard, M.T., Fleming, J.G., Weaver, R.J., Tanaka, S., Yu, L., Luettich, R.A., Dawson, C.N., Westerink, J.J., Wells, G., Lu, A., Vega, K., Kubach, A., Dresback, K.M., Kolar, R.L. , Kaiser, C., Twilley, R.R.. 2012. Surface trajectories of oil transport along the Northern Coastline of the Gulf of Mexico. *Cont. Shelf. Res.,* 41, 17–47.

Hughey, C., Galasso, S.A. & Zumberge, J.E.. 2007. Detailed compositional comparison of acidic NSO compounds in biodegraded reservoir and surface crude oils by negative ion electrospray Fourier transform ion cyclotron resonance mass spectrometry. *Fuel,* 86, 758-768.

Ijah, U.U.J. & Antai, S.P. 2003. Removal of Nigerian light crude oil in soil over a 12-month period. *Int. Biodeter. Biodegr.,* 51, 93-99.

Incardona, J., Carls, M., Day, H.L., Sloan, C., Bolton, J., Collier, T. & Scholz, N. Cardiac 2009. Arrhythmia Is the Primary Response of Embryonic Pacific Herring (Clupea pallasi) Exposed to Crude Oil during Weathering. *Environ. Sci. Technol.,* 43, 201–20.

Kelkar, D.S., Kumar, A.R. & Zinjarde, S.S. 2007. Hydrocarbon emulsification and enhanced crude oil degradation by lauroyl glucose ester. *Bioresource Technol.,* 98, 1505-1508.

Kobayashi, T., Murai, Y., Tatsumi, K. & Iimura, Y. 2009. Biodegradation of polycyclic aromatic hydrocarbons by Sphingomonas sp. enhanced by water-extractable organic matter from manure compost. *Sci. Total. Environ,* 407, 5805–5810.

Komukai-Nakamura, S., Sugiura, K, Yamauchi-Inomata, Y., Toki, H., Venkateswaran, K., Yamamoto, S., Tanaka, H. & Harayama, S. 1996. Construction of bacterial consortia that degrade Arabian crude oil. *J. Ferment. Bioeng.,* 82, 570-574.

Li, Z., Kepkay, P., Lee, K., King, T., Boufadel, M.C. & Venosa, A. 2007. Effects of chemical dispersants and mineral fines on crude oil dispersion in a wave tank under breaking waves. *Mar. Pollut. Bull.,* 54, 983-993.

Lindstrom, J.E. & Braddock, J.F.. 2002. Biodegradation of petroleum hydrocarbons at low temperature in the presence of the dispersant Corexit 9500. *Mar. Pollut. Bull.,* 4, 739-747.

López-Rodas, V. Carrera-Martínez,, D., Salgado, E., Mateos-Sanz, A., Báez, J.C. & E Costas, E. 2009. A fascinating example of microalgal adaptation to extreme crude oil contamination in a natural spill in Arroyo Minero, Río Negro, Argentina. *An. R. Acad. Nac. Farm.,* 75, 883-899.

Mariano, A.J. , Kourafalou, V.H. , Srinivasan, A., Kang, H., Halliwell, G.R., Ryan, E.H., Roffer, M.. 2011. On the modeling of the 2010 Gulf of Mexico Oil Spill. *Dynam. Atmos. Oceans*, 52, 322– 340.

Mckew, B., Coulon, F., Yakimov, M., Denaro, R., Genovese, M., Smith, M., Cindy, J., Osborn, A.M., Timmis, K. & McGenity, T. 2007. Evidence for in situ crude oil biodegradation after prestige oil spill. *Environ. Microbiol.*, 9, 1562-1571.

Medina-Bellver, J., Marin, P., Delgado, A., Rodriguez-Sanchez, A., Reyes, E., Ramos, J.L. & Marques, S. 2005. Efficacy of intervention strategies for bioremediation of crude oil in marine systems and effects on indigenous hydrocarbonoclastic bacteria. *Environ. Microbiol.*, 7, 773-779.

Mohajeri, L., Aziz, H., Isa, M. & Zahed, M. 2010. A statistical experiment design approach for optimizing biodegradation of weathered crude oil in coastal sediments. *Bioresource Technol.,* 101, 893-900.

Nikolopoulou, M. & Kalogerakis, N. 2008. Enhanced bioremediation of crude oil utilizing lipophilic fertilizers combined with biosurfactants and molasses. *Mar. Pollut. Bull.,* 56, 1855-1861.

Obayori, O., Adebusoye, S., Adewale, A., Oyetibo, G., Oluyemi, O., AMokun, R. & Ilori, M. 2009. Differential degradation of crude oil (Bonny Light) by four Pseudomonas strains. *J. Environ. Sci.,* 21, 243-248.

Obuekwe, C., Al-Jadi, Z. & Al-Saleh, E. 2009. Hydrocarbon degradation in relation to cell-surface hydrophobicity among bacterial hydrocarbon degraders from petroleum-contaminated Kuwait desert environment. *Int. Biodeter. Biodegr.,* 63, 273-279.

Pelletier, E., Delille, D. & Delille, B. 2004. Crude oil bioremediation in sub-Antarctic intertidal sediments: chemistry and toxicity of oiled residues. *Marine Environmental. Research,* 57, 311-327.

Rahman, K. Thahira-Rahman, J., Lakshmanaperumalsamy, P. & Banat, I.M. 2002. Towards efficient crude oil degradation by a mixed bacterial consortium. *Bioresource Technology,* 85, 257-261.

Reddy, C.M. & Quinn, J.G. 1999. GC-MS analysis of total petroleum hydrocarbons and polycyclic aromatic hydrocarbons in seawater samples after the North Cape oil spill. *Mar. Pollut. Bull.,* 38, 126-135.

Romero-Lopez, J., Lopez-Rodas,V., Costas, E.. 2012. Estimating the capability of microalgae to physiological acclimatization and genetic adaptation to petroleum and diesel oil contamination. *Aquat. Toxicol.,* 124–125, 227-237.

Saeki, H., Sasaki, M., Komatsu, K., Miura, A. & Matsuda, H. 2009. Oil spill remediation by using the remediation agent JE1058BS that contains a biosurfactant produced by Gordonia sp. strain JE-1058. *Bioresource Technol.,* 100, 572-577.

Stokstad, E. 2010. Dispersant Study Released by EPA, But Big Questions Remain. *Science,* Jun 30.

Trejo-Hernández, M.R., Ortiz, A., Okoh, A.I., Morales, D. & Quintero, D. 2007. Biodegradation of heavy crude oil Maya using spent compost and sugar cane bagasse wastes. *Chemosphere,* 68, 848-855.

Vega, F., Covelo, E., Reigosa, M. & Andrade, M.L.. 2009. Degradation of fuel oil in salt marsh soils affected by the Prestige oil spill. *J. Hazard. Mater,* 166, 1020-1029.

Venosa, A.D. & Zhu, X. 2003. Biodegradation of Crude Oil Contaminating Marine Shorelines and Freshwater Wetlands. *Spill Sci. Technol. B,* 8, 163-178.

Wang, Y., Zhou, Q., Peng, S., Ma, Q. & Niu, X. . 2009. Toxic effects of crude-oil-contaminated soil in aquatic environment on Carassius auratus and their hepatic antioxidant defense system. *J. Environ. Sci.,* 21, 612-617.

Winkelmann, M., Hunger, N., Hüttl, R. & Wolf, G. . 2009. Calorimetric investigations on the degradation of water insoluble hydrocarbons by the bacterium Rhodococcus opacus 1CP. *Thermochimica Acta,* 482, 12-16.

Yang, S., Jin, H., Wei, Z., He, R., Ji, Y., Li, X. & Yu, S. . 2009. Bioremediation of Oil Spills in Cold Environments: A Review. *Pedosphere,* 19, 371-381.

Zhang, G.,Wu, Y. Qian, X. & Meng, Q.. 2005. Biodegradation of crude oil by Pseudomonas aeruginosa in the presence of rhamnolipids. *J. Zhejiang Univ-Sc. B Science,* 6, 725-730.

Zinjarde, S.S. & Pant, A.A. 2002. Hydrocarbon degraders from tropical marine environments. *Mar. Pollut. Bull.,* 44, 118-121.

SECTION 4: COMPOSITE BIOREMEDIATION

In: Bioremediation: Processes, Challenges and Future Prospects ISBN: 978-1-62948-513-3
Editors: Jesús Bernardino Velázquez-Fernández et al.

Chapter 11

HYDROQUINONE: PROPERTIES AND MICROBIAL DEGRADATION PROCESSES

***Francisco J. Enguita*[1] *and Ana Lúcia Leitão*[2]**
[1] Instituto de Medicina Molecular,
Faculdade de Medicina da Universidade de Lisboa, Lisboa, Portugal
[2] Departamento de Ciências e Tecnologia da Biomassa,
Faculdade de Ciências e Tecnologia, Universidade Nova de Lisboa,
Quinta da Torre, Caparica, Portugal

ABSTRACT

Hydroquinone (1,4-dihydroxybenzene) is commonly used in human activities as well as in the rubber and food industry. Hydroquinone is frequently detected in aquatic ecosystems, as it could be partially removed or generated during the wastewater treatment process. This phenolic compound is considered an environmental pollutant and its removal is of great concern because of the high toxicity effect on living organisms. Hydroquinone can also be produced as a pathway metabolite during phenol, chlorophenol and benzene biotransformation. Furthermore, it can be auto-oxidized to form 1,4-benzoquinone, a product that has higher toxicity than the parent compound. The inherent versatility of microorganisms allows them to participate in pollutant transformations. Several microorganisms catalyze mineralization and/or hydroquinone transformation, being an attractive tool for removing this pollutant from the environment and reduce chemical toxicity. This review underscores the mechanisms of hydroquinone biotransformation and the role of microorganisms and their enzymes in this process, with special focus on fungi.

Keywords: Hydroquinone, bioremediation, microbial catabolic pathways, enzymes

INTRODUCTION

Rapid industrialization and the use of chemicals in many areas of our daily lives have resulted in the release of potentially toxic compounds into the water bodies, soils and atmosphere. In the last decade, environmental deterioration has been one of the largest concerns of science community and general public. As a consequence several reviews have been published describing environmental exposure and toxicity of hazardous chemicals, as well as the past, present, and future trends of remediation processes. If there are substances that are well documented, for instance benzene, PAH and heavy metals, there are others like hydroquinone that recent research on its possible toxic effect made it of particular concern (Yang et al., 2011; Hebeda et al., 2012; Lee et al., 2012; Shimada et al., 2012).

Hydroquinone, the most widely distributed representative of simple phenols, is a ubiquitous chemical in the environment, used in human and industry activities. For example, it is used as a developing agent in photography, dye intermediate, stabilizer in paints, varnishes, oils and motor fuels. It is also present in cigarette smoke. Additionally, hydroquinone could be an antioxidant in the rubber and food industry. From 1950s to 2001 hydroquinone has been used in commercially available cosmetic skin lightening formulations in European Union countries and since 1960s it was also marketed as a medical product. This dihydroxybenzene is also used in cosmetic formulations of products for coating finger nails and hair dyes (O'Donoghue, 2006). Meanwhile, hydroquinone occurs naturally as a conjugate with β-D-glucopyranoside in the fruit, bark and leaves of several plants, particularly the ericaceous shrubs such as bearberry, cranberry, among others. It has been isolated at low levels in coffee, red wine, wheat cereals and broccoli (Deisinger et al., 1996).

Hydroquinone is also the intermediate of diverse metabolic pathways during the biodegradation or chemical processes of several aromatics. It is a major benzene metabolite, which is produced after benzene biotransformation by peroxidase that convert phenol to hydroquinone, mainly in the lungs and liver and it is accumulated in the bone marrow (Subrahmanyam et al., 1991). On the other hand, advanced oxidation processes (APOs) of aromatic compounds particularly of phenol, yields several aromatic compounds, such as hydroquinone, catechol and resorcinol, as intermediate metabolites of its transformation. The formation of hydroquinone and ρ-benzoquinone at early stages of phenol oxidation increases the toxicity of phenol wastewaters, showing that these compounds were more toxic and less degradable than the original pollutant (Santos et al., 2004). This increment of toxicity was also detected in the oxidative degradation of hydroquinone under a supercritical condition (409.9 ºC and 24.5 MPa) and subcritical condition (359.9 ºC and 24.5 MPa), where the p-benzoquinone is an important intermediate (Thammanayakatip et al., 1998). Chemical processes like APOs processes have their inherent limitations, as for example, low rates of degradation, lower mineralization or severe operating conditions or even high costs of operation among others (Mahamuni and Pandit, 2006). Polymeric adsorbents, such as activated carbon cloth and granular activated carbon, are suitable for treatment of industrial effluents containing hydroquinone (Ayranci and Duman, 2005; Suresh et al., 2011). However, these adsorbents are only able to remove few milligrams of hydroquinone per gram of adsorbent, and regeneration process is not easy, which makes them a non-economic process. Biological transformations are therefore generally preferred for being considered as more economical and environmentally friendly. Microbial versatility is outstanding either in its tolerance for extreme conditions or by the capability to adapt its enzymatic system to the environment challenges (Tortella et al., 2005).

Although phenol derivatives biodegradation has been investigated in the past years, there is still considerable interest in the metabolic capacity of microorganisms that are able to use these aromatic compounds. Hydroquinone has been shown to be the major recalcitrant dihydroxybenzene. Meanwhile, there are several yeasts and yeastlike fungi that are able to grow on hydroquinone, such as *Cryptococcus* sp., *Trichosporon* sp., *Exophiala jeanselmei, Candida* sp., *Geotrichum klebahnii, Stephanoascus* sp., *Myxozyma geophila* (Middelhoven, 1993). Fungi like *Aspergillus fumigates, Candida parapsilosis, Tyromyces palustris, Gloeophyllum trabeum, Penicillium chrysogenum* and *Phanerochaete chrysosporium* are examples of microorganisms able to degrade hydroquinone (Jones et al., 1994; Eppink et al., 2000; Kamada et al., 2002; Leitão et al., 2007; Nakamura et al., 2012). However, few reports are available on biodegradation of hydroquinone by pure culture of bacteria under anaerobic conditions. Among them, the anaerobic degradation by a fermenting bacterium the strain HQGö1 (Szewzyk and Schink, 1989) and by a sulfate-reducing bacterium identified as a subspecies of *Desulfococcus multivorans* (Schnell et al., 1989).

We are aware that the environmental persistence of hydroquinone is not high, but this compound has relevance to aquatic systems, since it is considered to have high acute toxicity for aquatic organisms (see Table 1). Moreover, the transport and distribution of hydroquinone in water represents 99.6% (OECD, 2012). Industrial wastewater containing hydroquinone was conventionally treated by chemical methods using $FeSO_4$ and H_2O_2 to oxidize it into CO_2 and H_2O, meanwhile ferric sulfate is originated from $FeSO_4$ enabling recycled reactants. In alternative, the photocatalytic removal of aqueous hydroquinone by TiO_2/AC does not presented regeneration of catalyst or major technical problems. However, in order to achieve high removal efficiency, hydrogen peroxide should be added in correct proportion, and this is a clear obstacle of the process (Geng et al., 2008). Therefore, it is important to find alternatives to chemical methods; bioremediation seems to be a promising technology. Aspects related to hydroquinone transformation in terms of pathways and enzymes are the focus of our review.

PROPERTIES OF HYDROQUINONE

Hydroquinone, synonymous of 1,4-benzenediol, p-benzenediol or p-dihydroxybenzene, is an aromatic compound consisting of the benzene ring and two –OH groups at *para* position. It is available in the form of white crystals, but industrial use grades may be light grey or light tan. Contact with air and light causes oxidation and darkening of color. With a molecular weight of 110.11 g/mol, hydroquinone present several physical-chemical characteristics that are resumed as followed. The specific gravity is 1.332 g/mL at 15 ºC, the boiling point is 286 ºC, and the melting point is between 169-174 ºC. Its solubility in water ranged from 70-73 g/L at 25 ºC and the vapor pressure between 0.000018-0.000019 mm Hg at 25 ºC. The log *n*-octanol/water partition coefficient (log K_{ow}) is between 0.50-0.61. The pK_a values are 9.9 and 11.6. It is a reducing agent with a standard electrode potential of half reaction of 0.714 V (Song et al., 2006).

Hydroquinone is obtained from three chemical processes: 1) oxidation of aniline with manganese dioxide in acidic conditions, followed by reduction with iron dust in aqueous medium; 2) oxidation of phenol produce aromatic compounds, designated as phenolic compounds, one of these is hydroquinone; 3) alkylation of benzene with propylene to originate the *para* diisopropylbenzene isomer, beside others isomers, which is oxidized and produce the corresponding dihydroperoxide, that subsequently is treated with an acid to originate hydroquinone (Suresh et al., 2012).

Table 1.

Ecotoxicology	Species	Results	Reference
Toxicity to Marine Bacteria	*Vibrio fischeri* (also known as *Photobacterium phosphoreum)*	EC_{50} (15 min) = 0.041 mg/L	(Santos et al., 2004)
Acute/Prolonged Toxicity to Aquatic Invertebrates *Daphnia*	*Daphnia magna*	LC_{50} (24 h) = 0.090 mg/L LC_{50} (24 h) = 0.12 mg/L LC_{50} (24 h) = 0.050 mg/L	(OECD, 2012)
Toxicity to Aquatic Plants e.g. Algae	*Selenastrum capricornutum* *Pseudokirchneriella subcapitata* *Monoraphidium braunii* *Desmodesmus armatus* *Synechocystis* sp. *Nostoc* sp. *Microcystis aeruginosa*	LC_{50} (7 d) = 1.0-4.0 mg/L EC_{50} (48 h) = 8.9 mg/L EC_{50} (48 h) = 11.3 mg/L EC_{50} (48 h) = 1.5 mg/L EC_{50} (48 h) = 0.68 mg/L EC_{50} (48 h) = 0.21 mg/L EC_{50} (48 h) = 0.054 mg/L	(OECD, 2012) (Bahrs et al., 2013)
Acute/Prolonged Toxicity to Fish	*Pimephales promelas* *Brachydanio rerio*	LC_{50} (96 h) $\geq$ 0.40 mg/L LC_{50} (96 h)= 0.17 mg/L	(OECD, 2012)

EC_{50}, hydroquinone concentration inhibiting 50% of growth or luminescence properties
LC_{50}, hydroquinone concentration killing 50% of population

BIODEGRADATION AND BIOTRANSFORMATION OF HYDROQUINONE

Microbial degradation of hydroquinone has been reported by several research groups (Schnell et al., 1989; Szewzyk and Schink, 1989; Jones et al., 1994; Kamada et al., 2002). Hydroquinone is subject to abiotic and biotic transformations, depending on factors such as oxygen, light, microorganism(s) among others.

Hydroquinone was a product obtained from metabolism of several aromatic compounds. Example of them is the benzaldehyde and benzoic acid metabolism by the brown-rot basidiomycetes *Tyromyces palustris* and *Gloeophyllum trabeum.* The study conducted by Kamada and others showed that hydroquinone was metabolized but no formation of products was observed. The same authors reported the effective mineralization of aromatic compound by the brown-rot fungi using radioactive substrates (Kamada et al., 2002). The hydroxylated intermediate was also found as product in combined phenol and glucose cultures of *Penicillium chrysogenum* strain. During the metabolism of phenol, hydroquinone was accumulated in the early stages of incubation and disappeared after 80 h of culture, indicating that hydroquinone was a metabolic intermediate but it is not a dead-end product (Leitão et al., 2007). Jones et al. (1994) proposed the pathway for degradation of 4-ethylphenol by *Aspergillus fumigates* where hydroquinone was obtained by hydroxylation of 4-hydroxphenyl acetate. This aromatic compound undergoes further hydroxylation to form 1,2,4-trihydroxybenzene followed by ring fission substrate to produces maleylacetate (Jones et al., 1994).

Several aerobic bacteria are capable of utilizing hydroquinone as a product compound obtained from other substrate, involving a hydroquinone 1,2-dioxygenase to convert hydroquinone into 4-hydroxymuconic semialdehyde. Degradation of 4-chlorophenol via hydroquinone pathway has been reported for several strains belonging to the actinobacterium group. *Arthrobacter ureafaciens* CPR706 first eliminates the chloro-substituent to form hydroquinone. The CPR706 strain also degrades other *para*-substituted phenols, including 4-fluoro, 4-bromo, 4-iodo and 4-nitrophenol via the hydroquinone pathway (Bae et al., 1996). The hydroquinone pathway is also used in *para*-nitrophenol degradation of gram-negative bacteria such as *Moraxella* sp. (Spain and Gibson, 1991), *Pseudomonas* sp. strain WBC-3 (Zhang et al., 2009) and *Pseudomonas* sp. 1-7 (Zhang et al., 2012). Another example is the catabolism of 4-hydroxyacetophenone by *Pseudomonas fluorescens* ACB that proceeds through the initial formation of 4-hydroxyphenyl acetate and hydroquinone (Higson and Focht, 1990; Kamerbeek et al., 2001; Moonen et al., 2008). Meanwhile, a second pathway branch where established for 4-chlorophenol transformation into hydroquinone, which in turn was hydroxylated to originate hydroxyquinol (Nordin et al., 2005).

Degradation of hydroquinone under anaerobic conditions has been reported. Schnell et al. (1989) described a denitrifying bacterium capable to grow with hydroquinone (Schnell et al., 1989). In the same year, Szewzyk and Schink proposed the carboxylation of hydroquinone to gentisate as the first reaction in anaerobic degradation by a strictly anaerobic fermenting bacterium strain HQGö1. This strain converted hydroquinone to acetate and benzoate, and only traces of hydrogen were formed. Phenol was never found as intermediate or end product of hydroquinone catabolism by strain HQGö1 (Szewzyk and Schink, 1989). Later, Gorny and Schink (1994) proposed in their findings a pathway for hydroquinone degradation via reductive dehydroxylation of gentisyl-CoA by the fermenting strain HQGö1. Addition of $NaHCO_3$ was found to accelerate the rate of hydroquinone degradation by dense cell suspensions of HQGö1 (Gorny and Schink, 1994).

ENZYMES INVOLVED IN HYDROQUINONE DEGRADATION AND METABOLISM

Biodegradation of hydroquinone can be achieved either by an aerobic or by an anaerobic mechanism. Anaerobic metabolization of hydroquinone is a less extended process in nature, mainly restricted to a specific group of sulfate-reducing bacteria. It involves the conversion of hydroquinone to benzoate with an intermediate carboxylation, and activation of the products by their binding coenzyme A (Fig. 1). Cells can either employ benzoate as an anabolic brick or introduce the CoA-activated metabolites in the beta-oxidative pathway to be used as sources of energy.

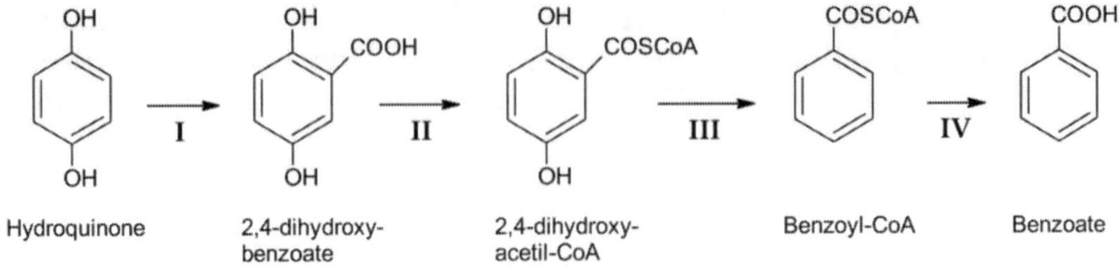

Figure 1. Anaerobic pathway for the metabolization of hydroquinone. I: hydroquinone carboxylase; II: hydroquinone Acetil-CoA transferase; III: benzoyl-CoA oxidoreductase; IV: benzoyl-CoA hydrolase.

In aerobic conditions, hydroquinone undergoes the beta-ketoadipate pathway through two different pathways (Fig. 2). The first pathway involves the initial hydroxylation of hydroquinone to 1,2,4-trihidroxy-benzene followed by a ring-fission reaction catalyzed by a 1,2-dioxygenase. The second pathway of hydroquinone degradation is less frequent. In this pathway, hydroquinone ring is directly cleaved by a specific hydroquinone 1,2-dioxygenase and the generated semialdehyde oxidized to maleyl-acetate. The first aerobic branch has been characterized in bacteria and fungi; meanwhile, the second is exclusive of prokaryotic organisms.

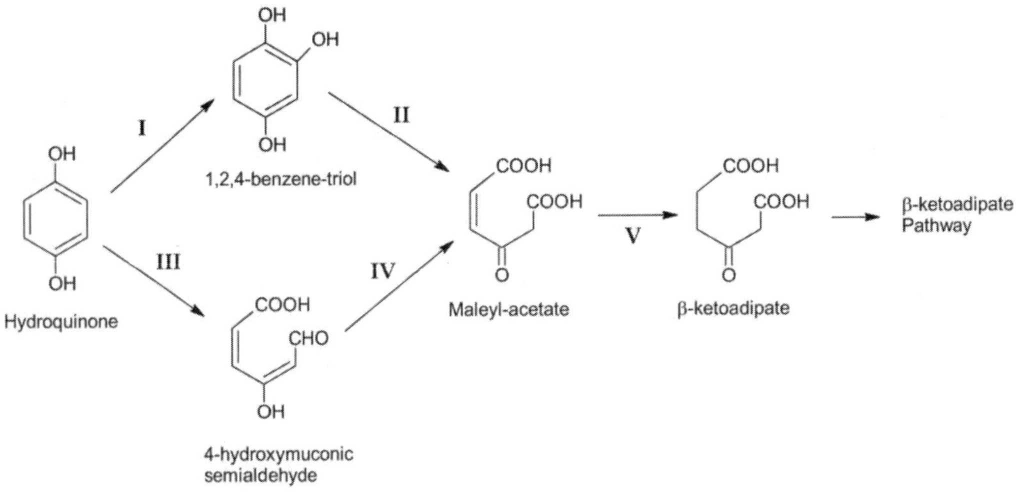

Figure 2. Two different branched pathways for the biodegradation of hydroquinone under aerobic conditions. I: hydroquinone hydroxylase; II: 1,2,4-trihidroxy-benzene 1,2 dioxygenase; III: hydroquinone dioxygenase; IV: 4-hydroxymuconic semialdehyde dehydrogenase; V: beta-ketoadipate oxidoreductase.

Aerobic Degradation of Hydroquinone

1. Direct Ring Fission Reaction: 1,2-Hydroquinone Dioxygenase

Hydroquinone 1,2-dioxygenases (HQDIOX) are mainly bacterial enzymes able to catalyze the direct phenolic ring fission using 1,4-dihydroxyphenols as substrates, and constituting a wide new family of aromatic ring-fission enzymes. Among all the HQDIOX we can distinguish two sub-families of enzymes: one constituted by monomeric non-heme iron containing enzymes (Xun et al., 1999), and another including two-component dioxygenases that were firstly characterized as a members of the cluster of genes responsible for the degradation of phenolic pollutants in *Pseudomonas* (Moonen et al., 2008).

Monomeric hydroquinone 1,2-dioxygenases were first characterized as a core catalytic members of the enzymatic system responsible for the degradation of some pollutants as pentachlorophenol (PCP). These single chain enzymes were first described in *Mycobacterium chrorophenolicum*, *Sphingobium chlorophenolicum* and *Novosphingobium* (Ohtsubo et al., 1999; Xun et al., 1999; Machonkin et al., 2010). In *S. chlorophenolicum* the HQDIOX is encoded by the *pcp*A gene, which translates a polypeptide of around 37 kDa. PcpA protein contains a non-heme Fe atom and has no homology with classical ring-fission enzymes such as catechol dioxygenase. The enzyme is able to catalyze the ring-opening of a wide range of substituted hydroquinones (Ohtsubo et al., 1999; Sun et al., 2011; Hayes et al., 2013).

Despite of the detailed knowledge of the PcpA enzyme from *S. chlorophenolicum*, other putative members of the family are present in several gram negative bacteria (Fig. 3).

The catalytic core of the PcpA enzyme is composed by a non-heme iron atom coordinated by two histidines and an aspartate (Fig. 4) as observed in the crystal structure of the PcpA enzyme from *S. chlorophenolicum* that has been recently determined (Hayes et al., 2013). The observed structure of PcpA suggested a potential catalytic mechanism, which will be distinct from the classical extradiol dioxygenases. In fact, the iron atom is located in a relatively hydrophobic location in the protein surface, which will facilitate the binding of the phenolic substrate. Interestingly, the 1,4-hydroxyl groups of the phenolic substrate seem essential for affinity, thus other phenols such as catechol, nitro-catechol, phenol, bromo and chloro-phenols did not show any apparent affinity for the PcpA protein (Hayes et al., 2013).

Another class of hydroquinone 1,2-dioxygenases has been characterized from *Sphingomonas* and *Pseudomonas* strains able to degrade phenolic compounds, with putative homologs also present in the genomes of *Photorhabdus*, *Burkholderia*, and *Variovorax* genera. This family of enzymes is completely unrelated with the PcpA-like ones. They are composed by two subunits encoded by two adjacent genes located in genomic clusters encoding proteins involved in the degradation of substituted aromatic compounds. In *Sphingomonas* sp., these genes are designated as *hdq*A and *hdq*B (Kolvenbach et al., 2011), and the corresponding homologs in *P. fluorescens* are *hap*C and *hap*D (Moonen et al., 2008). In *P. fluorescens*, the encoding genes are located in a cluster composed by *hapCDEFGHIAB*, responsible for the biodegradation of hydroxyl-acetophenone (Moonen et al., 2008). The purified enzyme from *P. fluorescens* is a heterotetramer with a quaternary structure of $\alpha_2\beta_2$ and a molecular weight of 115 kDa. The beta-subunit, encoded by *hap*D gene, has a molecular weight of 38 kDa, and contains a non-heme iron coordination site and is also responsible for the substrate binding. In spite of the lack of structural information, some details about the reaction mechanism of two-component hydroquinone hydroxylases are known.

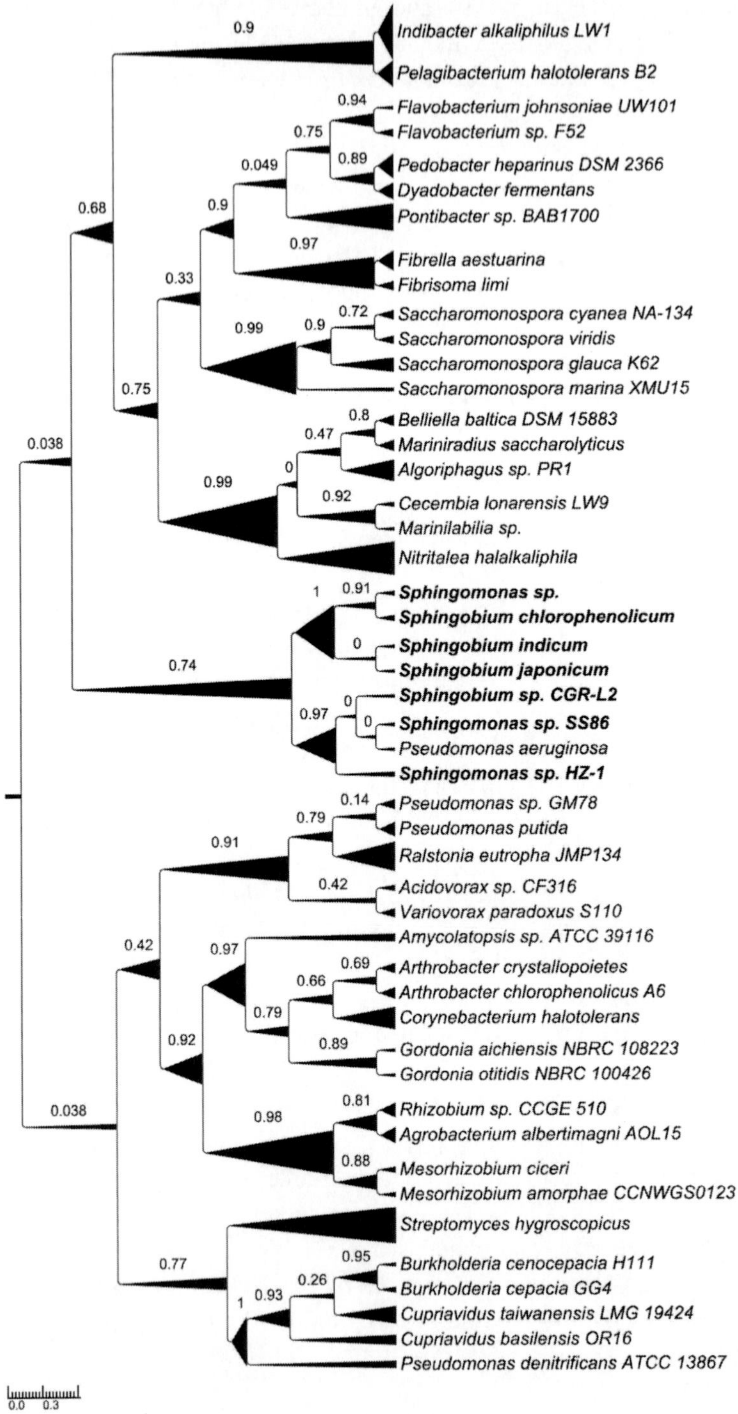

Figure 3. Phylogenetic chladogram of hydroquinone 1,2-dioxygenase sequences in diverse bacteria. The distances between nodes are depicted in the figure and the reference proteins from the geni *Sphingomonas* and *Sphingobium* represented in bold text.

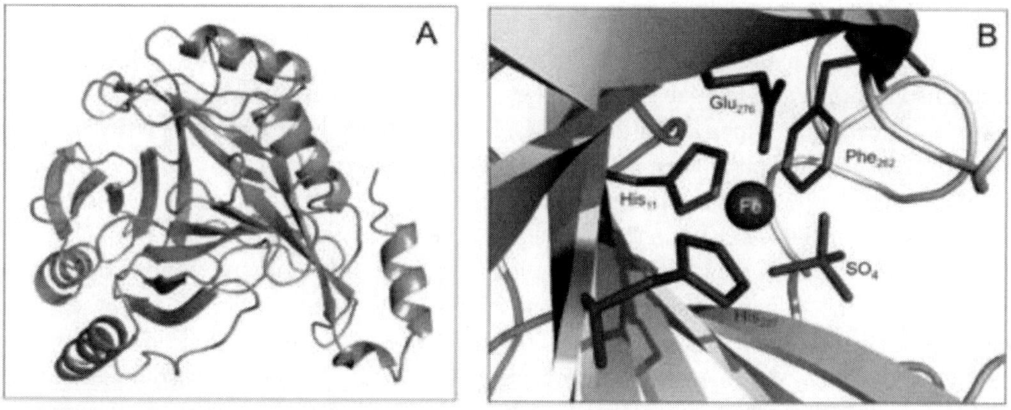

Figure 4. Structural features of the hydroquinone 1,2-hydroxylase from *S. chlorophenolicum*, as determined by X-ray crystallography (PDB code: 4HUZ). A: overall folding of the enzyme; B: close view of the coordinating residues and the Fe atom environment. The sulfate ion close to the iron atom is a component of the crystallization buffer.

The purified enzyme from *P. fluorescens* can act over a wide range of hydroquinone derivatives including 2-chloro-hydroquinone, 2-methyl-hydroquinone, 2-methoxy-hydroquinone and 2-ethyl-hydroquinone. The enzymatic activity over aliphatic derivatives with longer chains located in the *ortho* position of the benzene ring showed little or no detectable enzymatic activity, indicating the presence of a tight substrate pocket and steric clash interactions with substrates of higher molecular weights. Moreover, the enzyme is not inhibited my mono-phenols, but the activity is strongly decreased with the presence of cyanide compounds, as expected for an iron-dependent enzyme (Kamerbeek et al., 2001; Moonen et al., 2008). Apparently, the iron atom is in the core of the catalytic mechanism that could be essentially similar to the one postulated in the monomeric enzymes, however the role of the alpha small subunit of the enzyme remains elusive and will require further investigation.

2. 4-Hydroxymuconic Semialdehyde Dehydrogenase

In the microorganisms able to catalyze the direct cleavage of the hydroquinone ring, a second enzymatic activity will convert 4-hydroxymuconic acid into maleyl-acetate. The enzyme responsible for this activity has been described in *Pseudomonas* and *Sphingomonas* genera, and is encoded by genes associated to the degradation of *p*-nitrophenol and hydroquinone (Kolvenbach et al., 2011; Zhang et al., 2012). In *Pseudomonas*, the enzyme is encoded by the *hapE* gene, which encodes a protein of 487 aminoacids and a calculated molecular weight of 50 kDa. Other homologs of HapE protein are also present in *Burkholderia* sp., *Sphingomonas* sp., *Azospirillum amazonense*, and in *Brachymonas pretroleovorans*. The enzyme is an oxidoreductase that used NADP nucleotides as electron acceptors (Moonen et al., 2008; Vikram et al., 2012). It belongs to the NAD(P)-dependent aldehyde dehydrogenase superfamily, a group of enzymes with an important role not only in detoxification reactions but also in other metabolic pathways.

3. Hydroquinone Hydroxylase

Hydroquinone hydroxylase is a member of the family of phenol 2-monooxygenases. In comparison with hydroquinone dioxygenases, hydroquinone hydroxylases are less specific enzymes showing a wider substrate range. Hydroquinone hydroxylase activity has been extensively characterized in anamorphic yeasts as *Trichosporon cutaneum* (Kalin et al., 1992; Gerginova et al., 2007), *Candida* (Eppink et al., 2000; Vilimkova et al., 2008) and several species of gram negative bacteria (Izzo et al., 2011; Kolvenbach et al., 2011).

Phenol oxygenases are classified in two main groups; one family of monomeric enzymes and another one composed by multicomponent proteins. Interestingly, genes encoding monomeric hydroquinone hydroxylases are typically located in plasmid DNA, whereas multicomponent proteins are encoded by genes within the bacterial chromosome. In bacteria, hydroquinone hydroxylase activity has been associated to some monomeric metalloenzymes containing Fe^{2+}. In *Acinetobacter* and *Pseudomonas* the enzyme containg an iron-sulfur cluster of the type 2Fe-2S (Cadieux et al., 2002), responsible for the electron transfer from the substrate to the reduced cofactor NADH (Merimaa et al., 2006; Sazinsky et al., 2006).

Moreover in *Pseudomonas*, another family of phenol monooxygenases able to use hydroquinone as a substrate and belonging to the multicomponent group has been also described. Bacterial multicomponent monooxygenases (BMMs) consist of a protein complex with 250-300 kDa complex containing a dimeric hydroxylase (PHH) of the form $(\alpha\beta\gamma)_2$, a small regulatory protein (PHM) and a flavoprotein cointaining an iron-sulphur cluster (PHP), which acts as reductase supplying electrons to the hydroxylase component and consuming NADH as a cofactor. The hydroxylase component of the complex exerts its catalytic activity over phenolic substrates by the help of a di-iron center (Sazinsky et al., 2006; McCormick and Lippard, 2011; Tinberg et al., 2011). In spite of their structural complexity (Fig. 5), multicomponent phenol monooxygenases are extremely efficient enzymes, able to act over a wide range of substrates as nitro, amino and methyl-phenols and also halogenated phenols (McCormick and Lippard, 2011; Tinberg et al., 2011). They are typically inhibited by their substrates and also by inorganic anions such as nitrite, sulfate and phosphate, showing a slightly alkaline optimum pH.

The complexity of the multicomponent phenol monooxygenase is still far from being understood, since many additional protein components have been recently discovered. In fact genes encoding BMMs are located in a genomic cluster. Accessory components such PHK protein have been described as enhancers of the overall catalytic activity of the complex. PHK was assessed for its ability to interact with the active hydroxylase complex, being neither involved in the catalytic activity or required for the assembly of the apo-enzyme, suggesting that this component may be responsible for enhancing the incorporation of iron into the active site of the apo-hydroxylase (Izzo et al., 2011).

In eukaryotes, phenol hydroxylases with activity over hydroquinone are enzymes encoded by chromosomic genes. These enzymes have a molecular weight of about 150 kDa, and are functional dimers. In yeast as *Candida parapsilosis* and *Trichosporon cutaneum*, hydroquinone hydroxylase activity is induced when the cells are grown on mono or di-hydroxy benzoic acids as a sole source of carbon and energy (Eppink et al., 2000). Hydroquinone hydroxylase contains a tightly non-covalently bound FAD cofactor per each monomer, which is essential for the physiological reconstitution of the apoenzyme. These monooxygenases are able to catalyze the *ortho*-hydroxylation of a variety of substituted phenolic substrates including hydroquinone. The enzyme is able to use molecular oxygen and it is strictly dependent on NADPH as electron donor. In fact, in the absence of substrate, hydroquinone hydroxylase from *C. parapsilosis* showed a strong NADPH oxidase activity. Electrons produced from the oxidation of the substrates will be transferred to FAD cofactor thereafter. In *C. parapsilosis,* this flavin-dependent hydroxylase showed increased affinity over hydroquinone and phenol, if compared with catechol or resorcinol (Eppink et al., 2000).

Structural studies performed by X-ray crystallography in the *T. cutaneum,* hydroquinone hydroxylase showed that the protein has three clearly defined domains (Fig. 6).

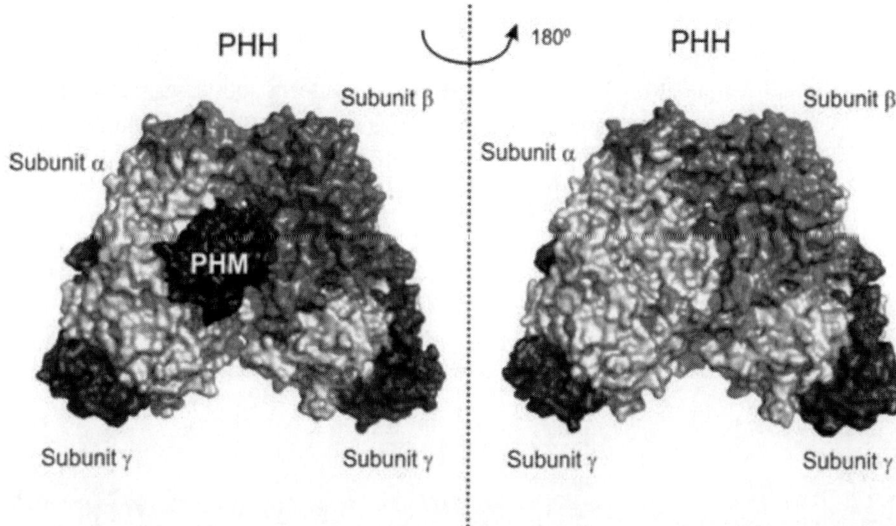

Figure 5. Surface representation of the overall tridimensional structure of the catalytic core PHH of the phenol monooxygenase from *Pseudomonas* sp. in complex with the regulatory subunit PHM (PDB code: 2INN). The catalytic core of the enzyme is composed by a dimer of three subunits αβγ (Sazinsky et al., 2006; McCormick and Lippard, 2011).

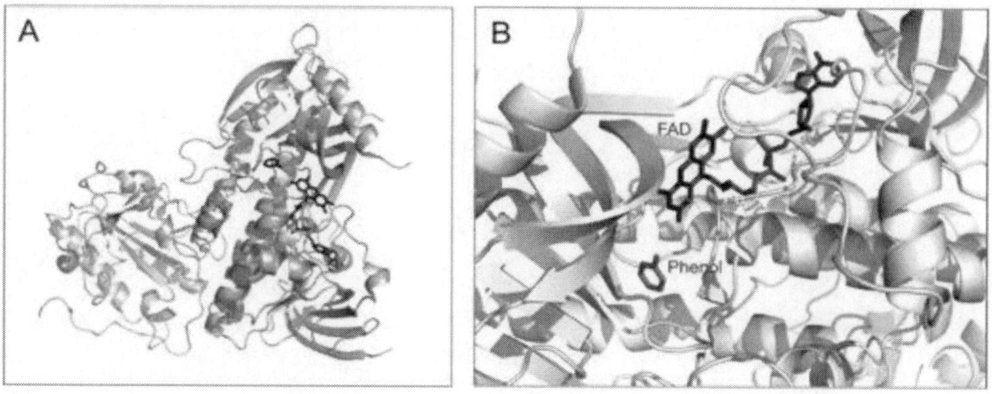

Figure 6. Structure of hydroquinone hydroxylase from *T. cutaneum* as determined by X-ray crystallography (PDB code: 1PN0). A: overall fold showing the structure of the three domains and the location of FAD cofactor and phenolic substrate in the catalytic pocket of the enzyme between domains 1 and 2; B: detailed view of the catalytic pocket, showing the phenolic substrate close to the FAD cofactor.

The first domain is a beta-sheet that forms the FAD binding site and also the substrate pocket close to the interface with the second domain. The third domain is composed by a thioredoxin-like fold involved in protein-protein interactions to constitute the apoenzyme dimer (Enroth et al., 1998). Eukaryotic phenol hydroxylases are able to act over simple phenols, but also over amino, halogenated and methyl-phenol derivatives with the only requirement of having a free *ortho* position in the phenolic ring.

4. 1,2,4-Trihidroxy-Benzene 1,2-Dioxygenase

Trihidroxy-benzene 1,2-dioxygenase is also known as hydroxyquinol 1,2-dioxygenase (1,2-HQD). 1,2-HQDs belong to the intradiol dioxygenase family and catalyze the oxidative ring cleavage of substituted 1,2-dihydroxy-benzenes. In bacteria and fungi, these enzymes are homodimers composed by two identical subunits of around 30-35 kDa, tightly packed in a very compact structure suggesting a possible cooperative mechanism between monomers.

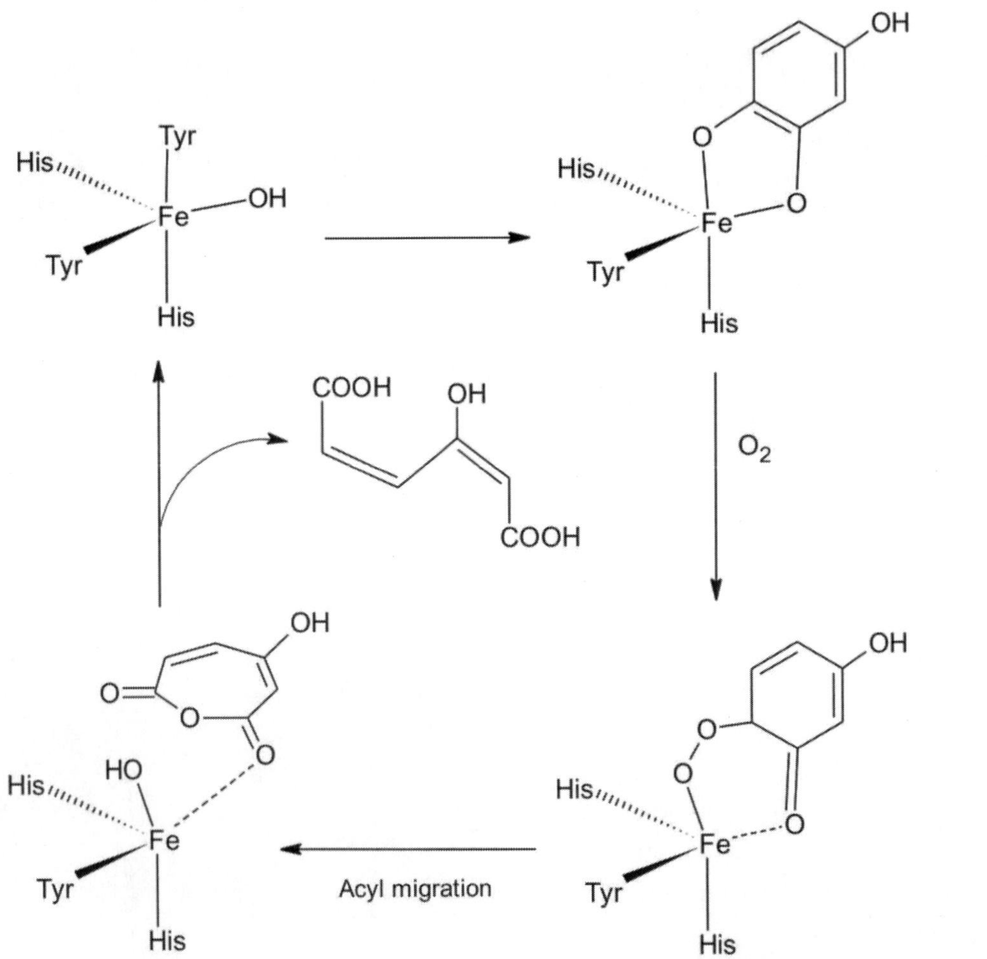

Figure 7. Proposed reaction mechanism for hydroxiquinol intradiol 1,2-dioxygenase. The enzyme catalyzed the ring fission reaction by generation of a seven-membered intermediate ring using molecular oxygen (Adapted from (Ferraroni et al., 2005)).

They have been described in gram negative bacteria (*Burkholderia cepacia*, *Azotobacter* sp., and *Ralstonia pickettii*), gram positive bacteria (*Nocardioides simplex* and *Arthrobacter* sp.), and fungi (*T. cutaneum* and *Phanerochaete chrysosporium*) (Asturias and Timmis, 1993; Eck and Belter, 1993; Bae et al., 1996; Daubaras et al., 1996; Ayoubi and Harker, 1998; Hino et al., 1998; Chang et al., 2003; Ferraroni et al., 2005; Wesche et al., 2005).

1,2-HQD are iron-dependent dioxygenases containing a non-heme penta-coordinated Fe(III) atom located close to the substrate binding pocket. This catalytic cavity for substrate binding comprises a tridimensional arrangement of beta-sheets together with a number of random coils rendering a small hydrophobic concavity that will place the substrate close to the iron atom. The possible catalytic mechanism for intradiol 1,2 cleavage has been recently proposed based on the crystallographic structure of 1,2-HQD from *N. simplex* (Ferraroni et al., 2005). The enzyme is able to catalyze the aromatic ring fission by a mechanism that involves the formation of an intermediate oxo-adduct and a seven-membered ring (Fig. 7). As reported in the tridimensional structure of the 1,2-HQD, the substrate specificity of the enzyme is expected to be controlled by the aromatic ring substituents (Fig. 8). In fact, 1,2-HQD is more similar to the Type I of intradiol dioxygenases, which showed an increased specificity in their substrates (Tao et al., 2004; Ferraroni et al., 2005).

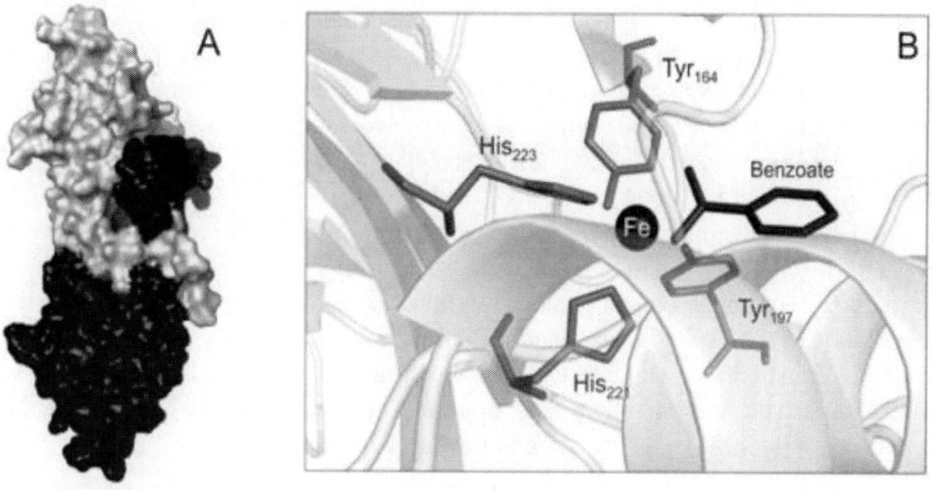

Figure 8. Tridimensional structure of the catalytic homodimer of Hydroxyquinol 1,2-dioxygenase from *N. simplex* as determined by X-ray crystallography (PDB code: 1TMX). A: global structure of the homodimer showing the tight packing between subunits; B: close look to the iron center, coordinating aminoacids and inhibitor (benzoate) binding to the iron atom.

Studies performed with diverse 1,2-HQD enzymes showed a decreased affinity and catalytic activity over catechol, pyrogallol and other substituted dihydroxybenzenes in comparison with 1,2,4-trihidroxybenzene. Interestingly, the sequence comparison of 1,2-HQDs among diverse microorganisms showed a divergent evolution pattern between this family of enzymes and other catechol dioxygenases. This indicates that the development of substrate specificity for hydroquinone was a very important step in the evolution of pathways for the efficient degradation of natural and xenobiotic compounds with a benzene nucleus (Murakami et al., 1997; Hatta et al., 1999; Chang et al., 2003).

Anaerobic Degradation of Hydroquinone

The chemical properties of hydroquinone are directly related with the intrinsic difficulty for its degradation under anaerobic conditions. In fact, the presence of two hydroxyl groups in *para* orientation within the benzene ring makes direct oxidative ring cleavage unlikely by any microorganism under these conditions. On the contrary, anaerobic hydroquinone-degrading microorganisms will engage a diverted pathway involving a carboxylation and activation with CoA to finish with the production of benzoate and the introduction of this compound into the classical anaerobic benzene metabolization pathways (Li et al., 2012; Valderrama et al., 2012). There are just a few reported examples in the literature describing anaerobic organisms able to degrade hydroquinone. They include sulfate-reducing bacteria from the genus *Desulfococcus* (Gorny and Schink, 1994; Gorny and Schink, 1994), and dehalogenating bacteria isolated from soil consortia together with filamentous fungi (Milliken et al., 2004; Milliken et al., 2004).

The reductive hydroquinone metabolization pathway starts with a carboxylation to produce gentisate (2,5-dihydroxybenzoate). This reaction is catalyzed by an uncharacterized carboxylase enzyme that is inducible by the presence of hydroquinone as a sole source of carbon and energy in anaerobic conditions (Gorny and Schink, 1994). Carboxylation of hydroquinone under these conditions is also stimulated by the addition of bicarbonate. After the synthesis of gentisate, this compound will be activated by the addition of a CoA group. This is a classical reaction step in the utilization of phenolic compounds by anaerobic microorganisms. The corresponding CoA-ligase involved, uses only Acyl-CoA as a donor, with no evidence of catalytic activity using acetyl or phenyl-CoA donors (Gorny and Schink, 1994; Gorny and Schink, 1994).

The most interesting reaction of the whole pathway is the reductive dehydroxylation of the gentisyl-CoA. This reaction is catalyzed by an oxygen-sensitive enzyme that removes both hydroxyl groups in one single step, and that is probably associated to cell membranes (Gorny and Schink, 1994; Gorny and Schink, 1994; Valderrama et al., 2012). However, further investigations are required to characterize this enzymatic activity. The gentisyl-CoA dehydroxylase will generate benzoate as a final product that will engage the anaerobic benzoate pathway for its final degradation towards beta-oxidation.

CONCLUSION

The wide use of hydroquinone as a main component in a myriad of industrial processes has led to the entry of this aromatic compound into the environment. Meanwhile, the microorganisms exposed to this pollutant evolved the capability to use it. The bioremediation technology explores microbial catabolism to convert organic pollutant, such as hydroquinone, into harmless substances or completely mineralized to CO_2 and H_2O. Therefore, this technology could be energy efficient and result in ecologically intact environments, being an important alternative to chemical methodologies. A major concern associated with the rising field of bioremediation technology is the applicability of laboratory research to actual large-scale contaminated field sites, and its inherent costs involved in the implementation. Due to their versatility in terms of low nutrient and oxygen demand and connected to the production of a high number of extracellular enzymes, fungi alone or in collaboration with bacteria display many of these features that could be important components of biotechnologies design to remediate contaminated water, soil and air. However, the majority of the studies on fungal performance have been conducted on control conditions of medium composition, temperature, redox potential, etc., which may not be significant when not all the natural environmental variables, including bacteria-fungi associations, are taken into consideration.

We have shown that several microorganisms and their biochemical potential can be of relevance for biotechnological application to remove hydroquinone. By understanding the biochemistry and genetic of hydroquinone catabolism it would be possible to enhance the capability of degradation of selected microorganisms, using recombinant DNA techniques, resulting in higher yields and less intermediates.

ACKNOWLEDGMENTS

The authors would like to thank Prof. C. Kang for sharing the 3D coordinates of PcpA protein before its public release.

REFERENCES

Asturias, J.A., Timmis, K.N., 1993. Three different 2,3-dihydroxybiphenyl-1,2-dioxygenase genes in the gram-positive polychlorobiphenyl-degrading bacterium *Rhodococcus globerulus* P6. *J. Bacteriol.* 175, 4631-40.

Ayoubi, P.J., Harker, A.R., 1998. Whole-cell kinetics of trichloroethylene degradation by phenol hydroxylase in a *Ralstonia eutropha* JMP134 derivative. *Appl. Environ. Microbiol.* 64, 4353-6.

Ayranci, E., Duman, O., 2005. Adsorption behaviors of some phenolic compounds onto high specific area activated carbon cloth. *J. Hazard Mater.* 124, 125-32.

Bae, H.S., Lee, J.M., Lee, S.T., 1996. Biodegradation of 4-chlorophenol via a hydroquinone pathway by *Arthrobacter ureafaciens* CPR706. *FEMS Microbiol. Lett.* 145, 125-9.

Bahrs, H., Putschew, A., Steinberg, C.E., 2013. Toxicity of hydroquinone to different freshwater phototrophs is influenced by time of exposure and pH. *Environ. Sci. Pollut. Res. Int.* 20, 146-54.

Cadieux, E., Vrajmasu, V., Achim, C., Powlowski, J., Munck, E., 2002. Biochemical, Mossbauer, and EPR studies of the diiron cluster of phenol hydroxylase from *Pseudomonas* sp. strain CF 600. *Biochemistry* 41, 10680-91.

Chang, H.K., Mohseni, P., Zylstra, G.J., 2003. Characterization and regulation of the genes for a novel anthranilate 1,2-dioxygenase from *Burkholderia cepacia* DBO1. *J. Bacteriol.* 185, 5871-81.

Daubaras, D.L., Saido, K., Chakrabarty, A.M., 1996. Purification of hydroxyquinol 1,2-dioxygenase and maleylacetate reductase: the lower pathway of 2,4,5-trichlorophenoxyacetic acid metabolism by *Burkholderia cepacia* AC1100. *Appl. Environ. Microbiol.* 62, 4276-9.

Deisinger, P.J., Hill, T.S., English, J.C., 1996. Human exposure to naturally occurring hydroquinone. *J. Toxicol. Environ. Health* 47, 31-46.

Eck, R., Belter, J., 1993. Cloning and characterization of a gene coding for the catechol 1,2-dioxygenase of *Arthrobacter* sp. mA3. *Gene* 123, 87-92.

Enroth, C., Neujahr, H., Schneider, G., Lindqvist, Y., 1998. The crystal structure of phenol hydroxylase in complex with FAD and phenol provides evidence for a concerted conformational change in the enzyme and its cofactor during catalysis. *Structure* 6, 605-17.

Eppink, M.H., Cammaart, E., Van Wassenaar, D., Middelhoven, W.J., van Berkel, W.J., 2000. Purification and properties of hydroquinone hydroxylase, a FAD-dependent monooxygenase involved in the catabolism of 4-hydroxybenzoate in *Candida parapsilosis* CBS604. *Eur. J. Biochem.* 267, 6832-40.

Ferraroni, M., Seifert, J., Travkin, V.M., Thiel, M., Kaschabek, S., Scozzafava, A., Golovleva, L., Schlomann, M., Briganti, F., 2005. Crystal structure of the hydroxyquinol 1,2-dioxygenase from *Nocardioides simplex* 3E, a key enzyme involved in polychlorinated aromatics biodegradation. *J. Biol. Chem.* 280, 21144-54.

Geng, Q., Guo, Q., Cao, C., Wang, L., 2008. Investigation into nanoTiO2/ACSPCR for decomposition of aqueous hydroquinone. *Ind. Eng. Chem. Res.* 47, 2561-68.

Gerginova, M., Manasiev, J., Shivarova, N., Alexieva, Z., 2007. Influence of various phenolic compounds on phenol hydroxylase activity of a *Trichosporon cutaneum* strain. *Z. Naturforsch C* 62, 83-6.

Gorny, N., Schink, B., 1994. Anaerobic degradation of catechol by *Desulfobacterium* sp. strain Cat2 proceeds via carboxylation to protocatechuate. *Appl. Environ. Microbiol.* 60, 3396-400.

Gorny, N., Schink, B., 1994. Complete anaerobic oxidation of hydroquinone by *Desulfococcus* sp. strain Hy5: indications of hydroquinone carboxylation to gentisate. *Arch Microbiol* 162, 131-35.

Gorny, N., Schink, B., 1994. Hydroquinone degradation via reductive dehydroxylation of gentisyl-CoA by a strictly anaerobic fermenting bacterium. *Arch. Microbiol.* 161, 25-32.

Hatta, T., Nakano, O., Imai, N., Takizawa, N., Kiyohara, H., 1999. Cloning and sequence analysis of hydroxyquinol 1,2-dioxygenase gene in 2,4,6-trichlorophenol-degrading *Ralstonia pickettii* DTP0602 and characterization of its product. *J. Biosci. Bioeng.* 87, 267-72.

Hayes, R.P., Green, A.R., Nissen, M.S., Lewis, K.M., Xun, L., Kang, C., 2013. Structural characterization of 2,6-dichloro-p-hydroquinone 1,2-dioxygenase (PcpA) from *Sphingobium chlorophenolicum*, a new type of aromatic ring-cleavage enzyme. *Mol. Microbiol.*, 88(3), 523-36.

Hebeda, C.B., Pinedo, F.J., Bolonheis, S.M., Ferreira, Z.F., Muscara, M.N., Teixeira, S.A., Farsky, S.H., 2012. Intracellular mechanisms of hydroquinone toxicity on endotoxin-activated neutrophils. *Arch .Toxicol.* 86, 1773-81.

Higson, F.K., Focht, D.D., 1990. Bacterial degradation of ring-chlorinated acetophenones. *Appl. Environ. Microbiol.* 56, 3678-85.

Hino, S., Watanabe, K., Takahashi, N., 1998. Phenol hydroxylase cloned from *Ralstonia eutropha* strain E2 exhibits novel kinetic properties. *Microbiology* 144 (Pt 7), 1765-72.

Izzo, V., Leo, G., Scognamiglio, R., Troncone, L., Birolo, L., Di Donato, A., 2011. PHK from phenol hydroxylase of *Pseudomonas* sp. OX1. Insight into the role of an accessory protein in bacterial multicomponent monooxygenases. *Arch. Biochem. Biophys* 505, 48-59.

Jones, K.H., Trudgill, P.W., Hopper, D.J., 1994. 4-Ethylphenol metabolism by *Aspergillus fumigatus*. *Appl. Environ. Microbiol.* 60, 1978-83.

Kalin, M., Neujahr, H.Y., Weissmahr, R.N., Sejlitz, T., Johl, R., Fiechter, A., Reiser, J., 1992. Phenol hydroxylase from *Trichosporon cutaneum*: gene cloning, sequence analysis, and functional expression in *Escherichia coli*. *J. Bacteriol.* 174, 7112-20.

Kamada, F., Abe, S., Hiratsuka, N., Wariishi, H., Tanaka, H., 2002. Mineralization of aromatic compounds by brown-rot basidiomycetes - mechanisms involved in initial attack on the aromatic ring. *Microbiology* 148, 1939-46.

Kamerbeek, N.M., Moonen, M.J., Van Der Ven, J.G., Van Berkel, W.J., Fraaije, M.W., Janssen, D.B., 2001. 4-Hydroxyacetophenone monooxygenase from *Pseudomonas fluorescens* ACB. A novel flavoprotein catalyzing Baeyer-Villiger oxidation of aromatic compounds. *Eur. J. Biochem.* 268, 2547-57.

Kolvenbach, B.A., Lenz, M., Benndorf, D., Rapp, E., Fousek, J., Vlcek, C., Schaffer, A., Gabriel, F.L., Kohler, H.P., Corvini, P.F., 2011. Purification and characterization of hydroquinone dioxygenase from *Sphingomonas* sp. strain TTNP3. *AMB Express* 1, 8.

Lee, J.S., Yang, E.J., Kim, I.S., 2012. Hydroquinone-induced apoptosis of human lymphocytes through caspase 9/3 pathway. *Mol. Biol. Rep.* 39, 6737-43.

Leitão, A.L., Duarte, M.P., Santos Oliveira, J., 2007. Degradation of phenol by a halotolerant strain of *Penicillium chrysogenum*. *Int. Biodeterior. Biodegradation* 59, 220-25.

Li, Y.N., Porter, A.W., Mumford, A., Zhao, X.H., Young, L.Y., 2012. Bacterial community structure and bamA gene diversity in anaerobic degradation of toluene and benzoate under denitrifying conditions. *J. Appl. Microbiol.* 112, 269-79.

Machonkin, T.E., Holland, P.L., Smith, K.N., Liberman, J.S., Dinescu, A., Cundari, T.R., Rocks, S.S., 2010. Determination of the active site of *Sphingobium chlorophenolicum* 2,6-dichlorohydroquinone dioxygenase (PcpA). *J. Biol. Inorg. Chem.* 15, 291-301.

Mahamuni, N.N., Pandit, A.B., 2006. Effect of additives on ultrasonic degradation of phenol. *Ultrasonics Sonochemistry* 13, 165-74.

McCormick, M.S., Lippard, S.J., 2011. Analysis of substrate access to active sites in bacterial multicomponent monooxygenase hydroxylases: X-ray crystal structure of xenon-pressurized phenol hydroxylase from *Pseudomonas* sp. OX1. *Biochemistry* 50, 11058-69.

Merimaa, M., Heinaru, E., Liivak, M., Vedler, E., Heinaru, A., 2006. Grouping of phenol hydroxylase and catechol 2,3-dioxygenase genes among phenol- and p-cresol-degrading *Pseudomonas* species and biotypes. *Arch. Microbiol.* 186, 287-96.

Middelhoven, W.J., 1993. Catabolism of benzene compounds by ascomycetous and basidiomycetous yeasts and yeastlike fungi. A literature review and an experimental approach. *Antonie Van Leeuwenhoek* 63, 125-44.

Milliken, C.E., Meier, G.P., Sowers, K.R., May, H.D., 2004. Chlorophenol production by anaerobic microorganisms: transformation of a biogenic chlorinated hydroquinone metabolite. *Appl. Environ. Microbiol.* 70, 2494-6.

Milliken, C.E., Meier, G.P., Watts, J.E., Sowers, K.R., May, H.D., 2004. Microbial anaerobic demethylation and dechlorination of chlorinated hydroquinone metabolites synthesized by basidiomycete fungi. *Appl. Environ. Microbiol.* 70, 385-92.

Moonen, M.J., Kamerbeek, N.M., Westphal, A.H., Boeren, S.A., Janssen, D.B., Fraaije, M.W., van Berkel, W.J., 2008. Elucidation of the 4-hydroxyacetophenone catabolic pathway in *Pseudomonas fluorescens* ACB. *J. Bacteriol.* 190, 5190-8.

Moonen, M.J., Synowsky, S.A., van den Berg, W.A., Westphal, A.H., Heck, A.J., van den Heuvel, R.H., Fraaije, M.W., van Berkel, W.J., 2008. Hydroquinone dioxygenase from *Pseudomonas fluorescens* ACB: a novel member of the family of nonheme-iron(II)-dependent dioxygenases. *J. Bacteriol.* 190, 5199-209.

Murakami, S., Kodama, N., Shinke, R., Aoki, K., 1997. Classification of catechol 1,2-dioxygenase family: sequence analysis of a gene for the catechol 1,2-dioxygenase showing high specificity for methylcatechols from Gram+ aniline-assimilating *Rhodococcus erythropolis* AN-13. *Gene* 185, 49-54.

Nakamura, T., Ichinose, H., Wariishi, H., 2012. Flavin-containing monooxygenases from *Phanerochaete chrysosporium* responsible for fungal metabolism of phenolic compounds. *Biodegradation* 23, 343-50.

Nordin, K., Unell, M., Jansson, J.K., 2005. Novel 4-chlorophenol degradation gene cluster and degradation route via hydroxyquinol in *Arthrobacter chlorophenolicus* A6. *Appl. Environ. Microbiol.* 71, 6538-44.

O'Donoghue, J.L., 2006. Hydroquinone and its analogues in dermatology - a risk-benefit viewpoint. *J. Cosmet. Dermatol.* 5, 196-203.

OECD SIDS, 2012. Hydroquinone (CAS No. 123-31-9). *UNEP Publications.*

Ohtsubo, Y., Miyauchi, K., Kanda, K., Hatta, T., Kiyohara, H., Senda, T., Nagata, Y., Mitsui, Y., Takagi, M., 1999. PcpA, which is involved in the degradation of pentachlorophenol in *Sphingomonas chlorophenolica* ATCC39723, is a novel type of ring-cleavage dioxygenase. *FEBS Lett.* 459, 395-8.

Santos, A., Yustos, P., Quintanilla, A., Garcia-Ochoa, F., Casas, J.A., Rodriguez, J.J., 2004. Evolution of toxicity upon wet catalytic oxidation of phenol. *Environ. Sci. Technol.* 38, 133-8.

Sazinsky, M.H., Dunten, P.W., McCormick, M.S., DiDonato, A., Lippard, S.J., 2006. X-ray structure of a hydroxylase-regulatory protein complex from a hydrocarbon-oxidizing multicomponent monooxygenase, *Pseudomonas* sp. OX1 phenol hydroxylase. *Biochemistry* 45, 15392-404.

Schnell, S., Bak, F., Pfennig, N., 1989. Anaerobic degradation of aniline and dihydroxybenzenes by newly isolated sulfate-reducing bacteria and description of *Desulfobacterium anilini*. *Arch. Microbiol.* 152, 556-63.

Shimada, A.L., Ribeiro, A.L., Bolonheis, S.M., Ferraz-de-Paula, V., Hebeda, C.B., Farsky, S.H., 2012. In vivo hydroquinone exposure impairs MCP-1 secretion and monocyte recruitment into the inflamed lung. *Toxicology* 296, 20-6.

Song, Y., Xie, J., Shu, H., Zhao, G., Lv, X., Xie, W., 2006. Calculation of standard electrode potential of half reaction for benzoquinone and hydroquinone. *Spectrochim. Acta A Mol. Biomol. Spectrosc.* 65, 333-9.

Spain, J.C., Gibson, D.T., 1991. Pathway for Biodegradation of p-Nitrophenol in a *Moraxella* sp. *Appl. Environ. Microbiol.* 57, 812-9.

Subrahmanyam, V.V., Kolachana, P., Smith, M.T., 1991. Hydroxylation of phenol to hydroquinone catalyzed by a human myeloperoxidase-superoxide complex: possible implications in benzene-induced myelotoxicity. *Free Radic. Res. Commun.* 15, 285-96.

Sun, W., Sammynaiken, R., Chen, L., Maley, J., Schatte, G., Zhou, Y., Yang, J., 2011. *Sphingobium chlorophenolicum* dichlorohydroquinone dioxygenase (PcpA) is alkaline resistant and thermally stable. *Int. J. Biol. Sci.* 7, 1171-9.

Suresh, S., Srivastava, V.C., Mishra, I.M., 2011. Adsorption of hydroquinone in aqueous solution by granular activated carbon. *J. Environ. Eng.* 137, 1145-57.

Suresh, S., Srivastava, V.C., Mishra, I.M., 2012. Adsorption of catechol, resorcinol, hydroquinone, and their derivatives: a review. *Int. J. Energy Environ. Eng.* 3, 32.

Szewzyk, U., Schink, B., 1989. Degradation of hydroquinone, gentisate, and benzoate by a fermenting bacterium in pure or defined mixed culture *Arch. Microbiol.* 151, 541-45.

Tao, Y., Fishman, A., Bentley, W.E., Wood, T.K., 2004. Oxidation of benzene to phenol, catechol, and 1,2,3-trihydroxybenzene by toluene 4-monooxygenase of *Pseudomonas mendocin*a KR1 and toluene 3-monooxygenase of *Ralstonia pickettii* PKO1. *Appl. Environ. Microbiol.* 70, 3814-20.

Thammanayakatip, C., Oshima, Y., Koda, S., 1998. Inhibition effect in supercritical water oxidation of hydroquinone. *Ind. Eng. Chem. Res.* 37, 2061-63.

Tinberg, C.E., Song, W.J., Izzo, V., Lippard, S.J., 2011. Multiple roles of component proteins in bacterial multicomponent monooxygenases: phenol hydroxylase and toluene/o-xylene monooxygenase from *Pseudomonas* sp. OX1. *Biochemistry* 50, 1788-98.

Tortella, G.R., Diez, M.C., Duran, N., 2005. Fungal diversity and use in decomposition of environmental pollutants. *Crit. Rev. Microbiol.* 31, 197-212.

Valderrama, J.A., Durante-Rodriguez, G., Blazquez, B., Garcia, J.L., Carmona, M., Diaz, E., 2012. Bacterial degradation of benzoate: cross-regulation between aerobic and anaerobic pathways. *J. Biol. Chem.* 287, 10494-508.

Vikram, S., Pandey, J., Bhalla, N., Pandey, G., Ghosh, A., Khan, F., Jain, R.K., Raghava, G.P., 2012. Branching of the p-nitrophenol (PNP) degradation pathway in *Burkholderia* sp. Strain SJ98: Evidences from genetic characterization of PNP gene cluster. *AMB Express* 2, 30.

Vilimkova, L., Paca, J., Kremlackova, V., Stiborova, M., 2008. Isolation of cytoplasmic NADPH-dependent phenol hydroxylase and catechol-1,2-dioxygenase from *Candida tropicalis* yeast. *Interdiscip. Toxicol.* 1, 225-30.

Wesche, J., Hammer, E., Becher, D., Burchhardt, G., Schauer, F., 2005. The bphC gene-encoded 2,3-dihydroxybiphenyl-1,2-dioxygenase is involved in complete degradation of dibenzofuran by the biphenyl-degrading bacterium *Ralstonia* sp. SBUG 290. *J. Appl. Microbiol.* 98, 635-45.

Xun, L., Bohuslavek, J., Cai, M., 1999. Characterization of 2,6-dichloro-p-hydroquinone 1,2-dioxygenase (PcpA) of *Sphingomonas chlorophenolica* ATCC 39723. *Biochem. Biophys. Res. Commun* 266, 322-5.

Yang, E.J., Lee, J.S., Yun, C.Y., Kim, I.S., 2011. The pro-apoptotic effect of hydroquinone in human neutrophils and eosinophils. *Toxicol. In Vitro* 25, 131-7.

Zhang, J.J., Liu, H., Xiao, Y., Zhang, X.E., Zhou, N.Y., 2009. Identification and characterization of catabolic para-nitrophenol 4-monooxygenase and para-benzoquinone reductase from *Pseudomonas* sp. strain WBC-3. *J. Bacteriol.* 191, 2703-10.

Zhang, S., Sun, W., Xu, L., Zheng, X., Chu, X., Tian, J., Wu, N., Fan, Y., 2012. Identification of the para-nitrophenol catabolic pathway, and characterization of three enzymes involved in the hydroquinone pathway, in *Peudomonas* sp. 1-7. *BMC Microbiol.* 12, 27.

In: Bioremediation: Processes, Challenges and Future Prospects ISBN: 978-1-62948-513-3
Editors: Jesús Bernardino Velázquez-Fernández et al.

Chapter 12

Endpoints and Bioassays to Assess Bioremediation Efficiency of Contaminated Soils

Dânia Elisa Christofoletti Mazzeo, *
Bruna de Campos Ventura-Camargo,
Lais Roberta Deroldo Sommaggio
and Maria Aparecida Marin-Morales
UNESP - Univ Estadual Paulista, Department of Biology,
Institute of Biosciences, Rio Claro, SP, Brazil

Abstract

The bioremediation process used to promote environmental decontamination, utilizes living organisms, mainly microorganisms or their enzymes, to clean up contaminated sites. It is considered an effective, low cost and low impact technology. In most cases, bioremediation mineralizes contaminants or reduces their toxicity. However, there are studies that show the formation of even more toxic metabolites than the original compound during such processes. Environmentally, soil is the destination of several pollutants derived from industrial discharge, accidents, spills and waste disposal. Due to its capacity to retain some types of substances, the dispersion process is more difficult to occur and many residues can persist in the soil. Bioassays are used to determine harmful biological effects of chemicals or mixtures in environmental samples and can be carried out with cells, tissues, organs or with the entire organism. Such tests can be performed with different species (e.g., bacteria, fungi, algae, plants and animals) to assess many endpoints (death, DNA damage, morphological and physiological effects and others), both in laboratorial and field conditions. Currently, the use of bioassays has been recommended to evaluate the bioremediation efficiency since they can predict a toxic response caused by possible metabolites originated in the process. Thereby, the aim of this review is to approach the main endpoints and bioassays that are used to estimate the

* Corresponding Author address: Av. 24-A, 1515, 13506-900 Rio Claro, SP, Brazil; Tel: +55 19 3526 4143; fax: +55 19 3534 0009. Email: daniamazzeo@gmail.com.

bioremediation effectiveness of contaminated soils. Furthermore, this work intends to emphasize the biological tests that are commonly used to monitor the bioremediation of contaminated soils, as well as assays that have appropriate responses, attempting to relate them to a particular type of contaminant or soil characteristic.

Keywords: Biodegradation, biological assays, toxic metabolites, phytoremediation, bioaugmentation, biopiles, landfarming

1. Introduction

Soil is the central component of terrestrial ecosystems and, due to its numerous functions; it plays an essential role for the maintenance of life. It is a source of nutrients and provides habitats for plants and animals (Yu et al., 2013). For humans, soil is essential to supply food and fibres (Glanz, 1995), and is the base of agriculture and animal productivity (Doran; Zeiss, 2000). Because of its intense microbial activity, soil is considered an ideal medium for decomposition of plant and animal residues and chemical transformation (Doran; Zeiss, 2000). Besides this, soil is also important to the protection of groundwater quality due to its filter property (Seabra, 2005).

However, the increase in the introduction of contaminants may interfere in the correct functioning of this environment. According to White and Claxton (2004), waste disposal; intentional or accidental discharge of industrial materials; fossil fuel combustion; pesticide use; solid waste burning; leaks and spills during transportation and storage of chemical products are among the main causes of soil contamination.

Once the xenobiotic enters soil, it can suffer different processes, such as volatilization, degradation, leaching, absorption by plants or animals and sorption (Haigh, 1996; Stokes et al., 2005). The physico-chemical properties of the contaminant (e.g. vapour pressure, solubility, chemical stability and others) and the characteristics of soil (e.g. pH, presence of organic material and microorganisms, redox potential and others) influence which of the mechanism will operate and if the contaminant will or will not persist in the soil (Haigh, 1996).

Thus, due to the frequent presence of xenobiotics and the persistence of specific contaminants in soil, besides the possibility to be source of contaminants for other environments, especially groundwater, the soil pollution is considered a global challenge and a serious environmental problem to be solved.

Different clean-up technologies can be applied to treat soil contamination. According to White and Claxton (2004), they can be classified in four types: physical (excavation, entombment and covering); chemical (washing, vacuum extraction, electro-reclamation for metal removal, oxidation, precipitation, reduction and adsorption to activated carbon); thermal (combustion); and biological (bioremediation). However, the use of bioremediation processes has been encouraged because they are considered effective, low cost and environmentally friendly (Russo et al., 2012). These processes use living organisms to degrade and mineralize pollutants, transforming them into less toxic compounds (Alexander, 1994; Eweis et al., 1998; Bamforth and Singleton, 2005). Nevertheless, the formation of toxic metabolites during the bioremediation has been reported by some studies (Alexander, 1999; Randerath et al., 1999; Phillips et al., 2000; Alexander et al., 2002).

Generally, the bioremediation effectiveness is evaluated by chemical analyses that verify the presence or absence of target contaminant, but such analyses do not predict the effects of possible intermediary substances formed during the process. Therefore, to assure the real bioremediation efficiency, biological assays must be used to monitor this process (Hund and Traunspurger, 1994).

According to Hansen et al. (2007) bioassays are tests that estimate deleterious biological effects of individual chemical compounds or a mixture of them, including environmental samples. To be considered efficient, bioassays require a quick and reliable methodology to characterize the extent of contamination and to measure contaminant bioavailability and toxicity (Maila and Cloete, 2005).

Maila and Cloete (2005) affirmed that the variability of soil properties must be considered when a bioassay is used in soil evaluation since these properties can interfere in the results. Thus, it is believed that further research should be done on soil bioindicators before they could be used to monitor contamination and bioremediation, by using more relevant ecological test species, by standardization of new assays or improvement of existing tests, and by the selection of relevant endpoints.

One of the most widely used bioassays to monitor soil bioremediation process is the Microtox® test. This method employs a bioluminescent bacterium (*Vibrio fischeri* or *Photobacterium phosphoreum*) as test organism. This assay measures the light intensity generated by the luminescent bacteria. A decrease in the produced bioluminescence means that the sample is toxic.

Seed germination and earthworm survival tests are also considered useful to be used for the same purpose due to their easy and simple methodology and moderate sensitivity to contaminants (Maila and Cloete, 2005).

Tests that assess parameters like reproduction represent important tools to evaluate ecosystem functions, since they allow predicting effects on population (Hansen et al., 2007). Generally, the main organisms employed are crustaceans and earthworms.

Biological assays that detect genotoxic and mutagenic effects have also been used to verify the success of bioremediation. These tests assess the possibility of a compound interact with the genetic material causing DNA breaks or losses and mutations that can lead to cell death or to carcinogenic processes.

Thus, this chapter focuses on studies that apply, besides chemical analyses, bioassays to monitor soil bioremediation process. Different types of bioremediation technologies associated with several bioassays are discussed along the text to highlight the importance of the use of the biological tests as a complementary tool to chemical analyses.

2. Different Bioremediation Processes Associated with Bioassays

The objectives of soil remediation processes are based on threshold levels of soil contaminants. Bioremediation has been successful in several applications for contaminated soils, either by pesticides, polycyclic aromatic hydrocarbons (PAHs) or many other chemical compounds. Although, during the different bioremediation processes, it is very provable to incur changes both in bioavailability of those chemicals and metabolites formation. Thereby, it is necessary to incorporate ecotoxicity assays in conjunction with chemical analysis to assess the effectiveness of the bioremediation processes (Frutos et al., 2012)

2.1. *In Situ* Bioremediation

The process of *in situ* bioremediation of contaminated soils aims to initiate or stimulate the elimination of contaminants by biological pathway in the field without soil extraction. In this process, different organisms that could be introduced in the soil and/or the indigenous soil biota are stimulated *in situ* attempting to increase the local biological activity for possibly reducing the toxicity of contaminated soils (Trapp, 2000 apud Frische and Höper, 2003).

A study performed by Frische and Höper (2003) monitored the *in situ* bioremediation process of topsoil contaminated with 2,4,6-trinitrotoluene (TNT) and other contaminants, due to military activities that occurred during the Second World War on a site situated in northern Germany. The authors evaluated changes within 6 experimental fields (without, with medium, and with high contamination by TNT) for a 17-months remediation period and compared the toxicity results of soil microbial and luminescent bacteria assays with chemical monitoring analyses. They observed a significant reduction of the soil contaminants in the treated fields after the 17 months remediation period, detected by the luminescent bacteria assay. On the other hand, the soil microbial assay, that analyzes the abundance and activity of the indigenous soil microbial community, did not minimize the soil contamination after the 17 months of bioremediation, but it evidenced the persistent adverse effects of chronic TNT contamination on the site-specific microbial community associated to the local carbon cycle in the soil. As in contrast to luminescent bacteria assays, the study did not reveal recovery of the soil at the end of the remediation period (17 months), there is a clear evidence for persistent adverse effects of chronic TNT contamination on the site-specific microbial community and the local carbon cycle in the soil.

Steliga et al. (2012) monitored the toxicity during and after *in situ* bioremediation of weathered drill wastes contaminated with petroleum hydrocarbons, located in an oil and gas plant in Grabownica (Poland). In this study, the *in situ* bioremediation process was implemented during three years and in three different phases: initial remediation (drainage), basic bioremediation (by inoculation with an indigenous bacteria-based biopreparation, as species that use hydrocarbons as the only source of carbon), and inoculation (with an indigenous bacteria-based biopreparation enriched with selected fungi species, which are able to degrade hydrocarbons). By the chemical analyses of petroleum hydrocarbons, it was detected the presence of aliphatic hydrocarbons, n-alkanes, isoprenoids as pristane and phytane, aromatic hydrocarbons (benzene, toluene, ethylbenzene and xylene - BTEX and PAH) and heavy metals in several soil samples. The basic bioremediation carried out with Actinomycetales bacteria showed that such microorganisms are able to degrade aliphatic and aromatic hydrocarbons. Other important conclusion was that the phases of inoculation with fungi species added to the bacterial consortium substantially increased the effectiveness of long-chain aliphatic and aromatic hydrocarbons biodegradation. All the toxicological tests used (Microtox®, Ostracodtoxkit™, Phytotoxkit™ and Ames) indicated that changes in toxicity during the *in situ* bioremediation were not connected with the decrease in the contents of petroleum contaminants, due to the formation of toxic metabolites. These assays also demonstrated that, after the *in situ* bioremediation process, non-toxicity (by the Microtox®,

Ostracodtoxkit™ and Phytotoxkit™ tests) and non-mutagenicity (by the Ames test) were found in the soil. Analyzing these results, the authors indicated that the studied soil after bioremediation could be used for agricultural purposes.

2.2. Natural Attenuation

Natural attenuation, also known as bioattenuation, refers to the decrease of the concentration of a certain contaminant in the soil as a result of a natural process, such as dilution, transformation, volatilization, sorption, run-off, volatilization, biodegradation and others (Yong and Mulligan, 2003; Megharaj et al., 2011). Nevertheless, according to Megharaj et al. (2011), in nutrient-poor soils and without the presence of degrading microorganisms, this process become slow and inadequate and require other technologies to improve this method.

Dawson et al. (2008) optimized the process of natural attenuation in a soil contaminated with BTEX (benzene, toluene, ethylbenzene and xylene), by amending with N and P. As the studied site had a long historical contamination by this mixture, it was not necessary the addition of BTEX-degraders microorganisms to the soil. The use of bioluminescent-based bacterial bioassays (constitutively expressed biosensor and BTEX-specific induced biosensor) indicated an increase in the toxicity for the first days of bioremediation, but after 38 days of the process, a reduction in the toxic potential of the samples was observed, corroborating the chemical analyses. Thus, the applied bioassays were able to detect the formation of toxic metabolites and alterations in the bioavailability of BTEX compounds.

2.3. Bioaugmentation/ Biostimulation

One of the most used bioremediation technologies is the bioaugmentation strategy. Bioaugmentation can be defined as the addition of microorganisms that are capable of biodegrading the target pollutant aiming to enhance and accelerate the biodegradation process. According to Lens et al. (2005) and Stroo et al. (2013), this strategy provides the enlargement of genes, enzymes and microbial diversity in the contaminated site to increase the variety of metabolic pathways that were not present in the indigenous population.

Another tool that is frequently used to promote a better biodegradation in contaminated soils is the biostimulation. This process supplies stimulating agents such as nutrients and oxygen to stimulate the growth of the natural microbial population present locally. According to Singh et al. (2011), factors such as adjusted pH, ideal C:N:P nutrient ratio in the medium, addition of biosurfactants or enzymes may contribute to the success of the biostimulation.

Xu and Lu (2010) assessed the efficiency of different biostimulation and bioaugmentation processes to bioremediate soils contaminated with crude oil under laboratory conditions. The authors selected an indigenous hydrocarbon-degrading bacterial consortium, using crude oil as the sole source of carbon, to promote bioaugmentation. To promote biostimulation peanut hull powder was used as a bulking agent and as a carrier material to immobilize bacteria cells. The Microtox® bioassay was used to evaluate the soil toxicity during the bioremediation processes. After 12 weeks of test, it was observed that bioaugmentation with immobilized cell on peanut hull powder soil was the most efficient treatment, removing 61% of crude oil. The performed bioassay indicated a high toxicity during the initial phase of the treatment probably due to the formation of toxic metabolites (e.g. aldehydes). However, a significant reduction in the toxicity of the samples was observed at the end, thus, proving the efficiency of this process.

A loamy sand soil contaminated with crude oil (38,300 mg/kg of total petroleum hydrocarbons) was bioremediated in laboratory by the combination of biostimulation and a modified Fenton oxidation (Gong, 2012). The toxicity was monitored along the process by the Microtox® bioassay. The efficiency of a treatment carried out with the addition of peanut hull and inorganic nutrients such as $(NH_4)_2SO_4$ and K_2HPO_4 (C:N:P ratio of 100:10:5) was compared with another treatment containing the same nutrients and the same bulking agent plus Fe(III) chelated with nitrilotriacetic acid (molar ratio of 1:1) to promote modified Fenton reactions. After 20 weeks, a decrease of 55.1% was observed for the biostimulation treatment and 88.9% for the combined treatment. The Microtox® test revealed an increase in the toxicity from week 2 until week 8, for both treatments. However, this toxic effect had a significant reduction after the chemical oxidation process, mainly for the last one. Thus, the use of toxicity test can contribute to determine the ideal treatment time and the most appropriated process to perform a more effective bioremediation.

Chemical and ecotoxicological assessments were carried out by Elgh-Dalgren et al. (2011) to evaluate the efficiency of two commercial methods to biodegrade PAH in an arsenic co-contaminated soil. The Daramend® and BioSan methods were chosen to study biostimulation and bioaugmentation, respectively. A non-ionic surfactant was added to the treatments to increase the PAH bioavailability. After 30 weeks, although a decrease in the PAH was observed for the majority of the treatments, none of them significantly reduced the contaminant concentrations. This was probably because the time used was not sufficient to biodegrade the contaminants at satisfactory levels. However, the microbial activity induced an arsenic remobilization in the soil, increasing the concentration of this metal in the available fraction, which is very harmful to the ecosystem. The Microtox® bioassay indicated that all the treatments induced a higher toxicity than the non-treated soil, mainly for the treatments that used surfactants to help in the biodegradation, possibly due to an increase in the PAH bioavailability. This study highlights the importance of using bioassays associated with bioremediation processes because besides identifying the presence of toxic metabolites they also provide information about changes in the bioavailability of the contaminants.

Spent mushroom compost was used to promote biostimulation and bioaugmentation in soils polluted with benzo[α]pyrene (Russo et al., 2012). The residual soil toxicity was evaluated by using Microtox® test with *V. fischeri* and by seed germination/root elongation tests with *Lepidium sativum*. Despite the reduction of 35% in the benzo[α]pyrene concentration by this treatment, the combination with ozone oxidation could contributed to a better efficiency, above 75%. Furthermore, the combination of these two processes also contributed to a decrease in soil toxicity.

Lettuce root elongation bioassay, growth inhibition and mortality of the ostracod *Heterocypris incongruens* were used to verify the efficiency of bioaugmentation and biostimulation treatments in the degradation of PAH-contaminated soils (Hamdi et al., 2007a). The authors added sewage sludge compost, decaying rice straw and/or known microorganisms PAH-degrader to soils contaminated with a mixture of anthracene, pyrene and benzo[α]pyrene to obtain a final concentration of 1000 mg of each PAH per kg of dry soil. After 120 days, it was observed that the use of decaying rice straw and bioaugmentation were the most efficient treatments, contributing to a decrease of 96% of the initial concentration of anthracene and pyrene. For the ecotoxicological assays, the ostracod showed a high sensibility to these compounds, since all the samples were toxic, causing 100% of mortality, while a decrease in the presence of PAHs in the samples also decreased the toxicity for lettuce. However, the addition of sewage sludge to the soil inhibited the seed germination, increasing the sample toxicity, even after 120 days of treatment.

Parameters such as low temperatures and lack of nutrients in the soil can jeopardize the microbial biodegradation. In an attempt to improve the biodegradation process of soils contaminated with petroleum hydrocarbons in high latitude environments, Coulon et al. (2005) determined the effects of adding nutrients and rising the temperature on the biodegradation of sub-Antarctic soils contaminated with diesel and crude oil. Despite the significant hydrocarbons degradation confirmed by chemical analyses, the toxicity evaluated by the modified Microtox® solid phase (STP) test remained almost the same after 180 days of bioremediation. Therefore, the authors suggested that as chemical and toxicity information do not always support each other, a bioassay to evaluate residual soil toxicity should be performed to confirm the bioremediation successfully.

Souza et al. (2013) studied the efficacy of the addition of sugar cane vinasse (by-product of the sugar industry) in a landfarming soil from a petroleum refinery in the reduction of genetic damages on *Allium cepa* meristematic cells. Although the use of this bulking agent in contaminated soils can favour the bioremediation to increase microbial activity, the authors concluded that the sugar cane vinasse contributed to an increase of chromosome breakages in this species, suggesting that the use of this compound associated with bioremediated organic residues can be harmful to the organism.

A study performed by Hu et al. (2012) evaluated the toxic and genotoxic potentials of a PAH-contaminated soil, before and after bioremediation using the chicken DT40 B-lymphocyte isogenic cell line and 15DNA-repair-deficient DT40 mutant cells. The soil used (66% sand, 28% silt, 6% clay, total organic matter of 16.6%) contained 14 of the 16 USEPA priority PAHs (not including acenaphthylene and indeno-[1,2,3-cd]pyrene) with a total concentration of 556 ± 50 ng/mg (dry mass basis, w/w). The aerobic bioreactor system used for the *ex situ* bioremediation of the soil was able to remove 69% of the total PAH. The *in situ* applied treatment, a biostimulated continuous-flow column system, removed 84% of the total PAH. The bioassays results indicated an increase in toxicity and genotoxicity for the bioreactor system over the course of a treatment cycle (7 days) and a decrease in these parameters for the long-term flow column system (2.5 years). So, the authors concluded that for PAH-contaminated soil, although different bioremediation strategies can reduce the contaminants, they can also lead to different results of toxicity, confirming the importance of using bioassays to verify the risks of bioremediation.

Another study also assessed the biodegradation of the 16 USEPA priority PAHs during 274 days and used the comet assay with RTL-W1 cells and the CALUX bioassay with H4IIE rat hepatoma cells to verify the toxic potential of samples from different periods (Andersson et al., 2009). A particulate material made from plant fibres and phosphate was added to the clayey contaminated soil. Although the chemical analysis indicated a small biodegradation after bioremediation, an increase in the biological activity was observed for both bioassays. The authors suggested that these results are due to the formation of toxic metabolites or the increase of PAHs bioavailability. So, bioassays can contribute to the choice for a safer bioremediation strategy, since the soil can become more toxic and harmful to the living organisms after this process.

Phillips et al. (2000) proved the importance of using more than one bioassay to monitor soil bioremediation. After adjusting the moisture content, temperature and nitrogen and/or phosphorus concentrations of soils contaminated with creosote, the authors used the Microtox® STP test, seed germination assay, earthworm survival test, the SOS-chromotest, Toxi-chromotest and a red blood cell (RBC) haemolysis assay to verify the success of bioremediation in a bench-scale microcosms. Chemical and biological analyses did not corroborate with each other, because an increase in toxicity was observed for earthworm, seed germination, RBC and Microtox® tests in spite of the biodegradation of total petroleum hydrocarbons. Moreover, it was observed that different bioassays can present different responses depending on the sensitivity of the organism used to a certain contaminant.

Different ecotoxicological endpoints (plant biomass production, earthworm reproduction, collembolans reproduction, microalgae growth and cladoceran reproduction) were used to verify the efficiency of a bioaugmentation with *Pseudomonas* sp ADP and a biostimulation with citrate in the improvement of the bioremediation of a soil contaminated with the atrazine herbicide (Chelinho et al., 2010). After 42 days of experiment, atrazine levels decreased to 98-99% of the initial concentration, proving the effectiveness of these combined strategies. For the ecotoxicological assays, after 10 days, no toxic effect was observed for the organisms tested.

2.4. Biosorption

Biosorption is also a widely used bioremediation strategy. This technique is based on the fact that a microbial biomass can bind and accumulate contaminants, removing them from the environment. The xenobiotics can bind to the cell walls, pigments and extracellular polysaccharides or accumulate inside the cells of bacteria, fungi and algae (Kotbra et al., 2011).

Srivastava and Thakur (2006) used the fungus *Aspergillus niger* to bioremediate a soil microcosm contaminated with different concentrations of chromium (250, 500, 1000, 1500 and 2000 mg/L). In addition, the seed germination inhibition test with wheat and the peroxidase enzyme assay with the leaves of wheat were used to verify the possible reduction on the toxicity of the samples. After 15 days, the authors observed a decrease of 75% for the lowest concentration and of 45% for the highest concentration, probably due to the bioaccumulation by the fungal mycelium. However, this reduction in the concentration of chromium in soil was not sufficient so that the values of seed germination, seedling length and peroxidase activity of the treatment became similar to the control.

2.5. Biopiles

An *ex situ* bioremediation methodology, known as biopile, mounted with excavation of contaminated soil, making it available in layers, promotes an improvement in the microbial anaerobic degradation (Wu and Crapper, 2009). Biopiles generally are from 2 to 4 metres high (Jørgensen et al., 2000), depending on the type of contaminated soil, space available and handling requirements (Kodres, 1999). Concentration of oxygen, soil humidity, concentration of nutrients and pH are controlled in the biopiles in order to optimize the growth and activity of the microorganisms (Kodres, 1999). Different agents such as straws and husks, wood chips, among other organic materials can be added to biopiles to increase their volume (Jørgensen et al., 2000).

Brooks et al. (1998) elaborated a pilot project to evaluate the efficiency of different methodologies based on their capacity in reducing total PAHs in soil, as well as analyze their toxic potentials. One of the methodologies chosen for the bioremediation was biopiles. In this experiment, the soil used was removed from a site contaminated with PAHs of creosote. Aiming to biomagnify the microbial population of this soil, cow manure (1%) was added to the biopile. The assessment and identification of the mutagenic agents in the soil, before and after the bioremediation, were estimated by fractionated extracts, subjected to the *Salmonella* histidine reversion assay. The authors also performed chromatographic chemical analyses of the samples (GC/MS). At the end of the experiment, it was observed a reduction in the concentrations of PAHs in the soils, when compared to the periods before and after the treatments. However, the authors observed an increase in the mutagenic activity, in relation to the control test. This result may be related with the presence of azaarenes, which are recognized mutagenic agents. This study proved the importance of the combination of bioassays with chemical analyses for a safer determination of the toxicity of environmental mixtures.

In order to evaluate the differences in the degradation of petroleum hydrocarbons, Delille et al. (2008) proposed several treatments for the bioremediation by biopiles. Soil samples from the Kerguelen Archipelago added with 180 g of diesel oil were used to make different biopiles. They were all exposed to different treatments, regarding ventilation, mechanical agitation, aggregation of organic nutrients and addition of surfactants. After one year of treatment, the authors performed analyses of the rates of the contaminants degradation (by chemical analyses - GC/MS) and toxicity assays with the luminescent bacterium *V. fischeri* (Microtox® SPT). In this study, it was observed almost total disappearance of the hydrocarbons for all the treatments. The researchers realized a reduction to less than 0.2% of total hydrocarbons for the biopiles treated with organic nutrients (fish compound), while the assays without this organic nutrient presented a reduction of less than 7%. This study showed a higher efficiency in the hydrocarbons biodegradation when an organic compound was added. Also, a higher amount of PAHs was observed in the biopiles that received the fish compound but did not receive physical treatment in relation to those that were ventilated or agitated. However, for several biopiles treated with organic compounds (fish compound), residual toxicity was observed as determined by the Microtox® assay. Thus, although the biopiles have been more efficient in the chemical reduction of the hydrocarbons with the addition of the fish compound, this nutrient contributed for an increase in the toxicity of the bioremediated sample, leading to an unsatisfactory result for this type of contaminant.

Plaza et al. (2005) studied two biopiles from stabilization ponds of a petroleum refinery in Czechowice, Poland. One of the analyzed biopiles, built in 1997, which had several active and passive aeration sections, was subjected to bioremediation process for four years and it was called biopile 1. The other biopile, built in 2001 (biopile 2), was treated passively for eight months. Thus, both biopiles received different biostimulation treatments. Then, hydrocarbons degradation rate was measured and the following bioassays were applied: Spirotox; Microtox®; Ostracodtoxkit F™; umu-test with S-9 activation; and assays with six species of higher plants (*Secale cereale, Lactuca sativa, Zea mays, Lepidium, Triticum vulgare, Brassica oleracea*). Toxicity of the soils contaminated with the bioremediated petroleum was evaluated by means of these bioassays. At each period, the authors observed a substantial decrease of the petroleum hydrocarbons in both treatments, being higher for biopile 1. A less toxic effect was also observed for biopile 1. The differences observed in this study can be assigned to the different treatment strategies employed for bioremediation, besides the exposure period of each treatment. The authors emphasized the importance of toxicological tests as environmental monitoring to evaluate the effectiveness of bioremediation technologies.

2.6. Landfarming

Landfarming is a medium to long term solid-phase remediation technology considered adequate for the treatment of soils contaminated by petroleum hydrocarbons, due to its containment, low cost, and high potential for success (HARMSEN, 1991; PEARCE; OELLERMANN, 1998). It is a soil bioremediation technology practiced by several oil refineries intended to reduce or eliminate hydrocarbons from petroleum sludge (Souza et al., 2009), using biological, chemical and physical natural processes (Pope and Matthews, 1993 apud Marin et al., 2005). According to these authors, the landfarming concept serves as a fundament for the design and operation of soil bioremediation technologies in many waste sites requiring cleanup.

A study developed by Mendonça and Picado (2002) monitored a coke oven soil belonging to a industrial site in operation located in Coina Estuary, Portugal, which is contaminated with high concentrations of PAHs (Picado et al., 2001), after the landfarming process, by chemical analysis and some ecotoxicological tests (Microtox® test, using *V. fischeri* for verifying bacterial toxicity; alga test, using *Pseudokirchneriella subcapitata* for measuring its growth inhibition; Daphnia test, for assessing the inhibition of the mobility of *Daphnia magna*; Mutatox test, analysis of the activation of the luminescence of *V. fischeri* to evaluate the samples genotoxicity; seed test, using *L. sativa* seeds to measure their inhibition of germination). In the beginning of the landfarming process, the aqueous elutriates of soil samples showed ecotoxic effects for *V. fischeri, D. magna* and *P. subcapitata*, but after two months of landfarming treatment it was observed a reduction in acute toxicity to *V. fischeri* and to *D. magna*, in addition to a gradual decrease of the soil chronic toxicity to the sensitive alga along the five months of treatment. Besides, toxic effects were verified for *L. sativa* only for the initial sample. Genotoxic effects were also detected in the initial soil sample, although after 2 months of landfarming treatment no genotoxic effects were registered. The toxic effects observed for all tested species in the beginning of the landfarming process were mainly related to the high concentrations of PAHs. The reduction of such effects, during the treatment, was associated to the biodegradation of low molecular weight PAH with two, three and four rings. All this demonstrated the effectiveness of the landfarming process as an efficient soil decontamination technology. In addition, according to the authors, the comparative analysis of these different ecotoxicity assays allowed to considerate that the alga test is the best procedure for the monitoring of soils polluted with PAHs.

Marin et al. (2005) evaluated the ability of landfarming to decrease or eliminate the total hydrocarbon content added to the soil with refinery sludge in an area with a semiarid climate

(Cantagena, Murcia, Spain), they also investigated the effect of this technique on the evolution of the microbial activity of the soil, using biological parameters (carbon fractions, microbial biomass carbon, basal respiration and ATP), determining biochemical parameters (enzymatic activities) and verifying the soil toxicity by the Microtox® assay after the landfarming technique. 80% of the hydrocarbons were eliminated after eleven months. Biological parameters of the samples of soils treated with landfarming technique showed higher values than the control soil during the first months, but these values were later reduced whereby the labile hydrocarbons were mineralized. The enzymatic activities (oxydoreductases and hydrolases) presented higher values in the soils impacted with the refinery sludge than in the control, although these values were reduced until levels similar to those of the control soil after eleven months. The percentage of inhibition of luminescent bacteria significantly decreased in the soils submitted to the landfarming process, which indicates that the water-soluble toxic compounds were totally cleaned after 9 months. All the results indicated that bioremediation of refinery sludge by landfarming technique is possible to be applied and good effectiveness in the tested conditions are obtained.

One research developed in Okayama, Japan, aimed to assess the ecotoxicity of a silt loam soil spiked with three PAHs at four levels (15, 75, 150, 300 mg ∑ 3 PAHs/Kg soil) after a long-term bioremediation process simulating landfarming, by using physico-chemical analysis, solid-phase bioassays and soil microbial activities. The bioremediation process consisted of irrigations in the presence (treated soil) or absence (control soil) of sewage sludge compost under greenhouse conditions. Fifteen months later, two PAHs (anthracene and pyrene) were almost completely degraded whereas benzo[α]pyrene persisted, especially in samples treated with sewage sludge compost. The application of this sludge was performed to evaluate the biowaste and stimulate PAHs biodegradation. Although, it caused soil salinization and pH reduction at the end of the bioremediation process, which consequently induced both strong phytotoxicity to lettuce as detected by the high inhibition of its germination. Also, it produced high rates of mortality to the ostracod *H. incongruens* in soils amended with sewage sludge compost. Moreover, in treated soils, the enzymatic activities of dehydrogenase and extracellular lignin peroxidase were not affected by the organic amendment and residual PAHs. Thereby, the authors concluded that the bioremediation efficiency was likely to be more limited by the bioavailability of PAHs rather than by the total number of PAH-degraders (Hamdi et al., 2007b).

Souza et al. (2009) assessed the genotoxicity of landfarming soil, before and after the addition of rice hulls using the *A. cepa* bioassay. They verified that the landfarming soil presented release of CO_2, besides high levels of total petroleum hydrocarbons and induced significant clastogenic and mutagenic effects in *A. cepa*, which were reduced after 108 days of biodegradation. Furthermore, landfarming treated with rice hulls promoted a considerable diminution in total petroleum hydrocarbons associated with the higher release of CO_2 and it caused the highest inhibition in the frequencies of mitotic indices and chromosome aberrations (mainly chromosomal adherence, bridges and breaks), micronuclei and nuclear buds. The authors inferred that the use of rice hulls accelerated the biodegradation effectiveness of landfarming, due to the increase of CO_2 release associated to the reduction of total petroleum hydrocarbons and the addition of rice hulls was able to reduce the landfarming genotoxicity. Therefore, such treatment could be a relevant indication to improve the efficiency of soils bioremediation processes and it may be a cost effective alternative for hydrocarbon bioremediation.

One study carried out by Frutos et al. (2012) evaluated the efficiency of two technologies (landfarming and slurry bioremediation) for the treatment of sludge (fine fraction with clay and silt) contaminated with diesel hydrocarbons from a soil washing process, by some chemical (gas chromatography) and ecotoxicity assays (survival test of the earthworm *Eisenia fetida*; seeds germination test with wheat, rape and red clover; microorganisms toxicity assays – soil respiration test and soil enzymatic activities; mobility test of *D. magna*; Algal (*Chlorella vulgaris*) growth inhibition test). Laboratory and pilot-real scales were studied, both for bioslurry (laboratory scale: microcosms – respirometry – and stirred reactors; pilot scale: air-lift pilot plant) and landfarming (laboratory scale: microcosms – respirometry – and mesocosms – columns+trays; demonstration scale: field – landfarming cell). The initial sample of sludge presented a total petroleum hydrocarbon content of 2243 mg/kg, which was decreased after both bioremediation treatments. In the case of bioslurry, there was a considerable reduction in the rates of total petroleum hydrocarbon of 57 and 65% in 28 days (laboratory and pilot scale, respectively). For landfarming process, there was a significant reduction in the levels of such hydrocarbons in six months (85% and 42%, for laboratory and real-scales, respectively). For earthworms, plants and microorganisms tested, the ecotoxic effects in the initial sludge sample were low, but after slurry and landfarming treatments, a decrease in such effects was observed, mainly in earthworm survival and in carbon mineralization at various concentrations tested. There was no relationship detected between the effects on seedling emergence and plant growth for untreated and treated sludge samples. No toxicity to aquatic organisms (algae and *Daphnia*) was observed in soil leachates, due to a strong adsorption onto the clay. Even though the results in real-scales of total petroleum hydrocarbons reduction efficiency were lower than that obtained at the laboratory scale, Frutos et al. (2012) indicated that the two bioremediation technologies studied, landfarming and slurry, are feasible alternatives for the recovery of sludges contaminated with hydrocarbons.

2.7. Soil Decomposer Animals

According to Haimi (2000), microorganisms are not the only responsible for degrading soil contaminants in bioremediation processes. Soil animals also have an important indirect role in these processes, because they are able to increase the microbial metabolic activities in the soil. In addition, they can be used as indicators of soil contamination before, during and after bioremediation. In addition to chemical analyses, soil animal studies are relevant to reveal associations between soil contamination levels and negative effects on local biota. For example, earthworms (*E. fetida, Eisenia andrei, Aporrectodea caliginosa* and *Lumbricus terrestris*) are considered soil decomposer animals highly accepted for use in simple assays of growth, reproduction and/or survival that evaluate effects of different chemical compounds on soil organisms. Other relevant soil decomposer organisms belonging to the Enchytraedis, Collembola, Mites, Isopods and Nematodes groups are also frequently used in the assessment of chemical toxicity and contaminated soils, by the analyses of reproduction and survival, besides the long life cycle analysis in the case of *Platynothrus peltifer* (Mites group), and the measurement of nutrient mineralization for *Porcellio scaber* and *Trichoniscus pusillus* (Isopods group). Despite the benefits that some studies show about minimizing the toxicity of contaminated soils using decomposer animals, it should be also considered that, in some cases, such animals may transfer toxicants accumulated in them to the above-ground food webs.

2.7.1. Vermicomposting

The vermicomposting is a type of bioremediation process that specifically uses earthworms, which eat, triturate, and digest organic wastes with the aid of aerobic and some anaerobic microflora, converting them into a much finer, humified and microbially active material (Ndegwa and Thompson, 2001; Maboeta and van Rensburg, 2003). Such process is able to reduce the toxicity of soils and sludges, because earthworms are capable of bioaccumulating heavy metals in their body tissues, as in chloragocytes, and their intestinal microflora that have the capacity of detoxifying some chemicals, like several pesticides (Srivastava et al., 2005). Wherefore, the earthworms could also provide an index of the bioavailability of heavy metals that might be present in the materials that are bioconverted, pointing out potential environmental hazards (Maboeta and van Rensburg, 2003). According to Ndegwa and Thompson (2001), the product generated after the vermicomposting is generally stable and homogenous, and may have reduced levels of contaminants.

Srivastava et al. (2005) evaluated the toxic, genotoxic and the phytotoxic potentials of three concentrations (2.5, 5 and 10%) of municipal sludge from a disposal site located on Lucknow-Kanpur, State of Uttar Pradesh, India, near to one automobile industry, by the *A. cepa* test, before and after the vermicomposting that used the species *E. fetida*. By Inductively Coupled Plasma Emission Spectrometry (ICP, Labtam Plasmalab 8440), the researchers determined some heavy metals (Cr, Cu, Ni e Pb) in the leachate of municipal sludge (10%), which were also detected in the leachate after the vermicomposting treatment, although showing lower values. Additionally, they observed that the municipal sludge before the vermicomposting process inhibited the mitotic index of the onion bulbs, plant weight, root elongation and chlorophyll contents in a concentration dependent manner. In contrast, the vermicomposting sludge induced similar values to the negative control. Only the highest concentration of the leachate tested (10%) without the vermicomposting induced significant morphological abnormalities and significant mitotic aberrations in the *A. cepa* roots, which became coiled or wavy and presented chromosomal bridges and fragments, unlike the leachate of the vermicomposting sludge. Therefore, the authors concluded that the municipal sludge tested was able to cause toxic, genotoxic and phytotoxic effects in *A. cepa*, which could be due to heavy metals or other organic compounds present in such sample. After the vermicomposting, this sludge induced a significant reduction of those negative responses, demonstrating that this bioremediation process can be employed as an important technique for recycling the municipal sludge, before depositing it in landfills or dumps.

Studies performed by Laine et al. (1995) apud Haimi (2000), corroborate the Srivastava et al. (2005) conclusions, by demonstrating the effectiveness of the vermicomposting to recuperate contaminated soils. Although, those authors had used other standard test species (*E. andrei*) that reduced the organic halogen compound levels in contaminated sawmill soils. On the other hand, Chang et al. (1997) showed that artificial soils treated by vermicomposting presented an increased toxicity when compared to the original lead-contaminated soil, due to the high salt levels originated from the bioremediation process with earthworms. Another study by Nedgwa and Thompson (2001) used a system consisting of two bioremediation processes – firstly the traditional thermophilic composting and after the vermicomposting with *E. fetida* – to degrade biosolids (activated sewage sludge) added with mixed paper-mulch as the carbon base. This system was able to reduce the stabilization time and ameliorate the biodegraded products quality, making it more stable and homogenous and with lower impact on the environment.

2.8. Phytoremediation

Phytoremediation is an emergent and low cost methodology, which makes use of vascular plants to remove or attenuate the effects of environmental pollutants (Cunningham and Berti, 1993; Rasking et al., 1994; Salt et al., 1998). Plants present a great capacity to absorb, adsorb, degrade and transform substances (Cunningham and Berti, 1993); therefore, they are considered efficient solar-driven pumping and filtering systems of pollutants (Cunningham et al., 1996). Associations established between microorganisms present in the soil and the root surfaces of plants can significantly increase the contact of the roots with the soil, helping in the remediation of contaminated environments (Cunningham and Berti, 1993). According to Salt et al. (1998), the phytoremediation technique can be applied in environments contaminated by organic and inorganic pollutants, present in several substrates. Plants can accumulate the pollutants in different parts of their biomass, such as, for example, in the aerial parts and roots. The roots can absorb or adsorb contaminants and can also decrease their bioavailability in the environment, besides enabling the volatilization of certain molecules or even removing them from the atmosphere around them (Salt et al. 1998).

Cofield et al. (2008) assessed the efficiency of the phytoremediation in the decontamination of soils contaminated by PAHs. The authors subjected samples of soils highly contaminated with PAHs of an old industrial complex of gas manufactured in Bedford, Indiana (USA) to the phyroremediation process for a period of 12 months. This was carried out with the species *Festuca arundinacea* and *Panicum virgatum*, besides performing two unvegetated controls, one unvegetated-unfertilized and the other unvegetated-fertilized. Bioassays were developed (3, 6, 9 and 12 months) to evaluate the survival of earthworms (*E. fetida*), survival of nematodes (*Caenorhabditis elegans*), germination of lettuce seeds (*L. sativa*) and microbial respiration. The authors observed that before the application of the phytoremediation process, the samples had toxic effects on all the bioassays (except for nematodes). After phytoremediation, total concentration of PAHs present in the soil decreased, and the concentration of many of the labile compounds became almost undetectable at the end of the bioassay. Moreover, there was an increase in the survival of earthworms, germination of lettuce and emission of CO_2. The results observed in the bioassays showed a correlation between the decrease in the soil toxicity at the end of the experiment and the decrease in the labile concentrations of the PAHs due to the Tenax-TA extraction procedure. This corroborates the importance of methodological associations to evaluate the residual toxicity, after remediation methodologies, as well as correlations of the bioassays with the presence of the substances in the samples.

Hamdi et al. (2012) conducted a study of phytoremediation with *Medicago sativa* (alfalfa) to remove PAHs present in contaminated soils. Besides the phytoremediation, the authors adopted a strategy of adding a bioresidue to the substrate (sewage sludge - SS) to increase/introduce new microorganisms to the indigenous microbiota. Also, a monitoring of the toxicity of the soil collected in a rice field (from the Research Institute for Bioresources – Japan) was carried out with Ostracoda (*H. incongruens*) and with *L. sativa* (lettuce germination test). To perform this study, besides a control test made with *in natura* soil, the soil was experimentally contaminated with a mixture of anthracene, pyrene and benzo[α]pyrene (proportion of 1:1:1) at four different concentrations (5, 25, 50 and 100 mg/kg dry soil).

Moreover, the authors also tested the efficiency of adding SS in the samples that presented the same conditions as the previous ones. Firstly, the soils were subjected to 15 months of bioremediation, simulating a landfarming process (absence of phytoremediation) to activate the microbial communities present there, aiming an improvement in the removal of PAHs. After this period, the soil samples were prepared to receive the alfalfa seeds, initiating the phytoremediation process, which was conducted in greenhouse for six months. Initially, the alfalfa germination rate in soils treated with SS was lower than in the untreated soils. However, at the end of the experimental period, this difference between the treatments was no longer significant, since, in both soils the plant growth was vigorous. No presence of PAHs was found in dried aerial parts of the plant, regardless the treatment performed. However, the phytoremediation had an important role, since it acted positively in the physicochemical properties of the soil, especially in those treated with SS, which were more acid and saline at the end of the 15 months exposure (simulating landfarming). After the phytoremediation period, anthracene was no longer detected in the treatments and pyrene was almost completely degraded. In this assessment, the benzo[α]pyrene was still persistent, but at concentrations below the initial levels.

According to the authors, this persistence was due to the benzo[α]pyrene irreversibly bound to the soil solid phase, which possibly is not bioavailable and does not represent a risk. The authors also alert for the possibility that the phytoremediation period have been insufficient or even to the inadequacy of the plant species (alfalfa) used in the process. Regarding the toxicity tests, the ecotoxicological data were directly correlated with the physicochemical alterations of the soil. After the phytoremediation period, survival of the ostracod increased, while phytotoxicity decrased. Thus, it confirms the suitability of both assays to reflect the direct toxicity of the soil.

Phytotoxicity tests can be useful to evaluate the persistence of PAHs in soils. These techniques can, eventually, be used to monitor the presence of these compounds in the soil. In this context, Maila and Cloete (2002) studied the phytoremediation of the species *Panacum maximum* (white buffalo grass) in the removal of these pollutants and the species *L. sativum* (garden cress) as a potential bioindicator. The soils employed in the study were: soil A (predominantly sandy loam soil taken from the CSIR site; Pretoria, SA); soil B (industrial soil taken from an industrial oil site; Secunda, South Africa and artificially contaminated with 1.2 g of PAH per kg of soil); soil C (industrial soil; Kwazulu-Natal, SA- containing 6% ilmenite, 0,4% rutile, 1 % zircon, 0,3% leucoxene, 1% magnetite and 1% kyanite); and soil D (white playpen sand bought from Lion Bridge SA Pty. Ltd.). The germination of *L. sativum* was performed in these different types of soil and the seeds were exposed to different concentrations of PAHs.

In order to establish the treatments, the authors made several combinations with the soils previously described with absence or presence of vegetation and with or without PAHs (1000 mg/L) rendering a total of 5 different treatments and a control test. The levels of seed germination and the fresh weight of *L. sativum* decreased with the increase in the concentration of PAHs. Phytoremediation by *P. maximum* effectively decreased the levels of

PAHs when compared with non-vegetated soil, suggesting a positive effect of the species to remove and detoxify PAHs.

Varun et al. (2011) evaluated the possibility of remediating environments contaminated with lead (Pb) with the plant species *Crinum asiaticum* (family Amaryllidaceae). The authors also assessed the efficacy of applying a bioassay to monitor the levels of this contaminant in the environment, using the germination index of the species *Hordeum vulgare* (barley). The authors prepared the samples, adding different concentrations of Pb (300, 600 and 1200 mg/L) to uncontaminated soils, resulting in three different treatments (T1, T2 and T3, respectively). The period of study was of four months and in 30-days intervals leaf samples of the plant and soil samples were collected both for the analysis of the Pb levels and to perform the bioassays. In the first periods of the experiment (30 and 60 days) similar results for the germination indices of the barley germination were obtained for the three treatments and the highest values were observed for the lower concentrations of the metal. At 90 days, the best germination persisted for the soils with lower levels of Pb, but the aspect of the seedlings was better when compared to the previous periods (which was curled, characteristic that indicates response to the toxicity of the environment). At the end of the experiment (120 days), the germination percentage improved significantly for all the treatments and it could be observed 90% of germination for T1 and T2, about 80% for T3 and 100% for the control. Furthermore, there was a reduction in the content of lead in the soil and an increase of it in the aerial part of the plant. Therefore, the species *C. asiaticum* can be considered an efficient biological material to remediate of soils contaminated with Pb. Also, *H. vulgare* seems to be a good bioindicator, characterizing itself as an important tool for the evaluation of the effectiveness of phytoremediation processes.

Pires et al. (2005) evaluated the ability of some plant species (*Cajanus cajan, Canavalia ensiformes, Dolichos lablab, Pannisetum glaucum, Estizolobium deeringianum, Estizolobium aterrimum* and *Lupinus albus*) to bioremediate soils contaminated with the herbicide tebuthiuron. For the experiment, Red-yellow Clay soil collected in secondary forest, plus four concentrations of tebuthiuron (0; 0.5; 1.0 and 1.5 Kg/ha) were used. As controls, soil samples with the same concentrations of the herbicide without the addition of plants were used. After 60 days of the phytoremediation, soils were fertilized, the bioindicator species *A. strigosa* (black oat) was sown and cultivated for another 60 days. After this period, height, dry weight of the aerial parts and intoxication symptoms of this plant were evaluated. The researchers observed that the plant height and the biomass of the aerial part of *A. strigosa* decreased as the levels of tebuthiuron increased. Lower phytotoxicity of the herbicide was observed when the soil was phytoremediated with the species *P. glaucum, L. albus* and *C. cajan* at a concentration of 0.5 kg/ha. For higher concentrations, the plants presented intoxication symptoms. However, lower effects were observed when the bioindicator was cultivated after the treatment with *C. ensiformes*. This suggests that, for this study, *C. ensiformes* was the best phytoremediator for tebuthiuron contaminated soils.

Table 1. List of the main researches carried out with several endpoints and bioassays to evaluate the efficiency of different bioremediation processes of contaminated soils

Bioremediation technologies	Contaminants	Bioassays	Endpoints	References
In situ bioremediation	TNT	Soil microbial assay Luminescent bacteria assay (*V. fischeri*)	Toxicity (CO_2-production rate) Toxicity (luminescence inhibition)	Frische and Höper (2003)
	Petroleum hydrocarbons (aliphatic hydrocarbons; n-alkanes; soprenoids; aromatic hydrocarbons) Heavy metals	Microtox® SPT (*V. fischeri*) Ostracodtoxkit™ (*H. incongruens*) Phytotoxkit™ (*Sorghum saccharatum, L. sativum, Sinapis alba*) Ames test (*Salmonella typhimurium*)	Toxicity (luminescence inhibition) Toxicity (mortality, growth inhibition) Toxicity (germination and growth inhibition) Mutagenicity (number of revertants)	Steliga et al. (2012)
Natural attenuation	BTEX	Bioluminescence-based bacterial bioassays	Toxicity (BTEX-specific induced biosensor, luminescence inhibition)	Dawson et al. (2008)
Bioaugmentation/ Biostimulation	Crude oil	Microtox® test (*P. phosphoreum*)	Toxicity (luminescence inhibition)	Xu and Lu (2010)
	Crude oil	Microtox® test (*P. phosphoreum*)	Toxicity (luminescence inhibition)	Gong (2012)
	PAHs and arsenic co-contamination	Microtox® (*V. fischeri*)	Toxicity (luminescence inhibition)	Elgh-Dalgren et al. (2011)
	Benzo[α]pyrene	Microtox® (*V. fischeri*) Seed germination and root elongation test (*L. sativum*)	Toxicity (luminescence inhibition) Toxicity (mortality, growth inhibition)	Russo et al. (2012)
	PAHs – anthracene, benzo[α]pyrene, pyrene	Root elongation test (*L. sativa*) Ostracod (*H. incongruens*)	Toxicity (growth inhibition) Toxicity (mortality)	Hamdi et al. (2007a)
	Diesel Crude oil	Microtox® SPT (*V. fischeri*)	Toxicity (luminescence inhibition)	Coulon et al. (2005)
	Petroleum hydrocarbons	Mitotic, nuclear and chromosomal aberrations assays (*A. cepa*)	Genotoxicity (chromosomal breaks, losses, bridges, adherences, multipolar anaphases, polyploidy cells, micronucleus and nuclear buds)	Souza et al. (2013)
	PAHs	DT 40 DNA damage response analysis (chicken DT40 B-lymphocyte isogenic cell line and 15DNA-repair-deficient DT40 mutant cells)	Toxicity (mortality) Genotoxicity (DNA damage)	Hu et al. (2012)
	PAHs	Comet assay (RTL-W1 cells) CALUX AhR agonist bioassay (H4IIE rat hepatoma cells)	Genotoxicity (DNA damage) Ah receptor (AhR) activating compounds	Andersson et al. (2009)

Table 1. (Continued)

Bioremediation technologies	Contaminants	Bioassays	Endpoints	References
Bioaugmentation/ Biostimulation	Creosote	Microtox® SPT (*P. phosphoreum*) Seed germination and root elongation test (*L. sativa*) Earthworm survival test (*E. fetida*) SOS-chromotest (*Escherichia coli*) Toxi-chromotest (*E. coli*) RBC haemolysis assay	Toxicity (luminescence inhibition) Toxicity (germination and growth inhibition) Toxicity (mortality) Genotoxicity (DNA damage) Toxicity (inhibition of β-galactosidase production) Toxicity (mortality)	Phillips et al. (2000)
	Atrazine herbicide	Earthworm reproduction test (*E.fetida*) Collembolans reproduction test (*Folsomia candida*) Cladoceran reproduction test (*D. magna*) Microalgae test (*P. subcapitata*) Biomass production test (*Avena sativa*)	Toxicity (reproduction) Toxicity (reproduction) Toxicity (reproduction) Toxicity (algal growth) Toxicity (biomass production)	Chelinho et al. (2010)
Biosorption	Chromium	Seed germination test (wheat) Peroxidase enzyme assay (wheat leaves)	Toxicity (germination and growth inhibition) Toxicity (peroxidase activity)	Srivastava and Thakur (2006)
Biopiles	PAHs	Ames test (*S. typhimurium*)	Mutagenicity (number of revertants)	Brooks et al. (1998)
	Petroleum hydrocarbons	Microtox® SPT (*V. fischeri*)	Toxicity (luminescence inhibition)	Delille et al. (2008)
	Petroleum hydrocarbons	Spirotox (*Spirostomum ambiguum*) Microtox® (*V. fischeri*) Ostracodtoxkit F^{TM} (*H. incongruens*) Umu-test with S9-activation (*S. typhimurium*) Seed germination and root elongation test (*S. cereale*, *L. sativa*, *Z. mays*, *L. sativum*, *T. vulgare*, *B. oleracea*)	Toxicity (mortality, bending and shortening of the cell) Toxicity (luminescence inhibition) Toxicity (mortality, growth inhibition) Genotoxicity (DNA damage) Toxicity (germination and growth inhibition)	Plaza et al. (2005)

Bioremediation technologies	Contaminants	Bioassays	Endpoints	References
Landfarming	PAHs	Microtox® test (*V. fischeri*) Alga test (*P. subcapitata*) *Daphnia* test (*D. magna*) Mutatox® test (*V. fischeri*) Seed test (*L. sativa*)	Toxicity (luminescence inhibition) Toxicity (growth inhibition) Toxicity (mobility inhibition) Genotoxicity (number of revertants) Toxicity (germination inhibition)	Mendonça and Picado (2002)
	PAHs	Microtox® test (*P.phosphoreum*)	Toxicity (luminescence inhibition)	Marin et al. (2005)
	PAHs	Phytotoxicity assay (*L. sativa*) Ostracod (*H. incongruens*)	Toxicity (growth inhibition) Toxicity (mortality)	Hamdi et al. (2007b)
	PAHs	Mitotic, nuclear and chromosomal aberrations assays (*A. cepa*)	Genotoxicity (chromosomal breaks, losses, bridges, adherences, multipolar anaphases, polyploidy cells, micronucleus and nuclear buds)	Souza et al. (2009)
	Diesel hydrocarbons	Earthworm survival test (*E.fetida*) Seed germination test (wheat, rape and red clover) *Daphnia* test (*D. magna*) Microbial assay Algal growth inhibition test (*C. vulgaris*)	Toxicity (mortality and weight) Toxicity (seedling emergence, wet weight) Toxicity (mobility inhibition) Toxicity (soil respiration inhibition, soil enzymatic activities inhibition) Toxicity (growth inhibition)	Frutos et al. (2012)
Vermicomposting	Municipal sludge (heavy metals; other organic compounds)	Mitotic index (*A. cepa*) Mitotic and chromosomal aberrations (*A. cepa*) Phytotoxicity assay (*A. cepa*)	Toxicity (mitotic index inhibition) Genotoxicity (chromosomal bridges, fragments and gaps) Phytotoxicity (plant weight, root elongation and chlorophyll contents)	Srivastava et al. (2005)
	Lead	Earthworm acute toxicity test (*L. terrestris*, *E. fetida*) Seed germination/Root elongation test (*L. sativa*, *A. sativa*) Mitotic index (*A. cepa*) Mitotic and chromosomal aberrations (*A. cepa*)	Toxicity (mortality) Toxicity (mortality, germination and growth inhibition) Toxicity (mitotic index inhibition) Genotoxicity (chromosomal aberrations)	Chang et al. (1997)
Phytoremediation	PAHs	Earthworm survival test (*E. fetida*) Nematodes survival test (*C. elegans*) Seed germination test (*L. sativa*) Microbial respiration	Toxicity (mortality and weight) Toxicity (mortality) Toxicity (germination inhibition) Toxicity (CO_2-production rate)	Cofield et al. (2008)

Table 1. (Continued)

Bioremediation technologies	Contaminants	Bioassays	Endpoints	References
Phytoremediation	PAHs – anthracene, benzo[α]pyrene, pyrene	OstracodToxKit F^{TM} (*H. incongruens*) Seed germination test (*L. sativa*)	Toxicity (mortality) Toxicity (germination inhibition)	Hamdi et al. (2012)
	PAHs – naphthalene, anthracene, phenanthrene, fluorine, pyrene	Seed germination and seedlings fresh weight tests (*L. sativum*)	Toxicity (germination inhibition and fresh weight reduction)	Maila and Cloete (2002)
	Lead	Seed germination test, root and shoot elongation test (*H. vulgare*)	Toxicity (germination and growth inhibition)	Varun et al. (2011)
	Tebuthiuron herbicide	Seed elongation and biomass test (*Avena strigosa*)	Toxicity (biomass, growth inhibition)	Pires et al. (2005)

CONCLUSION

According to our review of the literature, it is possible to conclude that different bioremediation processes have been extensively studied, proposing several technologies to be applied to remediate soil contamination. These technologies seem to be very efficient in the mineralization of pollutants commonly found in the soils, as petroleum hydrocarbons, pesticides, metals, complex mixtures and others. Furthermore, these bioremediation methods are considered a low cost and an environmentally friendly methodology. However, to better evaluate the effectiveness of bioremediation, especially regarding the formation of toxic intermediate metabolites and their bioavailability, we believe that some bioassays should be incorporated, before and after the bioremediation process. In table 1, we show some examples of endpoints and bioassays that are successfully used to verify the reduction of toxicity during different types of bioremediation.

REFERENCES

Alexander, M., 1994. *Biodegradation and Bioremediation.* Academic Press, 302p.

Alexander, M., 1999. *Biodegradation and bioremediation.* Academic Press, 453p.

Alexander, R.R., Tang, J., Alexander, M., 2002. Genotoxicity is unrelated to total concentration of priority carcinogenic polycyclic aromatic hydrocarbons in soils undergoing biological treatment. *J. Environ. Qual.* 31, 150-154.

Andersson, E., Rotander, A., 2009. AhR agonist and genotoxicant bioavailability in a PAH-contaminated soil undergoing biological treatment. *Environ. Sci. Pollut. R.* 16, 521-530.

Bamforth, S.M, Singleton, I., 2005. Bioremediation of polycyclic aromatic hydrocarbons: current knowledge and future directions. *J. Chem. Technol. Biotechnol* 80, 723-736.

Brooks, L.R., Hughes, T.J., Claxton, L.D., Austern, B., Brenner, R., Kremer, F., 1998. Bioassay-Directed Fractionation and Chemical Identification of Mutagens in Bioremediated Soils. *Environ. Health Perspect.* 106, 1435-1440.

Chang, L., Meier, J., Smith, M., 1997. Application of plant and earthworm bioassays to evaluate remediation of a lead-contaminated soil. *Arch. Environ. Contam. Toxicol.* 32, 166-171.

Chelinho, S., Moreira-Santos, M., Lima, D., Silva, C., Viana, P., André, S., Lopes, I., Ribeiro, R., Fialho, A.M., Viegas, C. A., Sousa, J. P., 2010. Cleanup of atrazine-contaminated soils: ecotoxicological study on the efficacy of a bioremediation tool with *Pseudomonas* sp. ADP. *J. Soil. Sediment.* 10, 568-578.

Cofield, N., Banks, M.K., Schwab, A.P., 2008. Lability of polycyclic aromatic hydrocarbons in the rhizosphere. *Chemosphere.* 70, 1644-1652.

Coulon, F., Pelletier, E., Gourhant, L., Delille, D., 2005. Effects of nutrient and temperature on degradation of petroleum hydrocarbons in contaminated sub-Antarctic soil. *Chemosphere.* 58, 1439-1448.

Cunningham, S.D., Anderson, T.A., Schwab, A.P., Hsu, F.C., 1996. Phytoremediation of soils contaminated with organic pollutants. *Adv. Agron.* 56, 55-107.

Cunningham, S.D., Berti, W.R., 1993. Remediation of contaminated soils with green plants: an overview. *In Vitro Cell. Dev.Biol.* 29, 207-212.

Dawson, J.J.C., Iroegbu, C.O., Maciel, H., Paton, G.I., 2008. Application of luminescent biosensors for monitoring the degradation and toxicity of BTEX compounds in soils. *J. Appl. Microbiol.* 104, 141-51.

Delille, D., Duval, A., Pelletier, E., 2008. Highly efficient pilot biopiles for on-site fertilization treatment of diesel oil-contaminated sub-Antarctic soil. *Cold Reg. Sci. Technol.* 54, 7-18.

Doran, J.W., Zeiss, M.R., 2000. Soil health and sustainability: managing the biotic component of soil quality. *Appl. Soil Ecol.*, 15, 3-11.

Elgh-Dalgren, K., Arwidsson, Z., Ribé, V., 2011. Bioremediation of a Soil Industrially Contaminated by Wood Preservatives—Degradation of Polycyclic Aromatic Hydrocarbons and Monitoring of Coupled Arsenic. *Water Air Soil Poll.* 214, 275-285.

Eweis, J.B., Ergas, S.J., Chang, D.P.Y., Schroeder, E.D, 1998. *Bioremediation Principles.* WCB/McGraw-Hill, 530p.

Frische, T., Höper, H., 2003. Soil microbial parameters and luminescent bacteria assays as indicators for *in situ* bioremediation of TNT-contaminated soils. *Chemosphere* 50, 415-427.

Frutos, F.J.G., Pérez, R., Escolano, O., Rubio, A., Gimeno, A., Fernandez, M.D., Carbonell, G., Perucha, C., Laguna, J., 2012. Remediation trials for hydrocarbon-contaminated sludge from a soil washing process: evaluation of bioremediation technologies. *J. Hazard. Mater.* 262-271.

Glanz, J.T., 1995. *Saving Our Soil: Solutions for Sustaining Earth's Vital Resource*. Johnson Books, 182p.

Gong, X.-B., 2012. Remediation of weathered petroleum oil-contaminated soil using a combination of biostimulation and modified Fenton oxidation. *Int. Biodeter. Biodegr.* 70, 89-95.

Haigh, S.D., 1996. A review of the interaction of surfactants with organic contaminants in soil. *Sci. Total Environ.* 185, 161-170.

Haimi, J. 2000. Decomposer animals and bioremediation of soils. *Environ. Pollut.* 107, 233-238.

Hamdi, H., Benzarti, S., Manusadžianas, L., Aoyama, I., Jedidi, N., 2007a. Bioaugmentation and biostimulation effects on PAH dissipation and soil ecotoxicity under controlled conditions. *Soil Biol. Biochem.* 39, 1926-1935.

Hamdi, H., Benzarti, S., Manusadžianas, L., Aoyama, I., Jedidi, N., 2007b. Solid-phase bioassays and soil microbial activities to evaluate PAH-spiked soil ecotoxicity after a long-term bioremediation process simulating landfarming. *Chemosphere* 70, 135-143.

Hamdi, H., Benzarti, S., Aoyama, I., Jedidi, N., 2012. Rehabilitation of degraded soils containing aged PAHs based on phytoremediation with alfafa (*Medicago sativa L.*). *Int. Biodeterior. Biodegrad.* 67, 40-47.

Hansen, P.-D., Blasco, J., DelValls, T.A., Poulsen, V., van den Heuvel-Greve, M., 2007. Biological analysis (Bioassays, Biomarkers, Biosensors). In: Barceló, D., Petrovic, M., (Eds.), *Sustainable Management of Sediment Resources: Sediment Quality and Impact Assessment of Pollutants*, pp. 131-161.

Harmsen, J., 1991. Possibilities and limitations of landfarming for cleaning contaminated soils. In: Olfenbuttel, R.F.H. (Ed.), *On-site bioremediation process for xenobiotic and hydrocarbons treatment*. Butterwort-Hetmann Publishimg, pp. 255-272.

Hu, J., Nakamura, J., Richardson, S.D., Aitken, M.D., 2012. Evaluating the effects of bioremediation on genotoxicity of polycyclic aromatic hydrocarbon-contaminated soil using genetically engineered, higher eukaryotic cell lines. *Environ. Sci. Technol.* 46, 4607-4613.

Hund, K., Traunspurger, W., 1994. Ecotox - Evaluation Strategy for Soil Bioremediation exemplified for a PAH - Contaminated Site. *Chemosphere*. 29, 371-390.

Jørgensen, K.S., Puustinen, J., Suortti, A.M., 2000. Bioremediation of petroleum hydrocarbon-contaminated soil by composting in biopiles. *Environ. Pollut.* 107, 245-254.

Kodres, C.A., 1999. Coupled water and air flows through a bioremediation soil pile. *Environ. Model. Softw.* 14, 37-47.

Kotrba, P., Mackova, M., Macek. T., 2011. *Microbial biosorption of metals*. Springer Science and Business Media, 329p.

Lens, P., Grotenhuis, T., Malina, G., Tabak, H., 2005. *Soil and Sediment Remediation: Mechanisms, Technologies and Applications*. IWA Publishing, 544p.

Maboeta, M.S., van Rensburg, L., 2003. Vermicomposting of industrially produced woodchips and sewage sludge utilizing Eisenia fetida. Ecotoxicol. *Environ. Saf.* 56, 265-270.

Maila, M.P., Cloete, T.E., 2002. Germination of *Lepidium sativum* as a method to evaluate polycyclic aromatic hydrocarbons (PAHs) removal from contaminated soil. *Int. Biodeterior. Biodegrad.* 50, 107-113.

Maila, M.P., Cloete, T.E., 2005. The use of biological activities to monitor the removal of fuel contaminants – perspective for monitoring hydrocarbon contamination: a review. *Int. Biodeter. Biodegr.* 55, 1-8.

Marin, J.A., Hernandez, T., Garcia, C., 2005. Bioremediation of oil refinery sludge by landfarming in semiarid conditions: Influence on soil microbial activity. *Environ. Res.* 98, 185-195.

Megharaj, M., Ramakrishnan, B., Venkateswarlu, K., Sethunathan, N., Naidu, R., 2011. Bioremediation approaches for organic pollutants: a critical perspective. *Environ. Int.* 37, 1362-1375.

Mendonça, E., Picado, A., 2002. Ecotoxicological Monitoring of Remediation in a Coke Oven Soil. *Environ. Toxicol.* 17, 74-79.

Ndegwa, P.M., Thompson, S.A., 2001. Integrating composting and vermicomposting in the treatment and bioconversion of biosolids. *Bioresource Technol.* 76, 107-112.

Pearce, K., Oellermann, R.A., 1998. Status and scope of bioremediation in South Africa. In: Sikdar, S.K., Irvine, R.L., (Eds), *Bioremediation: Principles and practice – Bioremediation technologies*, Technomic. Vol. III, pp. 155-182.

Phillips, T.M., Liu, D., Seech, A.G., Lee, H., Trevors, J. T., 2000. Monitoring bioremediation in creosote-contaminated soils using chemical analysis and toxicity tests. *J. Ind. Microbiol. Biot. 24*, 132-139.

Picado, A., Nogueira, A., Baeta-Hall, L., Mendonça, E., Rodrigues, M.F., Sàágua, M.C., Martins, A., Anselmo, A.M., 2001. Landfarming in a PAH-contaminated soil. *J. Environ. Sci. Health.* 36, 1579-1588.

Pires, F.R., Souza, C.M., Silva, A.A., Cecon, P.R., Procópio, S.O., Santos, J.B., Ferreira, L.R, 2005. Fitorremediação de solos contaminados com tebuthiuron utilizando-se espécies cultivadas para adubação verde. *Planta Daninha*. 23, 711-717.

Plaza, G., Nalecz-Jawecki, G., Ulfig, K., Brigmon, R.L., 2005. The application of bioassays as indicators of petroleum-contaminated soil remediation. *Chemosphere*. 59, 289-296.

Randerath, K., Randerath, E., Zhou,G.D., Suponpong, N., He, L.Y., Mcdonald, T.J., Donnelly, K.C., 1999. Genotoxicity of complex PAH mixtures recovered from contaminated lake sediments as assayed by three different methods. *Environ. Mol. Mutagen.* 33, 303-312.

Raskin, I., Kumar, P.B.A.N., Dushenkov, S., Salt, D.E., 1994. Bioconcentration of heavy metals by plants. *Curr. Opin. Biotechnol.* 5, 285-290.

Russo, L., Rizzo, L., Belgiorno, V., 2012. Ozone oxidation and aerobic biodegradation with spent mushroom compost for detoxification and benzo(a)pyrene removal from contaminated soil. *Chemosphere*. 87, 595-601.

Salt, D.E., Smith, R.D., Raskin, I., 1998. Phytoremediation. *Annu. Rev. Plant Physiol. Plant Mol. Biol.* 49, 643-668.

Seabra, P.N.C., 2005, Aplicação de biopilha na biorremediação de solos argilosos contaminados com petróleo. UFRJ: Tese em Ciências em Engenharia Química, 169 p.

Singh, A., Parmar, N., Kuhad, R.C., 2011. *Bioaugmentation, Biostimulation and Biocontrol.* Springer, Vol. 28, 364p.

Souza, T.S., Hencklein, F.A., Angelis, D.F., Gonçalves, R.A., Fontanetti, C.S., 2009. The *Allium cepa* bioassay to evaluate landfarming soil, before and after the addition of rice hulls to accelerate organic pollutants biodegradation. *Ecotox. Environ. Safe.* 72, 1363-1368.

Souza, T.S., Hencklein, F.A., de Angelis, D.F., Fontanetti, C. S., 2013. Clastogenicity of landfarming soil treated with sugar cane vinasse. *Ecotox. Environ. Safe.* 185, 1627-1636.

Srivastava, R., Kumar, D., Gupta, S.K., 2005. Bioremediation of municipal sludge by vermitechnology and toxicity assessment by *Allium cepa. Bioresource Technol.* 96, 1867-1871.

Srivastava, S., Thakur, I.S., 2006. Evaluation of bioremediation and detoxification potentiality of *Aspergillus niger* for removal of hexavalent chromium in soil microcosm. *Soil Biol. Biochem.* 38, 1904-1911.

Steliga, T., Jakubowicz, P. Kapusta, P., 2012. Changes in toxicity during *in situ* bioremediation of weathered drill wastes contaminated with petroleum hydrocarbons. *Bioresource Technol.* 125, 1-10.

Stokes, J.D., Paton, G.I., Semple, K.T., 2005. Behaviour and assessment of bioavailability of organic contaminants in soil: relevance for risk assessment and remediation. *Soil Use Manage.* 21, 475-486.

Stroo, H.F., Leeson, A., Ward, C.H., 2013. *Bioaugmentation for Groundwater Remediation.* Springer, 389p.

Varun, M., D'Souza, R., Kumar, D., Paul, M.S., 2011. Bioassay as monitoring system for lead phytoremediation through *Crinum asiaticum* L. *Environ. Monit. Assess.* 178, 373-381.

White, P.A., Claxton, L.D., 2004. Mutagens in contaminated soil: a review. *Mutat. Res.* 567, 227-345.

Wu, T., Crapper, M., 2009. Simulation of biopile processes using a hydraulics approach. *J. Hazard. Mater.* 171, 1103–1111.

Xu, Y., Lu, M., 2010. Bioremediation of crude oil-contaminated soil: comparison of different biostimulation and bioaugmentation treatments. *J. Haz. Mater.* 183, 395-401.

Yong, R.N., Mulligan, C.N., 2003. *Natural Attenuation of Contaminants in Soils*. CRC Press, 336p.

Yu, H., Li, F., Yu, W., Li, Y., Yang, G., Zhou, S., 2013. Assessment of organochlorine pesticide contamination in relation to soil properties in the Pearl River Delta, China. *Sci. Total Environ.* 447, 160-168.

In: Bioremediation: Processes, Challenges and Future Prospects ISBN: 978-1-62948-513-3
Editors: Jesús Bernardino Velázquez-Fernández et al.

Chapter 13

BIOREMEDIATION STRATEGIES FOR THE CLEAN-UP OF OIL CONTAMINATION IN AN ARID ENVIRONMENT: AN INSIGHT ON THE KUWAITI EXPERIENCE

Awatif Yateem*

Kuwait Institute for Scientific Research, Biotechnology Department, Kuwait

ABSTRACT

Two decades have passed since the State of Kuwait was subjected to one of the worst man-made environmental catastrophes. The destruction of the state's oil-producing facilities had led to the release of huge quantities of crude oil and the formation of oil lakes. The result was a massive oil contamination that covered in total around 25% of Kuwait's desert soil and formed an oil slick that was sedimented in the intertidal zone of the western coast of the Arabian/Persian Gulf. As the state griped and recovered from the crises, the Kuwaiti government gave priority to the cleanup research programs that utilized the facilities of different local environmental and oil sectors as well as a number of international organizations in collaborated research projects. The overall output of these projects was a comprehensive evaluation of the efficiency of physical, chemical, and biological remediation methods in treating and containing the oil contamination in soil, water and sediments as well as restoring and rehabilitating the severely damaged environment.

Biological remediation, or as generally known bioremediation, is one of the most appealing methods for restoring polluted sites by which hazardous contaminants are either broken down or stabilized as a result of the activities of living organisms. In Kuwait, different bioremediation strategies were tested and evaluated including windrow composting piles, land-farming, phytoremediation and rhizoremediation. However, the tested bioremediation technologies were governed with not only the nature and magnitude of oil pollution but also with the arid dry climate, the nutritionally poor nature of desert soil and the increased salinity in marine sediments environment. The major objective of this chapter is to provide a perspective on Kuwait's experience in developing bioremediation approaches that were applicable and proved to be successful under its harsh environmental conditions and arid climate.

* P.O. Box 24885, 13109, Safat, Kuwait; E-mail: ayateem@kisr.edu.kw.

Keywords: Biological remediation, windrow composting piles, land-farming, phytoremediation, rhizoremediation

INTRODUCTION

The State of Kuwait is an Arab nation located in the north-east of the Arabian Peninsula in western Asia and on the north-western shore of the Arabian/Persian Gulf. It is among the top 10 oil-producing countries with the second largest oil field in the world "Burgan" (EIA, 2011). Kuwait is an arid country with a harsh climate known for its dry hot summer and mild with limited rainfall winter. The soil of Kuwait is sandy and alkaline in nature with little organic matter (<1%), low contents of plant nutrients and high amounts of calcareous materials. Moreover, Kuwait's soil is also characterized with its low water-holding capacity (7%), high infiltration (50-100 cm/h) and high evaporation rate (3000 mm) (Abdal et al., 2002). Kuwait has nearly a 500 km coast on the Arabian/Persian Gulf that is considered as the main source of fresh water and electricity in the country. The coast is known for its elliptically shaped and shallow Kuwait Bay which protrudes from the Arabian/Persian Gulf in the westward direction at its northwestern corner (Al-Ghadban et al., 2001). This bay has a special importance since it is considered as unique nursing ecosystem for fish and shrimps. Most of the sediments in Kuwait's coast are sandy and muddy; in addition, the northern coast is bordered by extensive mudflats (Al-Ghadban et al., 2001).

GULF WAR ENVIRONMENTAL CATASTROPHE

As a result of the Gulf War that occurred in 1991, Kuwait's terrestrial and marine environment suffered considerable destruction. Around 700 oil wells in the Kuwaiti desert were damaged and the gushing oil formed over 300 oil lakes (Balba et al., 1996). The terrestrial contamination covered in total about 49 km^2 of Kuwait's desert with oil or heavy deposits of acidic and oily soot (Pilcher & Sexton, 1993) (Fig 1). The total volume of polluted desert soil was estimated to be around 20×10^6 m^3 and the oil penetrated to varying depth that ranged from 40 to 60 cm depending on the nature of the underlying soil (Radwan, 2008). Accordingly, some sites were defined as heavily contaminated containing up to 40% oil, while others which contained 5% oil or less were classified as lightly to moderately contaminated (Balba et al., 1996).

The Gulf water body also received its share of oil contamination which was estimated to be twenty times as much as the Exxon Valdez spill (Daniel and Poitevin, 1992). The release of 800,000 tons of crude oil in the sea formed a slick at the island adjacent to Al-Ahmadi oil terminal and close to the Kuwaiti coastlines (Cross, 1992). However, sea currents and south-east winds enhanced the movement of the oil slick and most of the crude oil became sedimented in the intertidal zone along more than 700 km of the western coast of the Arabian/Persian Gulf (Radwan, 2008; Daniel and Poitevin, 1992). Accordingly, only about 5% of the Kuwait's coastline was impacted by oil (Pilcher & Sexton, 1993; Al-Ghadban et al., 2001) (Figure 1).

The Kuwaiti government encouraged and supported research and development of appropriate technologies to prevent the aggravation of the problem and to restore the damaged environment. Local environmental and oil sectors such as Kuwait Institute for Scientific Research, Kuwait Foundation for the Advancement of Science, Kuwait University, Environmental Protection Authority, Kuwait Oil Company and Kuwait National Petroleum Company have directed their facilities and research programs to monitor the oil pollution and to lay the foundation for a full-scale remediation and rehabilitation plan of the contaminated sites. In addition, a group of international organizations have collaborated with the Kuwaiti sectors in joint research projects to contain and remediate the oil contamination in soil, sediments and water, as well as to restore the damaged environment. In this context, different remediation methods, including physical, chemical and biological, have been proposed and evaluated. The optimum technologies were selected to match site conditions, and in consideration of such factors as pollutants concentration, their components, and cost. The preliminary screenings identified that physical/chemical remediation technologies were best for treating heavily oil-contaminated soil, while bioremediation methods, due to their dependence on living organisms, were recommended for sites with light to moderate oil contamination (Balba et al., 1996).

Figure 1. Terrestrial and Marine Pollution in the State of Kuwait as a Result of Gulf War Environmental Catastrophe.

BIOREMEDIATION OF OIL-CONTAMINATED SITES

Bioremediation is the process in which microorganisms or plants are used to neutralize or breakdown and thereby detoxify contaminants. The success of bioremediation processes is dependent on a number of factors in the contaminated site. Among these factors are the nature of pollutants and their movement through the site structure. Chlorinated or nitrated compounds are known to be more difficult to degrade, while polycyclic aromatic hydrocarbons with high molecular weight usually have slower degradation rates. On the other hand, the soil and sediment structure and composition, the nutritional state of the contaminated site and its microbial composition are also important parameters of bioremediation processes. Accordingly, bioremediation can be stimulated by introducing sufficient air, moisture and nutrients, using proper soil amendment and agronomic techniques. Aeration can be improved by occasional soil turning and low moisture levels may be elevated by frequent irrigation. Fertilization, especially with nitrogen fertilizers and simple organic compounds, should be added to the contaminated site when poor in nutrients (Blackburn and Hafker 1993; Dua et al. 2002). The bioremediation process may also be enhanced by inoculation of oil-utilizing microorganisms in contaminated environment (Radwan, 2009). Moreover, whenever bioremediation rate slowdown and the contamination reach a tolerable concentration, some plant species were found capable of stabilizing pollutants and accumulating others or enhancing their degradation through their rhizospheric microflora.

In the view of the previous essential requirements, it is clear that applying bioremediation for the cleanup of the oil contamination in the Kuwaiti desert was not an easy job. The harsh and hot climate, the nitrogen poor soil and the scarcity of water altogether restrained microbial degradation of oil-hydrocarbons (Radwan, 2009). With this in mind, the target of bioremediation plans for treating oil-contaminated desert soil and marine sediments in Kuwait focused on simple and economical protocols that considered these restrictions and worked on improving them. Thus, a number of methods were tested and evaluated such as windrow composting piles, land-farming, phytoremediation and rhizoremediation and their advantages and disadvantages for application in arid environment were evaluated as will be discussed in the following sections.

BIOREMEDIATION OF MODERATELY CONTAMINATED DESERT SOIL USING WINDROW COMPOSTING PILES

In a joint research program between Kuwait Institute for Scientific Research and Japan Energy Center, oil-lake number 102 from Burgan oil field, which contained around 5000 m^3 of contaminated material, was selected for carrying out bioremediation and rehabilitation studies. The recoverable oil was collected from the 50 ha lake by Kuwait Oil Company, and an area of 133 x 75 m was demarcated and assigned for the project. In addition, a nearby location was set for carrying out bioremediation field studies (Balba et al., 1998).

In windrow composting study, moderately contaminated soil containing 3-5% of total petroleum hydrocarbons (TPH) was mixed with supplements and piled. The dimensions of each constructed pile were 3m wide, 20m in length and 1.5 m in height. The nitrogen poor soil was supplemented with inorganic fertilizer to provide a carbon: nitrogen ratio of 50:1. Given the dry nature of the desert soil in Kuwait, continuous supply of fresh water was provided equally for each pile to maintain soil moisture within the range of 5-10%. In addition, sufficient air was introduced by turning and mixing once a month using a front loader (Al-Daher et al., 2001). After 12 months treatment, around 70% reduction in TPH concentration was recorded which was used as an indication of active biodegradation within the piles (Al-Daher, et al., 2001). However, one of the drawbacks of composting soil pile method is that the aeration procedure used may promote partial evaporation of hydrocarbons and thus, maybe responsible for some of the reduction in TPH concentration (Benyahia, et al., 2005). Al-Daher et al. (2001) reported that the degradation rate was relatively enhanced in the autumn, winter and spring seasons compared to the hot and dry summer season. The study correlated this decrease in TPH degradation to the elevated ambient temperature (50-55°C) which resulted in rapid evaporation of water within the piles and thus negatively affected hydrocarbon degradation rates (Al-Daher, et al., 2001). Furthermore, the study noticed that no significant influence on TPH degradation rate was observed when inoculating the composting piles with a mixture of locally isolated hydrocarbon utilizing bacteria (Al-Daher, et al., 2001). Although the study did not provide an explanation for this situation which contradicted with most investigation in the field (Benyahia, et al., 2005; Radwan, 2009), it is possible that the added inocula did not survive or reproduce appreciably to influence the degradation of TPH due to unfavorable conditions specifically those related to temperature, moisture and oxygen. The study concluded that maintaining sufficient moisture and proper method of aeration are the key parameters that affect the efficiency of windrow composting pile method under the arid climate of Kuwait.

BIOREMEDIATION OF MODERATELY CONTAMINATED DESERT SOIL USING LAND-FARMING

The concept of land-farming depends on facilitating hydrocarbon degradation of contaminated soil, sediments or sludge by incorporating them with the upper surface of a clean soil and providing proper amendment to enhance microbial growth. Within the same joint research program between Kuwait Institute for Scientific Research and Japan Energy Center, land-farming method was tested to treat moderately contaminated soil taken from oil-lake number 102. In this study, the contaminated soil (3-5% TPH) was amended with inorganic fertilizer to give a carbon: nitrogen ratio of 50:1 and spread thinly (about 30 cm) over land-farming plots (40 x 40 m) (Balba et al., 1996). Sufficient aeration was maintained by periodical tilling while the moisture content was kept in an optimal range (8-10%) to enhance microbial degradation of oil hydrocarbons (Balba et al., 1996). Excellent biodegradation was obtained after 12 months treatment as around 80% of the original TPH concentration was reduced (Balba et al., 1998). However, the elevated ambient temperature in the summer reduced the moisture content to less than 5% within the experimental plots which slowdown TPH degradation rate (Al-Awadhi, et al., 1998). The study also evaluated the mobility of oil in the treated subsurface soil which is one of the main concerns when applying land-farming remediation process. The results revealed that no appreciable hydrocarbons or heavy metals leach was detected within the experimental plots that was attributed to the calcareous and alkaline nature of Kuwait's soil (Al-Awadhi, et al., 1998). An enhanced oil biodegradation was noticed in the first months after inoculating the contaminated soil with a booster of hydrocarbon utilizing microorganisms, but after six months their effect was not apparent. Therefore, the study conclude that an efficient and steady reduction in TPH concentration in the zone of incorporation within land-farming plots can be achieved by providing the optimum growth conditions for the indigenous soil microorganisms (Al-Awadhi, et al., 1998; Balba, et al., 1998).

PHYTO-RHIZOREMEDIATION OF TREATED OIL-CONTAMINATED DESERT SOIL

Phytoremediation by definition is the use of plants to remove, detoxify or immobilize environmental contaminants in a growth matrix (soil, water or sediments) through their natural biological, chemical or physical activities. This process can be successfully applied to uptake, contain and immobilize heavy metals when they are the source of contamination in certain sites. However, there is no evidence that plant roots may absorb water insoluble pollutants such as oil and oil products (Radwan, 2009). Usually, when phytoremediating oil-contaminated sites, plants utilize the activity of their rhizospheric microorganisms in a process known as rhizoremediation. Rhizospheric microorganisms can effectively degrade and utilize petroleum hydrocarbons in a contaminated matrix, and that is why the process is considered as the primary loss mechanism for these compounds. However, since plants have a direct effect on their rhizosphere and thus the inhabiting microorganisms, their role in rhizoremediation is quite important (Radwan, 2009). Practically, phytoremediation and rhizoremediation are found closely connected that one process is rarely applied without the other. However, both processes are used as a polishing step when the contamination becomes tolerable and non-toxic both for plants and their rhizospheric microflora. In the following sections we will discussed the findings of phyto-rhizoremediation experiments that were carried out as the next step in bioremediating oil-contaminated sites in Kuwait. The conducted experiments studied the potential of using selected species of native desert plants, crops and ornamental trees and grasses with their rhizospheric microflora to improve hydrocarbon degradation within bioremediated oil-contaminated desert soil.

Desert Plants

Shortly after the Gulf War environmental catastrophe, a group of wild plants typical of the Kuwaiti desert reappeared in some of the polluted areas with frequent annual increase (Radwan et al., 1998). A number of these desert plants were selected for phyto-rhizoremediation studies such as *Asthenatherum forsskalii, Cyperus conglomeratus, Launaea mucronata, Picris babylonica, Salsola imbricata, Senecio glaucus* and *Stipagrostis plumosa.* It was observed that some of these plants, although growing in black, polluted sand, frequently possessed white clean roots rich in oil-utilizing microorganisms (Radwan et al., 1995). However, it was difficult to vegetate desert plants or transfer them to other contaminated locations, since most of them are seasonal plants that grow periodically and are strongly adapted to their desert environment (Yateem et al., 2000).

Leguminous Plants

The nutritionally poor nature of the soil in Kuwait and the ability of leguminous plants to enrich the rhizosphere with fixed atmospheric nitrogen via nodule bacteria directed some of the early research in the field of phyto-rhizoremediation towards studying their potential in restoring oil-contaminated desert soil in Kuwait. Among the different legume plant species, the potential of using *Vicia faba* (broad beans) in phyto-rhizoremediation of oil-contaminated desert soil in Kuwait was intensively studied (Table 1). A team from Kuwait University carried out a number of research studies on this particular legume specie. The researchers were encouraged with the findings of the studies conducted by Radwan et al (1998; 2000) and Malallah et al. (1996), which demonstrated that *V. faba* could tolerate and survive in desert soil contaminated with 10% to 12.5% (w/w) crude oil. Radwan et al. (2000) also proved that the excised roots of *V. faba* with their intact rhizosphere were capable of attenuating n-octane, phenanthrene and crude oil when shaken into sterile desert soil extract containing these hydrocarbons. Moreover, the study observed that the amount of hydrocarbons eliminated were greater with roots of plants previously raised in oily soil than with roots of plants raised in clean soil which was attributed to the enrichment and adaptation of the of hydrocarbon utilizing microorganisms in the rhizosphere of *V. faba*.

El-Nawawy et al. (1999) and Yateem et al. (2000) studied the utilization of residual TPHs in the rhizosphere in *V. faba* in comparison to another legume, *Medicaga sativa* (alfalfa). The oil-contaminated soil used in the study was bioremediated for 18 months using land farming until the oil concentration dropped to reach around 1% (w/w) before being cultivated with the selected leguminous plants for 28 wk under field environmental conditions. The rhizosphere effect of the tested plants was expressed as rhizosphere: soil ratio (R/S) which was calculated by dividing the numbers of microorganisms in the rhizosphere soils by their numbers in the non-rhizosphere soils. Both legumes demonstrated similar TPH degradation rate as around 36.6% and 35.8% of the residual TPH was degraded in the rhizosphere of *V. faba* and *M. sativa* respectively compared to only 13% in the control non-vegetated soil. In general, TPH reduction was coupled with an increase in the R/S ratio of the hydrocarbon utilizing bacterial (HUB) counts associated with the two legumes which attributed TPHs degradation to the metabolic activity of hydrocarbon utilizing rhizospheric microorganisms. However, *V. faba* recorded higher R/S ratios of HUB than *M. sativa* (Yateem et al., 2000).

The ability of *V. faba* to tolerate and enhance the degradation of relatively high concentration of oil via their rhizospheric microflora encouraged further assessments of their ability to accumulate oily hydrocarbons pollutants. In the study conducted by Nageswara Rao et al. (2007), the roots of *V. faba* were analyzed to detect the absorbed aliphatic (AHs) and poly-aromatic hydrocarbons (PAHs) after 6 week growth in 12.5% (w/w) oil-contaminated soil. The obtained results demonstrated the ability of *V. faba* roots to absorb the AHs fraction of the crude oil in the polluted experimental soil. However, PAHs were not absorbed which the study related it to the possibility of the insolubility and toxicity of the available concentration of PAHs to the tested plant. Similar data was also reported by El-Nawawy et al. (1996) that investigated the uptake of elements, heavy metals, TPHs and PAHs by *V. faba* grown in land-farmed oil-contaminated soil (1 % w/w residual oil). The data indicated that no accumulation of the tested parameters was noticed in *V. faba* plant and heavy metals concentration did not exceed permissible limits for plants. Nevertheless, it is recommended that crop plants used in phyto-rhizoremediation experiments should be handled as toxic and not used as food or fodder (Radwan, et al., 2000).

Two elaborate studies were conducted by Radwan et al. (2007) and Dashti et al. (2009) on hydrocarbon utilization by nodule bacteria and plant-growth-promoting rhizo-bacteria (PGPR) associated with *V. faba*. In the study conducted by Radwan et al. (2007), quantitative gas liquid chromatography analysis revealed the ability of representative species of nodule bacteria and PGPR in utilizing crude oil plus a wide range of AHs and PAHs. Further, intact roots with nodules of *V. faba* significantly reduced hydrocarbon levels; this emphasized the importance of legume nodules not only in rhizoremediation but also as a biotechnological material to bioremediate oily liquid wastes. Dashti et al. (2009), on the other hand, proved that the surfaces of root nodules of legume plants may colonize HUB which contributes together with the rhizospheric HUB in the rhizoremediation of oil-contaminated soils. In this study, different HUB strains were isolated from the bacterial films spreading around root nodules collected from *V. faba* grown in desert soil mixed with 0.5% (w/w) weathered crude for 8 wk. The isolated HUB exhibited the ability to consume crude oil and representative pure AHs and PAHs and grew well in nitrogen-free medium.

Table 1. Summary of *Vicia faba* data in phyto-rhizoremediation experiments

Legume Plant	Tested Oil	Conc. % (w/w)	Experiment Duration (wk)	Reduction %*	Rhizosphere Effect of HUB (R/S Ratio)
Vicia faba (Broad Beans)	Crude	1.0	12	0.13[2]	26.5[1]
		5.0	12	0.82[2]	626.3[1]
		10.0	12	ND	647.1[1]
	Weathered	0.5	8	58[3]	ND
	Residual	1.0	28	23.6[4]	6.4[4]

ND= Not determined.

R/S Ratio= Numbers of HUB in the rhizosphere soil divided by their numbers in the non-rhizosphere soil.

* Calculated by subtracting the reduction recorded in the control non-planted treatment.

1. Radwan et al. (1998).
2. Radwan et al. (2000).
3. Dashti et al. (2009).
4. Yateem et al. (2000).

HUB in the rhizosphere of leguminous plants other than *V. faba* was investigated by Al-Awadhi et al. (2009) and Sorkhoh et al. (2010). In both studies, the roots of *Pisum sativum* (peas) and *Phaseolus vulgaris* (bean) were analyzed for the total numbers of HUB harbored in rhizospheric soil samples. In his study, Al-Awadhi et al. (2009) grew the selected legumes for 4wk in desert soil mixed with 0.5% (w/w) weathered crude oil; while, Sorkhoh et al. (2010) used soil mixed with 1% (w/v) crude oil, and grew the plants for 8 wk in the greenhouse. In both studies, a number of HUB were isolated and demonstrated a good potential to biodegrade crude oil, AHs and PAHs in a nitrogen free medium. It was therefore suggested that these strains of HUB may be sustainable regarding nitrogen nutrition. In spite of the similarities in the selected legume plants, the identities of the most frequently cultivable HUB differed tremendously between the two studies. It is our opinion however; that the type of the oil used to prepare the artificial oil-contaminated soil (crude or weathered) and its concentration may be the reason for such unexplained variation.

Grasses

In the studies conducted by Yateem et al. (2000; 2007) the potential of using selected species of grasses in phyto-rhizoremediation of oil contaminated soils in Kuwait was assessed under different experimental conditions. The grasses used in these studies were *Lolium perenne* (Yateem et al., 2000), *Zoysia tenuifolia* and *Paspalum vaginatum* (Yateem et al., 2007). The original TPH in the bioremediated soil used for vegetating *Lolium perenne* was 11300 ppm from which 24% was reduced compared to 13% in the control un-vegetated soil after 7 months duration of the study (Yateem et al., 2000). *L. perenne* exhibited a normal growth in the tested environment and an enhanced HUB growth as the calculated log counts of colony forming unit (CFU) of HUB /g of rhizosphere soil was 4 compared to 3 in the non-vegetated control. On the other hand, both *Zoysia tenuifolia* and *Paspalum vaginatum* grasses were tested in a bioremediated soil in which the recorded TPH did not exceed 2000 ppm a situation that may facilitated the growth of HUB as their calculated log counts of CFU were as high as 6 and 7 per gram of rhizosphere soil associated with *Z. tenuifolia* and *P. vaginatum* respectively. In this study, a number of aerobic mesophilic and thermotolerant Gram-negative and positive hydrocarbon utilizing bacteria, actinobacteria and fungi were isolated from the rhizosphere of both grasses. The study proved that the isolated hydrocarbon-utilizing microorganisms were affected by the grass species as *P. vaginatum* demonstrated more diverse assemblage of HUB among which some were known for their nitrogen-fixing capabilities (Yateem et al., 2007). The study also concluded that although the hydrocarbon degradation capabilities of the isolated microorganisms were variable, their action as a concert was outstandingly efficient.

Ornamental Trees

Trees have been successfully used in phyto-rhizoremediation applications of sites polluted with oil, heavy metals, pesticides and PAHs (Komives and Gullner, 2006; Pilon-Smits, 2005; Pulford and Watson, 2003). They have the advantage over grasses and other smaller plants in such applications due to their long penetrating roots that can easily reach subsurface pollutants (Komives and Gullner, 2006).

In Kuwait, limited studies assessed the use of trees in phyto-rhizoremediation applications. The trees were chosen based on their ability to withstand the arid dry climate of Kuwait as well as to tolerate and successfully grow in a bioremediated soil (Suleiman and Bhat, 2003). Al-Surrayai et al, (2009) investigated the reduction in TPH, PAH, and heavy metals concentrations in the rhizosphere soil of *Conocarpus lancifolius* that was used to phytoremediate soil contaminated with 800 ppm residual TPH. After three successive growing seasons, 85.7% of TPH concentration was reduced and the concentrations of the detectable PAHs were less than 0.02 ppm. Samples were also collected from the roots and the vegetative parts of the tree and were analyzed for heavy metals accumulation. The results demonstrated that *C. lancifolius* was able to accumulate high levels of chromium (Cr), vanadium (V), and nickel (Ni) in its roots, whereas Zn was accumulated in the leaves. The original oil-contaminated soil had high levels of heavy metals (Balba et al., 1996) and in spite of the different bioremediation methods applied, only 10 to 20% reduction was detected in the concentration of some heavy metal while the concentration of most remained unchanged (Al-

Daher, et al., 2002). Accordingly, the findings of this study were considered important for the long-term management of bioremediated soil.

Yateem et al, (2008), investigated the efficiency of hydrocarbon utilization by microorganisms associated with *C. lancifolius* and *Ficus infectoria* rhizosphere. The selected trees were grown in bioremediated soil with residual TPH that range from 800 to 1200 ppm. In this study, a number of HUB were isolated from the rhizosphere soil of both trees using culture-based methods which implies that such HUB may not necessarily be the dominant degraders. Among the different bacterial isolates, the study reported that *Rhodococcus equi*, which was isolated from *C. lancifolius* rhizosphere, demonstrated a pronounced degradation of a mixture of AHs and PAHs. Similarly, both *Arthrobacter globiformis* and *Mycobacterium fortuitum*, that were isolated from the rhizosphere of *F. infectoria*, were also distinguished with their efficient hydrocarbon degradation. The study was among a few that explored the role of rhizospheric fungi in hydrocarbon degradation. The obtained results demonstrated the efficient degradation ability of *Aspergillus versicolor*, which was isolated from the rhizosphere of *C. lancifolius*, and *Fusarium oxysporium*, which was isolated from the rhizosphere of *F. infectoria*, of the tested hydrocarbons especially PAHs. However, the highest degradation percentage of the tested hydrocarbons was noticed when a mixture of both HUB and fungi was used.

BIOREMEDIATION OF OIL-CONTAMINATED SEDIMENTS

One of the first signs of living that reappeared in some of the oil-contaminated coasts in Kuwait was the mats of blue green algae (Sorkhoh, et al., 1992). The microbiological analysis of these mats revealed that they harbor a number of immobilized heterotrophic oil-utilizing bacteria (Radwan, 2008). This fact was attributed to the ability of blue green algae to provided rather favorable living conditions for these bacteria especially those related to oxygen and nitrogen availability (Radwan, 2008). Such early observations directed most of the research in the field of restoring oil-contaminated coasts and sediments in Kuwait towards the self-cleaning attenuation as valuable and encouraging bioremediation systems.

The study conducted by Radwan et al. (1999) was among the first documenting the self-cleaning potential of the Kuwaiti coast due to the existence of oil-utilizing bacteria. This preliminary study identified a number of oil-utilizing bacteria mostly belonging to the genera *Acinetobacter* and *Micrococcus* from selected location along the Kuwaiti coast. The overall conclusion of the study was that oil pollution should be attenuated along the coast rather than in water body.

Other studies that were carried out by Al-Mailem et al. (2010_a) and Al-Saleh et al. (2009) focused on the diversity of oil utilizing bacteria along the Kuwaiti coasts. Actually, Kuwait Bay is known for its hyper saline coasts with salinity values that reach as high as 38-43% (Bakri and Kittaneh, 1998), which suggests that indigenous bacteria should be halotolerant, if not halophilic, to survive such an extreme environment. This fact was utilized by Al-Mailem et al. (2010_a) to study the hydrocarbon degradation ability of extreme halophilic archaeal strains isolated from selected hyper-saline coast in Kuwait, and the effect of the increased salinity on their biodegradation potential. The isolated strains of *Haloferax*, *Halobacterium* and *Halococcus* genera demonstrated enhanced oil and hydrocarbons degradation that was not negatively affected with the increased salinity. Therefore, the study concluded that these halophilic archaea could contribute to self-cleaning and bioremediation of oil-contaminated hypersaline environments.

Al-Saleh et al. (2009), on the other hand, studied the predominant culturable oil-utilizing bacteria in sediment samples collected from seven different sampling locations along the Kuwaiti coast. The isolated oil-utilizing bacteria mostly belonged to the genera *Pseudomonas*, *Bacillus*, *Staphylococcus*, *Kocuria*, *Acinetobacter* and *Micrococcus*. Using 16S-RFLP analysis, the genus diversity of the isolated oil-degrading bacteria was assessed, and the results revealed that four phylotypes of *Pseudomonas sp* and seven phylotypes of *Bacillus sp* were identified. The study suggested that such a result implies the high degree of diversity of oil-degrading bacteria at the strain level, but low diversity at the genus level.

PHYTO-RHIZOREMEDIATION OF OIL-CONTAMINATED SEDIMENTS

The potential of applying phyto-rhizoremediation for treating oil-contaminated sediments in Kuwait was explored using perennial marine vegetation naturally inhabiting hyper saline coastal areas of the Arabian /Persian Gulf such as *Halonemum strobilaceum* and *Avicennia marina*. In the study conducted by Al-Mailem et al. (2010_b), the potential of using *H. strobilaceum* in phyto-rhizoremediation of oil-polluted hypersaline coast was investigated. By analyzing the HUB associated with its roots, genera affiliated to *Halobacterium sp*, *Halococcus sp*, *Brevibacillus borstenlensis*, *Pseudoalteromonas ruthenica* and *Halomonas sinaensis* were isolated. Most of the isolated rhizospheric bacteria demonstrated a powerful hydrocarbon-attenuation even in a nitrogen-free media and tolerated a wide range of salinity. Therefore, the study concluded that *H. strobilaceum* could be a valuable tool for phytoremediation of oil-polluted hypersaline environments via rhizosphere technology.

Yateem and Al-Sharrah (2011), assessed the ability of another intertidal vegetation *A. marina*, that is commonly known as gray mangroves, to facilitate hydrocarbon degradation in polluted sediments by enhancing halotolerant rhizospheric HUB. Different HUB were isolated from the rhizosphere of *A. marina* among which *Bacillus subtilis ssp subtilis* was distinguished with its efficient degradation capabilities of aliphatic hydrocarbons as well as PAHs plus its toleration to the increased salinity of the marine sediments environment. The study proved the importance of this rhizospheric *B. subtilis ssp subtilis* (GenBank JQ396173) as none of the local studies identified its isolation from marine sediments from any location along Kuwait's coasts (Radwan, 2008; Al-Saleh et al., 2009; Al-Mailem et al., 2010). Additionally, although *B. subtilis ssp subtilis* is a common rhizobacteria, this study is considered to be the first to document its association with mangroves rhizosphere (Bayoumi and El-Nagar, 2009; Sahoo and Dhal, 2009; Bharathkumar, et al., 2008; Kathiresan and Selvam, 2006). The ability of *A. marina* to enhance the growth of such an effective HUB implies its important role in phyto-rhizoremediation applications of oil-contaminated saline sediments.

CONCLUSION

Planning a successful bioremediation protocol to clean and treat oil-contaminated sites in Kuwait was affected greatly with the nature and magnitude of oil pollution, the arid climate plus the physical and chemical characteristics of the polluted sites. The tremendous oil contamination which occurred in the Kuwaiti desert varied in concentration, yet bioremediation could be only applied to treat sites with moderate to light levels of contamination. Applied methods of bioremediation as windrow composting piles and land-farming, though proved to be effective in treating oil contamination, required intensive labor and heavy field equipments to maintain sufficient moisture and aeration which added to the maintenance cost of these applications (Al-Daher, et al., 2001; Balba, et al., 1998).

Figure 2. The Bio-soil Park (5,600 m^2) at Al-Ahmadi Area in the State of Kuwait.

On the other hand, applying phyto-rhizoremediation as a polishing step to treat residual oil pollution in the bioremediated soil has several advantages. The process is rather simple, economical and aesthetically pleasing (Komives and Gullner, 2000). In fact, when comparing phyto-rhizoremediation with other remediation technologies it appears that it is quicker but more expensive than natural self-cleaning attenuation and, conversely, slower but less expensive than most engineering techniques and traditional bioremediation methods. In Kuwait, some of the early studies were directed toward utilizing desert plants in phyto-rhizoremediation. However, such studies were terminated due to the difficulties in controlling the growth of such annual plants. The nutrient poor nature of the polluted desert soil tempted researchers in the field to utilize the symbiotic nitrogen fixing capabilities of legume plants in phyto-rhizoremediation experiments. The findings of these studies proved the nitrogen fixing capabilities of a number of HUB associated with the rhizosphere of some legumes, as well as the hydrocarbon utilizing ability of nitrogen fixing bacteria colonizing the surfaces of their root nodules. Experiments using ornamental trees had an extra advantage over other plants tested as they can be used for landscape plantation, a situation that is thought to be extremely suitable for such an arid environment. In fact, the findings of some of the phyto-rhizoremediation experiments were utilized to establish a bio-soil park in which the top 60-cm layer was replaced with bioremediated soil (TPH ~ 0.56%) and vegetated with different ornamental shrubs and trees (Figure 2) (Al-Zalzaleh and Shabbir, 2004).

In general, the marine environment in Kuwait was less impacted with the oil-pollution compared to the terrestrial environment; therefore, most of the studies recommended treating oil-contaminated marine sediments by self cleaning attenuation which is achieved by indigenous HUB. Still, other studies proved that the rhizosphere of some marine plants harbors unique and yet effective HUB which proposed the application of phyto-rhizoremediation as an aiding tool in treating oil-contaminated marine sediments.

REFERENCES

Abdal, M., Suleiman, M., Al-Ghawas, S. (2002). Variation of water retention in various soils of Kuwait. *Arch. Agron. Soil Sci.* 48, (1) 25-30.

Al-Awadhi, H., El-Nemr, I., Mahmoud, H., Sorkhoh, N.A., Radwan, S. (2009). Plant-associated bacteria as tool for the phytoremediation of oily nitrogen-poor soils. *Int. J. Phytorem.* 11, 11-27.

Al-Awadhi, N., Balba, M.T., Al-Daher, R., El-Nawawy, A., Yateem, A. (1998). Evaluation of bioremediation technologies for restoration of damaged desert soil in Kuwait. In: S. A. S. Omar, R. Misak and D. Al-Ajmi (Eds.), Sustainable Development in Arid Zones, pp. 93-108. The Netherland, A.A. Balkema, Rotterdam.

Al-Daher, R., Al-Awadhi, N., Yateem, A., Balba, M.T., El-Nawawy, A. (2001). Compost soil piles for treatment of oil-contaminated soil. *Soil Sediment Contam.*10 (2), 197-209.

Al-Daher, R., Al-Awadhi, N., Yateem, A., Balba, M.T., Chino, H., and Tsuji, H. (2002). Remediation and rehabilitation of oil lake beds in Kuwait: Field-scale demonstration. In: Y.A. Al-Shayji, J.S. Sidhu, M. Saleem and K. Guerinik (Eds.), Biotechnology Applications for the Arid Regions. pp. 269-279. Kuwait, Kuwait Institute for Scientific Research Press.

Al-Ghadban, A.N., Al-Majed, N., Al-Muzaini, S. (2001). The state of marine pollution in Kuwait: Northern Arabian Gulf. In: M. Al-Sarawi, M. S. Massoud (Eds.), The Impact of Environmental Pollution on Development in the Gulf Region, pp. 97-131. Kuwait, Four Films Press.

Al-Mailem, D.M., Sorkhoh, N.A., Al-Awadhi, H., Eliyas, M., Radwan, S.S. (2010_a). Biodegradation of crude oil and pure hydrocarbons by extreme halophilic archaea from hypersaline coasts of the Arabian Gulf. *Extremophiles* 14,321-328.

Al-Mailem, D.M., Sorkhoh, N.A., Marafie, M., Al-Awadhi, H., Eliyas, M., Radwan, S.S. (2010_b). Oil phytoremediation potential of hypersaline coasts of the Arabian Gulf using rhizosphere technology. *Bioresour. Technol.* 101 (15), 5786-5792.

Al-Saleh, E., Al-Drobiova, H., Obuekwe, C. (2009). Predominant culturable crude oil-degrading bacteria in the coast of Kuwait. *Int. Biodeterior. Biodegrad.* 63(4), 400-406.

Al-Surrayai, T., Yateem, A., Al-Kandari, R., Al-Sharrah, T., and Bin-Haji, A. (2009). The use of *Conocarpus lancifolius* trees for the remediation of oil-contaminated soils. *Soil Sediment Contam* 18(3), 354-368.

Al-Zalzaleh H, Shabbir G (2004) Effect of bioremediated soil on growth of different plant species. Kuwait J. Sci. Eng. 31(1): 107-118.

Bakri, D., Kittaneh, W. (1998). Physicochemical characteristics and pollution indicators in the intertidal zone of Kuwait: Implications for benthic ecology. *Environ. Manage.* 22(3), 415-424.

Balba, M.T., Al-Awadhi, N., Al-Daher, R., Chino, H., Tsuji, H. (1996). Remediation and Rehabilitation of Oil-Lake Beds in Kuwait: 1. Bioremediation of Oil-Contaminated Soil. In: N. Al-Awadhi, M.T. Balba, C. Kamizawa (Eds.), Restoration and Rehabilitation of the Desert Environment, pp. 21-40. The Netherland, Elsevier Science Press.

Balba, M.T., Al-Daher, R., Al-Awadhi, N., Chino, H., Tsuji, H. (1998). Bioremediation of oil-contaminated desert soil: The Kuwaiti experience. *Environ. Int.* 24 (1-2), 163-173.

Bayoumi, R., El-Nagar, A. (2009). Safe control methods of petroleum crude oil pollution in the mangrove forests of the Egyptian Red Sea coast. *J. Appl. Sci. Res.* 5(12), 2435-2447.

Benyahia, F., Abdulkarim, M., Zekri, A., Chaalal, O., Hasanain, H. 2005. Bioremediation of crude oil contaminated soils a black art or an engineering challenge. *Process Saf. Environ. Prot.* 83(B4): 364-370.

Bharathkumar, S., Paul, D., Nair, S. (2008). Microbial diversity of culturable heterotrophs in the rhizosphere of salt march grass *Porteresia coarctata* (Tateoka) in mangrove ecosystem. *J. Basic Microbiol.* 48, 10-15.

Blackburn, J. W., Hafker, W.R. (1993). The impact of biochemistry, bioavailability and bioactivity on the selection of bioremediation techniques. *Trends Biotechnol.* 11,328-333.

Cross, A.M. (1992). Monitoring marine oil pollution using AVHRR data: Observation of the coast of Kuwait and Saudi Arabia during January 1991. *Int. J. Remote Sensing* 13(4), 781-788.

Daniel, P., Poitevin, J. (1992). A numerical study on movements of an oil silk on the sea surface in the Persian Gulf from 25 January 1991 till 1 February 1992. First Thematic Conference on Remote Sensing For Marine and Coastal Environments. New Orleans, Louisiana, USA. pp. 249- 260.

Dashti, N., Khanafer, M., El-Nemr, I., Sorkhoh, N., Ali, N., Radwan, S. (2009). The potential of oil-utilizing bacterial consortia associated with legume root nodules for cleaning oily soils. *Chemosphere* 74,1354-1359.

Dua, M., Sethunathan, N., Johri, A.K. (2002). Biotechnology and bioremediation: successes and limitations. *Appl. Microbiol. Biotechnol.* 59,143-152.

EIA, Energy Information Administration Reports (2011) http://www.eia.gov/countries /country-data.cfm?fips=KU

El-Nawawy, A.S., Al-Daher, R., Yateem, A. (1999). Plant-soil interaction during bioremediation of oil contaminated soil. In: P. Kostecki, M. Behbehani (Eds), Assessment and Remediation of Oil Contaminated Soils, pp. 218-224., Kuwait, Arab School of Science and Technology, New Age International (P) Limited Publishers.

El-Nawawy AS, Al-Daher R, Yateem A, Al-Awadhi N (1996) Bioremediation of oil-contaminated soil in Kuwait: II- Enhanced landfarming for bioremediation of oil-contaminated soil. In: Moo-Young M, Anderson WA, Chakrabarty AM (eds) Environmental Biotechnology: Principles and Application, Kluwer Academic Publishers, The Netherland. pp. 249-258.

Kathiresan, K., Selvam, M.M. (2006). Evaluation of beneficial bacteria from mangrove *soil. Bot. Mar.* 49(1), 86-88.

Komives, T., Gullner, G. (2006). Dendroremdiation: The use of trees in cleaning-up polluted soils. In: M. Thomas (Ed.), Phytoremediation-Rhizoremediation, pp. 23–31. The Netherlands, Springer Verlag Press.

Malallah, G., Afzal, M., Gulshan, S., Abraham, D., Kurian, M., Dhami, M.S.I. (1996) *Vicia faba* as a bioindicator of oil pollution. *Environ. Pollut.* 92(2), 213-217.

Nageswara Rao, C.V., Afzal, M., Malallah, G., Kurian, M., Gulshan, S. (2007). Hydrocarbon Uptake by Roots of *Vicia faba* (Fabaceae). *Environ. Monit. Assess.* 132,439-443.

Pilcher, C.W.T., Sexton, D.B. (1993). Effects of the Gulf War oil spills and well-head fires on the avifauna and environment of Kuwait. *Sandgrouse* 15, 6-17.

Pilon-Smits, EAH. (2005). Phytoremediation. *Ann. Rev. Plant Biol.* 56, 15-39.

Pulford, I.D., Watson, C. (2003). Phytoremediation of heavy metal-contaminated land by trees- a review. *Environ. Int.* 29, 529-540.

Radwan, S. (2009). Phytoremediation for Oily Desert Soils. In: A. Singh, R. C. Kuhad, O. P. Ward (Eds.), Advances in Applied Bioremediation, Soil Biology 17, pp. 279-298. Berlin Heidelberg, Springer-Verlag Press.

Radwan, S. (2008). Microbiology of Oil-Contaminated Desert Soils and Coastal Areas in the Arabian Gulf Region. In: P. Dion, C.S. Nautiyal (Eds.), Microbiology of Extreme Soils, Soil Biology 13, pp. 275-298. Berlin Heidelberg, Springer-Verlag Press.

Radwan, S. S., Dashti, N., El-Nemr, I., Khanafer, M. (2007). Hydrocarbon utilization by nodule bacteria and plant growth-promoting rhizobacteria. *Int. J. Phytorem.* 9, 475-486.

Radwan, S. S., AI-Awadhi, H., EI-Nemr, 1.M. (2000). Cropping as a phytoremediation practice for oily desert soil with reference to crop safety as food. *Int. J. Phytorem* 2(4), 383-396.

Radwan, S.S., Al-Hasan, R.H., Al-Awadhi, H., Salamah, S., Abdullah, H.M. (1999). Higher oil biodegradation potential at the Arabian Gulf coast than in the water body. *Mar Biol* 135,741-745.

Radwan, S.S., AI-Awadhi, H., Sorkhoh, N.A., EI-Nemr, 1.M. (1998). Rhizospheric hydrocarbon-utilizing microorganisms as potential contributors to phytoremediation for the oily Kuwaiti desert. *Microbiol. Res.* 153, 247-251.

Radwan, S.S., Sorkhoh, N.A., EI-Nemr, I. (1995). Oil biodegradation around roots. *Nature* 376, 302.

Sahoo, K., Dhal, N.K. (2009). Potential microbial diversity in mangrove ecosystems: A review. *Indian J. Mar. Sci.* 38(2), 249-256.

Sorkhoh, N., Al-Hassan, R., Radwan, S., Hopner, T. (1992). Self-cleaning of the Gulf. *Nature* 359,109.

Sorkhoh, N.A., Ali, N., Salamah, S., Eliyas, M., Khanafer, M., Radwan, S.S. (2010). Enrichment of rhizospheres of crop plants raised in oily sand with hydrocarbon utilizing bacteria capable of hydrocarbon consumption in nitrogen free media. *Int. Biodeterior. Biodegrad.* 64, 659-664.

Suleiman, M.K., Bhat, N.R. (2003). Performance of Ornamental Plants in Bioremediated Soil. *Arid Land Res. Manage.* 17, 169-176.

Yateem, A., Al-sharrah, T. (2011) Enhanced Hydrocarbons Degradation in the Rhizosphere of Mangrove Plants by a Halophilic *Bacillus subtilis subtilis* Strain. *Soil Sediment Contam.* 20,261-273.

Yateem, A., Al-Sharrah, T., Bin-Haji, A. (2008). Investigation of microbes in the rhizosphere of selected trees for rhizoremediation of hydrocarbon-contaminated soils. *Int. J. Phytorem* 10(4), 1-10.

Yateem, A., Al-Sharrah, T., Bin-Haji, A. (2007) Investigation of microbes in the rhizosphere of selected grasses for rhizoremediation of hydrocarbon-contaminated soils. *Soil Sediment Contam* 16(3), 269-280.

Yateem, A., Balba, M.T., El-Nawawy, A.S., Al-Awadhi, N. (2000). Plants-Associated microflora and the remediation of oil-contaminated soil. *Int. J. Phytorem.* 2(3),183-191.

SECTION 5: DERIVATIVE BIOREMEDIATION

In: Bioremediation: Processes, Challenges and Future Prospects ISBN: 978-1-62948-513-3
Editors: Jesús Bernardino Velázquez-Fernández et al.

Chapter 14

ANNONACEOUS ACETOGENINS AS BIOFILM FORMATION PROMOTERS: CHALLENGES IN INCREASING THE EFFICIENCY OF PAHS BIODEGRADATION PROCESSES

***Adriana Neske*[1*], *Elena Cartagena*[1], *Alicia Bardón*[1,2] and *Eduardo Alberto Parellada*[1]**

[1] Instituto de Química Orgánica, Facultad de Bioquímica, Química y Farmacia. Universidad Nacional de Tucumán, Tucumán, Argentina
[2] INQUINOA–CONICET, Tucumán, Argentina

ABSTRACT

As a strategy for bioremediation, biofilm forming bacteria might be employed assisted by natural products, like annonaceous acetogenins (ACG), isolated only from plants of the Annonaceae family. The present chapter provides evidence that ACGs can act as biofilm formation promoters by a mechanism of stress without reducing the amount of bacteria and thus, improving efficiency of PAHs biodegradation.

Keywords: Annonaceous acetogenins, bacterial biofilm, naphthalene degradation

INTRODUCTION

Polycyclic Aromatic Hydrocarbons (PAHs) As Priority Pollutants for Remediation

The interest in the remediation mechanism and environmental fate of PAHs is their ubiquitous distribution, their low bioavailability and high persistence in soil, and their

[*] Corresponding Author address: Instituto de Química Orgánica, Facultad de Bioquímica, Química y Farmacia. Universidad Nacional de Tucumán; Email: aneske@fbqf.unt.edu.ar.

potentially deleterious effect on human health (Johnsen et al., 2005). Many PAHs are considered to be environmental pollutants that can have a detrimental effect on the flora and fauna of affected habitats, resulting in the uptake and accumulation of toxic chemicals in food chains and, in some instances, in serious health hazard and/or genetic defects in humans. On the basis of their abundance and toxicity, 16 PAHs have been identified by the US Environmental Protection Agency (EPA) as priority pollutants for remediation (Figure 1).

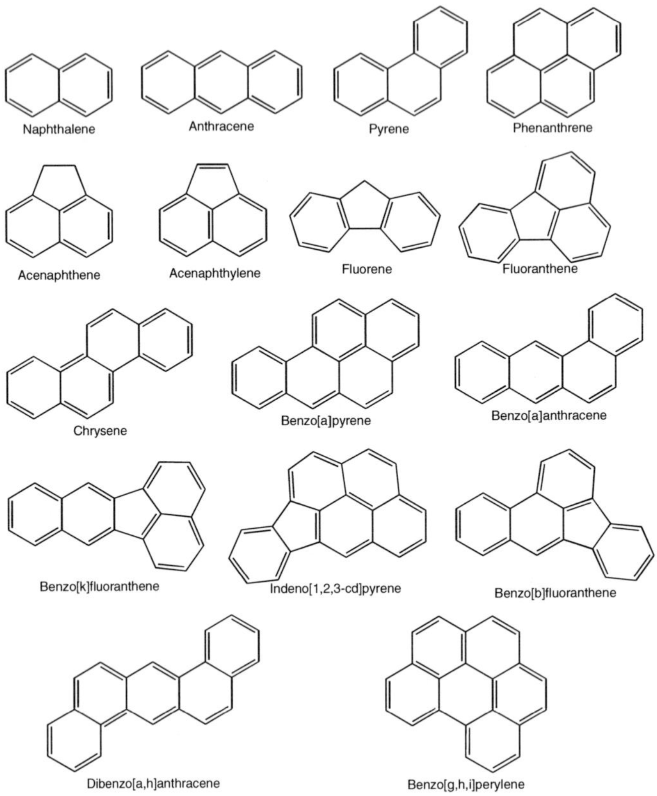

Figure 1. Chemical structures of 16 PAHs selected as priority pollulants for remediation by EPA.

Chemistry and Properties of PAHs

PAHs are a large group of organic compounds with two or more fused aromatic rings in linear, angular or cluster arrangements whose biochemical persistence arises from dense clouds of π-electrons on both sides of the ring structures, making them resistant to nucleophilic attack. They have a relatively low solubility in water, high hidrophobicities, and low volatilities, except small compounds like naphthalene. PAHs show long half-lives in geological media. In aerobic sediment, for example, half lives range from 3 weeks for naphthalene up to 300 weeks for benzo[a]pyrene. PAHs are regarded as persistant organic pollulants in the environment. This persistence is increasing with ring number and condentation degree (Henner et al., 1997).

Most of the PAHs with low vapour pressure in the air are adsorbed on particles. When dissolved in water or adsorbed on particulate matter, PAHs can undergo photodecomposition when exposed to ultraviolet light from solar radiation. PAHs may also be degraded by some microorganisms in the soil (WHO, 2000).

Sources of PAHs

PAHs are formed mainly as a result of pyrolytic processes, especially the incomplete combustion of organic materials during industrial and other human activities, such as processing of coal and crude oil, combustion of natural gas, including for heating, combustion of refuse, vehicle traffic, cooking and tobacco smoking, as well as in natural processes such as carbonization. There are several hundred PAHs; the best known is benzo[a]pyrene.

Health Effects of PAHs

According to the experimental results, the most significant health effect to be expected from inhalation exposure to PAHs is a high risk to lung cancer. In the past, chimney sweeps and tar workers were dermally exposed to substantial amounts of PAHs leading to skin cancer in many of them. Epidemiological studies in coke-oven workers, coal-gas workers and employees in aluminum production plants provide sufficient evidence of the role of inhaled PAHs in the induction of lung cancer. A significant high value of lung cancer mortality was found in coke-oven workers (WHO, 2000; Hemminki, 1993).

Naphthalene as Model PAHs for Remediation

Naphthalene, the first member of the PAH group, is a common micropollutant in potable water and serves as a model for understanding the properties of a large class of environmentally prevalent PAHs (Liu et al., 2001; Samanta et al., 2002).

The toxicity of naphthalene has been well documented and cataractogenic activity has been reported in laboratory animals (Goldman et al., 2001; Mastrangela et al., 1997). Naphthalene binds covalently to molecules in liver, kidney and lung tissues, thereby enhancing its toxicity; it is also an inhibitor of mitochondrial respiration (Falahatpisheh et al., 2001). Acute naphthalene poisoning in humans can lead to haemolytic anaemia and nephrotoxicity. In addition, dermal and ophthalmological changes have been observed in workers occupationally exposed to naphthalene (Samanta et al., 2002).

Microbial Degradation of PAHs

Contamination of soils, groundwater, sediments, surface water, and air with hazardous and toxic chemicals is one of the major problems facing the industrialized world today. The need to remediate these sites has led to the development of new technologies that emphasize the destruction of the recalcitrant compounds rather than the conventional approach of disposal. Bioremediation, the use of microorganisms or microbial processes to degrade environmental contaminants, is among these new technologies.

To be successful, bioremediation methods depend on having the right microbes in the right place with the optimum environmental factors. The right microbes are bacteria or fungi, which have the physiological and metabolic capabilities to degrade the pollutants. Bioremediation offers several advantages over conventional techniques such as land filling or incineration. Bioremediation can be done on site, it is often less expensive and site disruption is minimal, it eliminates waste permanently, and has greater public acceptance, counts with regulatory encouragement, and it can be coupled with other physical or chemical treatment methods (Boopathy, 2000).

The first step in the microbial degradation of PAHs is the action of dioxygenase, which incorporates atoms of oxygen at two carbon atoms of a benzene ring of a PAH resulting in the formation of *cis*-dihydrodiol (Kanaly et al., 2000), which undergoes re-aromatization by dehydrogenases to form dihydroxylated intermediates. Dihydroxylated intermediates subsequently undergo ring cleavage and form TCA-cycle intermediates (Sabaté et al., 1999).

A wide variety of PAHs are abundant in nature due to incomplete combustion of organic matter. The PAHs from extraterrestrial matter are also oxidized and reduced thanks to prevalent astrophysical conditions, and result in the formation of various organic molecules, which are the basis of early life on primitive earth. Moreover, the microorganisms (naturally occurring or genetically engineered) can mineralize toxic PAHs into CO_2 and H_2O (Samanta et al., 2002).

Bacterial Biofilm. A Promising Bioremediation Strategy of PAHs

Biofilm associated microorganisms play a crucial role in terrestrial and aquatic nutrient cycling and environmental contaminant biodegradation (Li et al., 2009). Biofilm-mediated bioremediation is safer and more efficient than that produced by planktonic organisms. This is because biofilm cells have a better chance for adaptation and survival (especially during periods of stress) since they are protected within the matrix (Decho, 2000).

Biofilm systems are especially suitable for the treatment of recalcitrant compounds because of their high microbial biomass and ability to immobilize compounds. Bioremediation is also facilitated by enhanced gene transfer among biofilm organisms and by the increased bioavailability of pollutants like PAHs for degradation as a result of bacterial chemotaxis. Strategies for improving bioremediation efficiency include genetic engineering to improve strains and chemotactic ability, the use of mixed population biofilm and optimization of physico–chemical conditions (Singh et al., 2006).

What Is Biofilm?

Biofilms are sessile microbial communities whose cells are anchored to a substrate or to an interface of other bacteria that produce and exhibit an altered phenotype with respect to their growth rate and gene transcription. These bacteria are embedded in a matrix of extracellular polymeric substances that they produce (Rodney et al., 2002).

Small Signals Molecules Regulate the Bacterial Communal Behavior

A variety of bacteria employs quorum sensing (QS) to coordinate communal behavior. QS consists in the regulation and coordinated expression of genes in response to cell density (Gonzalez and Keshavan, 2006). This phenomenon is caused by the production, release and detection of small signal molecules called type 1 autoinducers (AI-1). In Gram negative bacteria they are *N*-acyl-homoserine lactones (AHLs) (Bassler and Losick, 2006). AHLs play an important role in the regulation of biofilm formation in some bacteria (Waters and Bassler, 2005) like *Pseudomonas aeruginosa*.

Several bacterial phenotypes essential for the successful establishment of symbiotic, pathogenic, or commensal relationships with eukaryotic hosts, including motility, exopolysaccharide production, biofilm formation, and toxin production, are often regulated by quorum sensing. Interestingly, eukaryotic organisms like plants produce quorum-sensing-interfering (QSI) compounds that have a positive or negative influence on the bacterial signaling network (Gonzalez and Keshavan, 2006).

Natural Products from Annonaceae Plants As Biofilm Formation Promoters

Annonaceous Acetogenins

Annonaceous acetogenins (ACGs) constitute a series of natural products isolated from Annonaceae species (Zeng et al., 1996; Zafra-Polo et al., 1996, 1998; Cavé et al., 1997; Alali et al., 1999; Tormo et al., 1999) that are widely distributed in tropical and sub-tropical regions. ACGs exhibit a broad range of biological properties such as cytotoxic, antitumoural, antiparasitic, pesticidal, antimicrobial and immunosuppresive activities. Mechanism of action studies have shown that ACGs are the most potent inhibitors of the mitochondrial respiratory chain complex I. The biogenetic pathway of such unusual secondary metabolites is supposed, since no systematic studies have been carried out so far. Starting from a very long chain fatty acid, introduction of the terminal γ-lactone appears first, then, by oxidation of the unsaturated units present, followed by opening and closing reactions, the tetrahydrofuran (THF) and tetrahydropyran (THP) rings are introduced.

Until 2005, the number of ACGs was 417, 40 of them being linear, 16 epoxy, 206 mono-THF, 146 bis-THF, 1 tri-THF, and 8 other ACGs belonging to the THP group (Bermejo et al., 2005). The general structures of ACGs have 32 or 34 carbons, including a long chain alkyl group on one end and a mono unsaturated or saturated γ-lactone on the other end (Alali et al., 1999; Bermejo et al., 2005). Several oxygenated functions, such as hydroxyl, ketone, epoxide, THF and THP, may be present, as well as double and triple bonds (Figure 2).

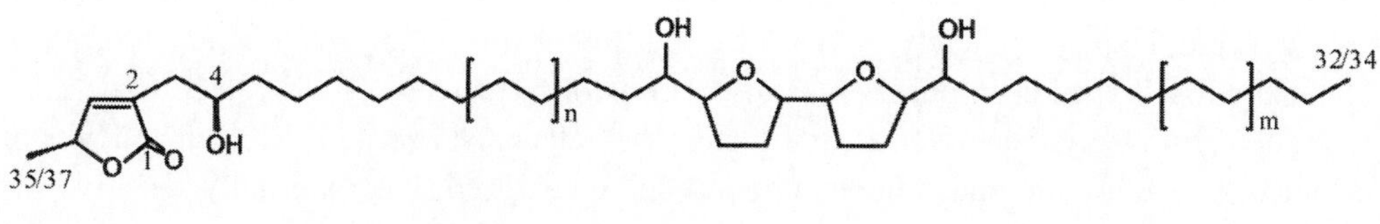

Figure 2. General structure of ACGs.

Extraction, Isolation and Purification of ACGs

The classic extraction of ACGs from plants is carried out by successive solvent extractions with increasingly polar solvents, or by liquid/liquid partition from an initial alcoholic extract. The separation of ACGs is then performed by chromatography on silica gel or by preparative HPLC with refractive index (RI) and ultraviolet (UV) detectors, or by mass spectrometry (MS), with or without separating the individual components by liquid chromatography (LC) (Zeng et al., 1996; Cavé et al., 1997; Alali et al., 1999). Positional isomeric and epimeric ACGs have been successfully separated by preparative HPLC. Countercurrent chromatography (CCC or CPC) has also been used for isolating ACGs. This method is highly efficient since large amounts of mixtures and crude extracts, when purified by this method, have yielded pure compounds. Again, positional isomers, epimers and homologous ACGs have been successfully separated by CPC (Bermejo et al., 2005).

Structural Elucidation of ACGs

The structural elucidation of ACGs has already been summarised in previous reviews (Zeng et al., 1996; Cavé et al., 1997; Alali et al., 1999). This consists of analysis of the mass spectra to determine the exact molecular formula, followed by Electronic Ionization Mass Espectrometry (EI-MS) and Fast Atom Bombardment (FAB-MS), which allow to determine the position of the functional groups on the alkyl chain. Elucidation of the relative stereochemistry of the stereogenic centres by detailed analysis of the ^{1}H- and ^{13}C-NMR spectra is now straightforward, thanks to the possible comparisons of the spectra between those of the natural products and synthetic models. To determine the absolute configurations, the advanced Mosher ester methodology has allowed several authors to determine unambiguously the absolute configuration of the carbinols present in the vicinity of THF rings, and thus to assign the whole configuration of the THF units. For isolated carbinols, the use of 2-NMA (2-napthylmethoxy acetic acid) esters is of great interest, since the influence of the aromatic rings can extend to five carbon–carbon bonds (Duret et al., 1997).

Concerning the absolute configuration of the terminal γ-lactone, several methods have been used and described in the literature, but they all required the degradation of the natural ACGs in order to obtain lactic acid derivatives, which can be analysed by chromatography. Recently, it has been proposed to analyse ACGs by ^{1}H-NMR spectroscopy, in the presence of a chiral solvating agent (CSA), and deduce the absolute configuration of the terminal γ-lactone by analysing the chemical shift differences of the carbinolic proton of the lactone in the presence of the (*R*)- and (*S*)-CSA (Latypov et al., 2002). Circular dichroism has also been used for the determination of the absolute configuration of the terminal γ-lactone (Gawronski and Wu, 1999).

ACGs are poor chromophores, so they can be tentatively identified as any peak in a chromatogram that has only one λ_{max}, at or about 210 nm. May be used an HPLC with a C-18 (octadecyl silica, or ODS) semipreparative scale column and RI detection with a methanol gradient as mobile phase to analyze ACGs from different species (Bermejo et al., 2005).

Once an ACG has been purified, Fourier-tramsform infrared (FTIR), NMR and MS are important methods and complementary techniques of analyzing it. FTIR can tell whether functional groups are present in the compound being analyzed. It is especially sensitive for –OH groups. FTIR produces relatively broad peaks that are characteristic of functional groups, such as the OH at about 3400 cm^{-1}, CH_2 at about 2922 cm^{-1}, a THF ring at about 2850 cm^{-1} and a butenolide at about 1744 cm^{-1} (Bermejo et al., 2005).

ACGs can be identified by characteristic peaks in the NMR spectrum. An unsaturated methylated γ-lactone will produce ^{13}C-NMR peaks at about 173.9, 134.3, 148.0, 77.4 and 19.2 ppm due to carbons 1 (O–C=O), 2 (C=CH), 33 (C=CH) and 34 (CH), respectively, and ^{1}H-NMR peaks at 7.0 (H-33, doublet), 5.00 (H-34, quartet and doublet C=CH) and 1.41 ppm (H-35, doublet, 7.0 Hz). The –CH_2– that is attached to the lactone (C 3), produces a ^{1}H-NMR peak at 2.25 (triplet, J = 7.0 Hz) and a ^{13}C-NMR peak at 25.0 ppm (Melot et al., 2009). A saturated γ-lactone will produce ^{13}C-NMR peaks at about 34.4, 44.2, 79.0 and 178 ppm, due to the CH CH_2 CH and O–C=O of carbons 2, 3, 4 and 1, respectively (Gallardo and Aragon, 1998). The THF ring will produce ^{1}H-NMR peaks at or near 1.65 and 1.98 ppm and ^{13}C-NMR peaks at about 28.8 and 29.7 ppm (Melot et al., 2009). Those ACGs that have an oxirane will produce ^{1}H-NMR peaks at or near 2.90 and 2.93 ppm and ^{13}C-NMR peaks at about 56.8 and 57.3 ppm (Melot et al., 2009). Some of them are connected to a –CH_2-CH_2-CH=CH, with ^{1}H-NMR peaks at about 1.61, 2.22, 5.39 and 5.41 and ^{13}C-NMR peaks at about 27.9, 24.3, 128.1 and 130.8 ppm (Melot et al., 2009). It should be noted that the exact chemical shifts depend on the deuterated solvent in which the sample is dissolved, its concentration and moisture content.

Often, MS can provide the molecular weight and the molecular formula. NMR can tell the number of hydrogens and carbons that are present, and can also be used to determine the conformation of ACGs (Latypov et al., 2002). Since ACGs have –OH groups, they were identified by their MS characteristic fragmentation patterns (Gu et al., 1997). They lose waters from molecular ions and sodium adducts (FAB-MS) of molecular ions. This is abbreviated by the following for the loss of the first three waters: $[M + H - H_2O]^+$, $[M + H - 2H_2O]^+$, $[M + H - 3H_2O]^+$, $[M + Na - H_2O]^+$, $[M + Na - 2H_2O]^+$ and $[M + Na - 3H_2O]^+$.

METHODS

Towards Understanding the Mode of Action of Natural ACG on the Stimulation of Bacterial Biofilm and Naphthalene Degradation

The first indication that acetogenins might be capable of stimulating bacterial biofilm formation arose from assays performed by Cartagena et al. (2007). At that time, the authors tested various plant sesquiterpene lactones by their ability to inhibit biofilm formation of a pathogenic strain of *Pseudomonas aeruginosa*. Among those compounds were some natural acetogenins isolated from *Annona cherimolia* seeds. One of them, squamocin, unexpectedly, was able to significantly stimulate the biofilm production of the strain. Thereafter, 7 PAH-degrading strains were tested for their attachment capability and biofilm development on polystyrene microplates using ACGs as biofilm promoters. From 7 strains tested, 3 showed natural formation of bacterial biofilm that increases in the presence of annonacin-A, laherradurin and cherimolin-2 (159, 64.0 and 41.5%, respectively) (Parellada et al., 2009). In this way, different ACGs stimulate the biofilm formation in environmental bacterial strains. This observation has prompted researchers to propose ACGs and the PAHs-degrading strains for future trials.

Effect of the ACGs on the Degradation of Naphthalene Mediated by a PAHs Degrading Bacteria, *Pseudomonas plecoglossicida* J26

The effects of the ACGs, squamocin and laherradurin on naphthalene degradation and biofilm formation of P. plecoglossicida J26 were evaluated in batch planktonic cultures with developing biofilm, and by biofilm attached on glass beads (Parellada et al., 2012). This strain was isolated from intertidal sediments of Patagonia, Argentina by selective enrichment with naphthalene as its sole carbon and energy source. Its naphthalene-degrading capacity was demonstrated by Riva Mercadal et al. (2010). Previously described by Nishimori, it has a respiratory but not a fermentative metabolism (Nishimori et al., 2000) and exhibits biofilm formation capacity (Li et al., 2009).

Naphthalene Degradation in Batch Planktonic Cultures with Developing Biofilm

According to Parellada et al. (2012), squamocin (2.5 $\mu g.ml^{-1}$) significantly stimulated biofilm formation of *Pseudomonas plecoglossicida* J26. At 6 h of incubation, the stimulation was 67% compared to control. The biomass percentage was similar until 12 h of incubation.

Laherradurin (2.5 μg.ml^{-1}) also significantly stimulated biofilm formation. At 6 h of incubation, the stimulation was 48% compared to control. Squamocin and laherradurin were able to increase the initial naphthalene degradation rate (0 to 6 h). In presence of squamocin, 43% of initial naphthalene was degraded during the first 6 h (naphthalene μM concentration dropped from 767.10±8.82 to 439.60±6.11). In presence of laherradurin, 43% of initial naphthalene was degraded during the first 6 h (naphthalene μM concentration dropped from 771.70±2.00 to 441.84±3.15). Only 15% of naphthalene was degraded in the control flasks without ACG (naphthalene μM concentration dropped from 771.70±1.43 to 653.23±3.15). From experimental data, naphthalene concentration vs time were fitted to lineal curves using the least squares method. Naphthalene consumption rate has been increased three-fold in the presence of ACGs. This fact could be employed to increase biofilm formation and efficiency of PAH bioremediation.

Naphthalene Degradation by Biofilm Attached on Glass Beads

Twelve hours of incubation yielded the consumption of 47% of initial naphthalene in the control (naphthalene μM concentration dropped from 1031.44±0.01 to 519.44±0.01) while the biofilm produced (compared to control) in presence of squamocin (145%) has consumed 61% (naphthalene μM concentration dropped from 993.91±0.01 to 384.56±0.01) and the biofilm produced in presence of laherradurin (120%) has consumed 54% (naphthalene concentration dropped from 1031.44±0.01 to 479.05±0.01 μM).

Therefore, this experimental evidence supports the use of laherradurin and squamocin as biofilm formation promoters that could make more efficient, safer and durable naphthalene bioremediation processes.

Bacterial Stress and Biofilm Formation

Parellada et al. (2011), studied ACG mode of action to stimulate the biofilm formation of *P. plecoglossicida* J26. They studied the effect of various stressors (naphthalene, octanol, HCl, and NaCl) on cell growth, biofilm formation and autoinducer production of *P. plecoglossicida* and compared with the effect of the ACG squamocin to establish its mode of action on biofilm formation. All stressors that inhibited growth stimulated autoinducer production while squamocin was growth stimulant at concentrations above 2.5 μg.ml^{-1}. Despite structural similarities, squamocin is not an autoinducer agonist. It indirectly stimulates autoinducer production and increases *P. plecoglossicida* J26 cell growth.

Finally, squamocin is indirectly involved in quorum sensing mechanism by inducing a stress related increase in AI production for a given incubation time. Therefore, the exacerbation of biofilm formation is due to increased production of AI-1.

REFERENCES

Alali, F.Q., Liu, X.X. and McLaughlin, J.L. 1999. Annonaceous Acetogenins: Recent Progress. *J. Nat. Prod.* 62, 504-540.

Bassler, B.L., Losick, R. 2006. Bacterially speaking. *Rev. Cell.* 125, 237-246.

Bermejo, A., Figadère, B., Zafra-Polo, M.C., Barrachina, I., Estornell, E. and Cortes, D. 2005. Acetogenins from Annonaceae: recent progress in isolation, synthesis and mechanisms of action. *Nat. Prod. Rep.* 22, 269-303.

Boopathy, R. 2000. Factors limiting bioremediation technologies. *Biores.Technol.* 74, 63-67.

Cartagena, E., Álvarez Colom, O., Neske, A., Valdez, J.C., Bardón A. 2007 Effects of plant lactones on the production of biofilm of *Pseudomonas aeruginosa. Chem. Pharm. Bull.* 55, 22-25.

Cavé, A., Figadère, B., Laurens, A. and Cortes, D. 1997. Acetogenins from Annonaceae, in Progess in the Chemistry of Organic Natural Products, eds. W. Herz, G. W. Kirby, R. E. Moore, W. Seglich and Ch. Tamm, Springer-Verlag,Wien. 70, pp. 81-288.

Decho, A.W. 2000. Microbial biofilms in intertidal systems: An overview. Continent. *Shelf Res.* 20, 1257-1273.

Duret, P., Waechter, A.I., Margraff, R., Foucault, A., Hocquemiller, R. and Cavé, A. 1997. High-speed contercurrent chromatography: a promising method for the separation of the Annonaceous acetogenins. *J. Liq. Chrom. Rel. Technol.* 20, 627-635.

Falahatpisheh, M.H., Donnelly, K.C., Ramos, K.S. 2001. Antagonistic interactions among nephrotoxic polycyclic aromatic hydrocarbons. *J. Toxicol. Environ. Health.* 62, 543-560.

Gallardo, T., Aragon, R. 1998. Acetogenins from *Annona glabra* seeds. *Phytochem.* 47, 811-816.

Gawronski, J. and Wu, Y.C. 1999. Anote on the determination of absolute configuration of acetogenins by circular dichroism. *Pol. J. Chem.* 73, 241-243.

Goldman, R., Enewold, L., Pellizzari, E., Beach, J.B., Bowman, E.D., Krishnan, S.S., Shields, P.G. 2001. Smoking increase carcinogenic polycyclic aromatic hydrocarbons in human lung tissue. *Cancer Res.* 61, 6367-6371.

Gonzalez, J.E., Keshavan N.D. 2006. Messing with bacterial quorum sensing. *Microbiol. Mol. Biol. Rev.* 70, 859-875.

Gu, Z.M., Zhou, D., Wu, J., Shi, G., Zeng, L., McLaughlin, J.L. 1997. Screening for Annonaceous acetogenins in bioactive plant extracts by liquid chromatography/mass spectrometry. *J. Nat. Prod.* 60, 242-8.

Hemminki, K. 1993. DNA adducts, mutations and cancer. *Carcinogenesis.* 14, 2007-2012.

Henner, P., Schiavon, M., Morel, J.L., Lichtfouse, E. 1997. Polycyclic aromatic hydrocarbons (PAH) occurrence and remediation methods. *Analusis.* 25, 9-10.

Johnsen, A.R., Wick, L.Y., Harms, H. 2005. Principles of microbial PAH-degradation in soil. *Environ. Poll.* 133, 71-84.

Jolad, S. D.; Hoffmann, J. J.; Schram, K. H.; Cole, J. R.; Tempesta, M. S.; Kriek, G. R.; Bates, R. B. 1982. Uvaricin, a new antitumor agent from *Uvaria accuminata* (Annonaceae). *J. Org. Chem.* 47, 3151-3153.

Kanaly, R.A., Harayama, S. 2000. Biodegradation of high-molecular-weight polycyclic aromatic hydrocarbons by bacteria. *J. Bacteriol.* 182, 2059-2067.

Latypov, S., Franck, X., Jullian, J.C., Hocquemiller, R. and Figadère, B. 2002. NMR determination of absolute configuration of butenolides of Annonaceous type. *Chem. Eur. J.* 8, 5662-5666.

Li, M.Y., Zhang, J., Lu, P., Xu, J.L., Li, S.P. 2009. Evaluation of biological characteristics of bacteria contributing to biofilm formation. *Pedosphere.* 19, 554-561.

Liu, K., Han, W., Pan, W.P., Riley, J.T. 2001. Polycyclic aromatic hydrocarbon (PAH) emissions from a coal fired pilot FBC system. *J. Hazard. Mater.* 84, 175-188.

Mastrangela, G. 1997. Polycyclic aromatic hydrocarbons and cancer in man. *Environ. Health Perspect.* 104, 1166-1170.

McLean, K.H., Winson, M.K., Fish, L., Taylor, A., Chhabra, S.R., Camara, M., Daykin, M., Swift J.H.S., Bycroft, B.W., Stewart Gordon, S.A.B., Williams, P. 1997. Quorum sensing and *Chromobacterium violaceum*: Exploitation of violacein production and inhibition for the detection of *N*-acyl homoserine lactones. *Microbiol.* 143, 3703-3711.

Melot, A., Fall, D., Gleye, C., Champy, P. 2009. Apolar Annonaceous Acetogenins from the Fruit Pulp of *Annona muricata*. *Molecules.* 14, 4387-4395.

Nishimori, E., Kita-Tsukamoto K. and Wakabayashi, H. 2000. *Pseudomonas plecoglossicida* sp. nov., the causative agent of bacterial haemoirrhagic ascites of ayu, *Plecoglossus altivelis. Int. J. Sys. Evol. Micr.* 50, 83-89.

Parellada, E., Rodriguez, A., Cartagena, E., Ferrero, M., Neske, A., Bardón, A. 2009. Annonaceous acetogenins stimulants of biofilm formation in aromatic hidrocarbon degrading bacteria. *Electron Microscopy and Cell Biol.* 33, A179.

Parellada, E.A., Ramos, A.N., Ferrero, M., Cartagena, E., Bardón, A., Valdez, J.C., Neske, A. 2011. Squamocin mode of action to stimulate biofilm formation of *Pseudomonas plecoglossicida* J26, a PAHs degrading bacterium. *Int. Biodeter. Biodegr.* 65, 1066-1072.

Parellada, E.A., Ramos, A.N., Ferrero M., Cartagena E., Bardón, A., Neske, A. 2012. Effect of the annonaceous acetogenins, squamocin and laherradurin, on the degradation of naphthalene mediated by *Pseudomonas plecoglossicida* J26. *Int. Biodeter. Biodegr.* 72, 82-87.

Riva Mercadal, J.P., Isaac, P., Siñeriz, F., Ferrero, M.A. 2010. Indigo production by *Pseudomonas* sp. J26, a marine naphthalene-degrading strain. *J. Basic Microb.* 50, 290-293.

Rodney, M., Donlan, L., Costerton, J.W. 2002. Biofilms: Survival mechanisms of clinically relevant microorganisms. *Clin. Microbial. Rev.* 15, 167-193.

Sabaté, J., Grifoll, M., Viñas, M., Solanas, A.M. 1999. Isolation and characterization of a 2-methylphenanthrene utilizing bacterium: identification of ring cleavage metabolites. *Appl. Microbiol. Biotechnol.* 52,704-712.

Samanta, S.K., Singh, O.V., Jain, R.K. 2002. Polycyclic aromatic hydrocarbons: environmental pollution and bioremediation. *Trends Biotechnol.* 20, 243-248.

Santos Lima, L.A.R., Pimenta L.P.S., Boaventura M.A.D. 2010. Acetogenins from *Annona cornifolia* and their antioxidant capacity. *Food Chem.* 122, 1129-1138.

Shimada, H., Grutzner, J.B., Kozlowski, J.F., Mc Laughlin, J. 1998. Membrane conformations and their relation to cytotoxicity of asimicin and its analogues. *Biochemistry US.* 37, 854-886.

Singh, R., Paul, D., Jain R.K. 2006. Biofilms: implications in bioremediation. *Trends Microbiol.* 14, 389-397.

Tormo, J.R., Gallardo, T., González, M.C., Bermejo, A., Cabedo, N., Andreu, I. and Estornell, E. 1999. Annonaceous acetogenins as inhibitors of mitochondrial complex I. *Curr. Top. Phytochem.* 2, 69-90.

Van Hamme, J.D., Ajay S., Ward, O.P. 2003. Recent advances in petroleum microbiology. *Microbiol. Mol. Biol. R.* 4, 503-549.

Waters, C.M., Bassler, B.L. 2005. Quorum sensing: Cell-to-cell communication in bacteria. *Annu. Rev. Cell. Dev. Biol.* 21, 319-346.

WHO Regional Office for Europe, Copenhagen, Denmark. 2000. PAHs Air Quality Guidelines - Second Edition. Chapter 5.9.

Zafra-Polo, M.C., González, M.C., Estornell, E., Sahpaz, S. and Cortes, D. 1996. Acetogenins from Annonaceae, Inhibitors of Mitochondrial Complex I. *Phytochemistry*. 42, 253-271.

Zafra-Polo, M.C., Figadère, B., Gallardo, T., Tormo, J.R. and Cortes, D. 1998. Natural Acetogenins from Annonaceae, synthesis and mechanisms of action. *Phytochemistry*. 48, 1087-1117.

Zeng, L., Ye, Q., Oberlies, N.H., Shi, G., Gu, Z.M., He, K. and McLaughlin, J.L. 1996. Recent Advances in Annonaceous Acetogenins. *Nat. Prod. Rep*. 13, 275-306.

In: Bioremediation: Processes, Challenges and Future Prospects ISBN: 978-1-62948-513-3
Editors: Jesús Bernardino Velázquez-Fernández et al. © 2014 Nova Science Publishers, Inc.

Chapter 15

BIOREMEDIATION BY RHAMNOLIPIDS PRODUCED BY ENVIRONMENTAL ISOLATES OF *PSEUDOMONAS AERUGINOSA*

Milena G. Rikalovic[1], *Miroslav M. Vrvic*[1] *and Ivanka M. Karadzic*[2*]
[1]Faculty of Chemistry, University of Belgrade, Belgrade, Serbia
[2] Department of Chemistry, Faculty of Medicine, University of Belgrade, Belgrade, Serbia

ABSTRACT

The efficacy of biodegradation of organic pollutants by *Pseudomonas* largely depends on its ability to produce biosurfactants, which, aside from tensioactive properties, were found to have the capacity to bind heavy metal ions. In addition to low molecular weight biosurfactants, *Pseudomonas aeruginosa* strains often produce extracellular polysaccharides with biosurfactant properties. Unlike synthetic surfactants, surfactants obtained by microbial biosynthesis are readily biodegradable and are particularly suitable for use in bioremediation. Microbial surfactants are low-toxic or nontoxic products, stable in a wide range of environmental conditions such as temperature, pH, pressure and salinity, and as such, they have potential for commercial use in bioremediation. Biosurfactants, in particular rhamnolipids, have been studied and shown to be highly efficacious in biodegradation of organics and biodegradation of organics in metal-organic co-contaminated systems, as well as being metal flushing and organic flushing agents. Microorganisms tolerant to organic and metallic pollutants are of special importance for environmental technologies. Particularly, microorganisms isolated from polluted environments, where natural selection has occurred have been recognized as a powerful tool for solving difficult problems in environmental protection to detoxify and degrade contaminants, and to clean up polluted environments. Microorganisms recovered from polluted sites are classified as environmental isolates in contrast to clinical and type cultures from culture collections.

* Visegradska 26, 11000 Belgrade; Email: ikaradzic@med.bg.ac.rs.

The most persistent organic contaminants exhibit low water solubility, and hence, their bioavailability can often be improved by addition of biosurfactants. The role of rhamnolipids in biodegradation of organic pollutants by environmental isolates of *P. aeruginosa* is discussed in this chapter. Several strains isolated from polluted areas were characterized by an extraordinary capacity to produce structurally diverse rhamnolipids that greatly facilitated biodegradation of organic contaminants. Rhamnolipids enhance desorption of heavy metals from soils, as well the remediation of heavy metals and organic compounds from co-contaminated sites. A particular advantage of bioremediation of co-contaminated soil is the possibility of *in situ* production of biosurfactants using the organic contaminants as substrates, which would lead to the remediation of both contaminants along with great reduction of the remediation cost.

The considerable potential for the use of biosurfactants as environmentally compatible soil-washing agents has resulted in demand for the improved production of biosurfactants. Three main postulates in rhamnolipid production on a large scale are: high yield, low cost and the ecologic aspect. Productivity of biosurfactants by environmental isolates of *P. aeruginosa* recovered from hydrocarbon-polluted sites and the multiple factors responsible for rhamnolipid production, such as bacterial strain, carbon sources, culture conditions and isolation procedures, are summarized. Conditions for improved production of rhamnolipids, especially use of low cost substrates (distillery and crude whey waste, orange peelings, molasses, sunflower and olive frying oil, soap stock, oil refinery waste) are summarized. The prevalence of particular rhamnolipid congeners according to their synthesis conditions and surface activity has been analysed to define the conditions for their production by suitable *P. aeruginosa* strains isolated from specific environments.

Keywords: Rhamnolipids, *Pseudomonas aeruginosa*, environmental strain, application, bioremediation, hydrocarbon, heavy metals

BIOSURFACTANTS

Biosurfactants are surface-active substances synthesized by living cells, mostly by microorganisms. They contain hydrophobic and hydrophilic moieties that confer the ability to accumulate between fluid phases, thus reducing surface and interfacial tension at the surface and interface, respectively (Mythusamy et al., 2008; Van Hamme et al., 2006). Biosurfactants are categorized by their chemical composition, molecular weight, physico-chemical properties, mode of action and microbial origin (Pacwa-Płociniczak et al., 2011). On the basis of their molecular weight, they are divided into low-molecular-mass biosurfactants, such as: glycolipids, phospholipids and lipopeptides, and into high-molecular-mass biosurfactants containing amphipathic moieties, such as: polysaccharides, proteins, lipopolysaccharides, lipoproteins or complex mixtures of these biopolymers (Pacwa-Płociniczak et al., 2011; Rosenberg and Ron 1999).

Unlike synthetic surfactants, surfactants obtained by microbial biosynthesis are readily biodegradable and particularly suitable for use in bioremediation. Microbial surfactants are usually considered as a low-toxic or nontoxic products, stable in a wild range of environmental conditions such as temperature, pH, pressure and salinity, making them applicable for broad commercial use in bioremediation (Muthusamy et al., 2008; Pacwa-Płociniczak et al., 2011). Biosurfactants may participate in the solubilization of organic compounds, as a critical step that limits the biodegradation process (Górna et al., 2011), and in the removal of heavy metals from even multi-metal contaminated soil (Juwarkar et al., 2008).

Pseudomonas aeruginosa - Primary Producers of Biosurfactants

The genus *Pseudomonas* encompasses, arguably, the most diverse and ecologically significant group of bacteria on the planet. Members of the genus are found in large numbers in all of the major natural environments (terrestrial, freshwater and marine) and also form intimate associations with plants and animals. This universal distribution suggests a remarkable degree of physiological and genetic adaptability (Spiers et al., 2000). Diversity within the genus is not limited to physiological traits. The diversity of phenotypes is also reflected at the genetic level, and evidence is mounting to suggest that the diversity of genome architecture (of both chromosomes and accessory genetic elements) is of a particular significance.

Fingerprinting studies performed by many workers have revealed a remarkable degree of restriction fragment polymorphism among strains of a species and even among strains that are closely related on phenotypic grounds (Spiers et al., 2000). *Pseudomonas* spp. were found to be the most common producers of biosurfactants (Górna et al., 2011), with *P. aeruginosa* being preeminent rhamnolipid (RL) biosurfactant-producing strains (Abdel-Mawgoud et al., 2010; Górna et al., 2011).

Pseudomonas includes 156 species initially recognized with nomenclatural standing, although 48 of these species have now been reclassified and are now considered to be basonyms or synonyms of species placed in other genera (Moor 2006). Also, 1803 full length sequences of 16S rRNA of strains of *Pseudomonas* species are available in the public databases (Bodilis et al., 2012).

Genome sizes of *Pseudomonas* vary widely, ranging from 3±7 Mbp for *P. stutzeri* to 7±1 Mbp for *P. aeruginosa*, with the genome sizes of *P. fluorescens* SBW25, *P. syringae* pv. *phaseolicola* and *P. putida* KT2440 being 6±7Mbp, 5±6 Mbp and 6±1 Mbp, respectively, while approximately 5500 putative genes have been detected. Genomes of *Pseudomonas* species are among the largest of the bacterial genomes that have been sequenced to date and the G+C contents of these genomes are among the highest: *P. aeruginosa*, 66.6%; *P. putida*, 61.5%; *P. syringae*, 58.4%; and *P. fluorescens*, 60.0% (Moor 2006). In addition, studies of multiple isolates from a single species have shown substantial genome size polymorphism; *P. stutzeri* spans from 3±7 to 4±7 Mbp and *P. aeruginosa* spans from 5±2 to 7±1Mbp (Spiers et al., 2000). Interestingly, this extensive diversity appears not to be reflected at the level of DNA sequence, as might be expected. For example, sequence analysis of housekeeping loci from 19 environmental and clinical isolates of *P. aeruginosa* showed low sequence diversity (Kiewitz and Tummler 2000; Spangenberg et al., 1998).

Consistent with a large genome size notable in several *P. aeruginosa* strains: 5683 genes (6264404 bp) for *P. aeruginosa* PAO1, 6062 (6601757 bp) for *P. aeruginosa* LESB58, 5965 (6537648 bp) for *P. aeruginosa* PA14 and 6369 (6588339 bp) for *P. aeruginosa* PA7, their high environmental adaptability has been proposed (Stover et al., 2000; Winsor et al., 2011). It was suggested that the *P. aeruginosa* genome size and complexity reflected evolutionary adaptation allowing it to grow in diverse environments, containing the highest proportion of regulatory genes observed for a bacterial genome and a large number of genes involved in the catabolism, transport and efflux of organic compounds. These regulatory genes presumably modulate the diverse genetic and biochemical capabilities of this bacterium in changing environmental conditions.

While some species of the genus *Pseudomonas* (e.g. *P. aeruginosa*) are homogeneous taxonomic units that are relatively easily differentiated and identified, other species include complex subdivisions, such as the *P. fluorescens* and *P. putida* biovars, the *P. syringae* and *P. marginalis* pathovars, or the *P. stutzeri* genomovars (Moore 2006).

P. aeruginosa, as all species that belong to genus *Pseudomonas*, is metabolically diverse and able to utilize more than 80 organic substrates as a carbon source. Due to this metabolic diversity, *P. aeruginosa* strains have potential for adaptation, survival and growth in a wide range of environmental conditions (Yahr and Parsek 2006).

Numerous studies confirmed that *P. aeruginosa*, due to its catabolic potential, is able to efficiently degrade a vast number of different hydrocarbons (Hasanuzzaman et al., 2007; Ko and Lebeault, 1999; Wongsa et al., 2004; Zhang and Miller 1994). The efficacy of organic pollutants biodegradation by *Pseudomonas* depends largely on its ability to form biosurfactants, which, aside from tensioactive properties, were found to have the capacity to complex heavy metal ions (Singh 2004). Furthermore, the fact that this species is capable of producing RL biosurfactants has significantly contributed to increasing the number of environmental clean-up studies (Górna et al., 2011; Noordman and Janssen 2002; Zhang and Miller 1994, 1995).

Environmental Isolates of *P. aeruginosa*

Biochemical decomposition of harmful pollutants to stable end products by indigenous and exogenous bacteria and fungi has been used as an effective and environmentally friendly strategy for bioremediation. Microorganisms tolerant to organic and metallic pollutants are of special importance for environmental technologies. Particularly, microorganisms isolated from polluted environments, where natural selection has occurred, have been recognized as a powerful tool for solving difficult problems in environmental protection to detoxify and degrade contaminants, and to clean up polluted environments (Satyanarayana 2012).

Their potential to metabolize chemical pollutants in the environment makes *Pseudomonas* species suitable for use as bioremediation agents able to remove heavy metal pollutants and numerous toxic organic compounds. Along with efficient heavy metal removal, *Pseudomonas* has an apparent advantage as an organic pollutant degrader, since it can secrete hydrolytic enzymes and biosurfactants - RLs and tolerate a high content of hydrocarbons or their derivatives. These kinds of microorganisms appear to be rare in the environment. Microorganisms recovered from the polluted sites are classified as environmental isolates in contrast to clinical and type cultures from culture collections.

The clinical and environmental isolates of *P. aeruginosa* were found to form a coherent taxonomic group with a difference related to utilization of hydrocarbons as a sole carbon source (Foght et al., 1996).

Sequencing of the 16S rRNA gene and analytical profile index (API)-tests have appeared as the predominant strategies for characterisation of newly isolated *P. aeruginosa* strains (Aparna et al., 2012; Górna et al., 2011; Nie et al., 2010; Rikalovic et al., 2013). Taxonomical and biochemical characterisation of novel *P. aeruginosa* strains indicate that the 16RNA sequences of different *P. aeruginosa* isolates show a high percentage of sequence identity between different strains (regardless of the strain origin), and that biochemical tests gave similar results for clinical and environmental strains. In the study of Nie et al. (2010), 16S RNA sequence of a newly isolated strain, *P. aeruginosa* NY3, showed high similarity to the 16S rRNA genes from *P. aeruginosa* strains which were all environmental strains. Five environmental isolates of *P. aeruginosa* showed similarities in biochemical tests with the clinical strain *P. aeruginosa* ATCC 27853 (Rikalovic et al., 2013) and sequence identity with different environmental and clinical strains. Also, the 16S RNA sequence of strain *Pseudomonas* sp. from the study of Aparna et al. (2012) showed a high percentage of sequence identity with *P. aeruginosa* strains isolated from contaminated environments and from host organisms.

From 1990 to 2012, according to Google Scholar, the number of published articles related to biosurfactants from environmental isolates of *P. aeruginosa* was significantly greater (4628) compared to investigations on biosurfactants from clinical isolates (1707). The number of published articles about the application of RLs in biomedicine (835) and bioremediation (2023) also implies the fact that bioremediation has been more studied of late. The driving force for the intensive investigation into RL in bioremediation seems to be based on the great progress in analytical and screening methods, which has resulted in the opportunity to more rapidly isolate RL producers originating from different environments like contaminated soil, water or industrial waste.

RHAMNOLIPIDS

Rhamnolipids are glycolipids, which belong to the class of low molecular weight biosurfactants (Van Hamme et al., 2006). These glycosides are composed of one or two L-rhamnose units and one or two units of 3-hydroxy fatty acid. Variations in lipid component contribute to biodiversity of RLs (Abdel-Mawgoud et al., 2010). Some authors classified RLs into four homologue groups (Figure 1): RL1 - mono-rhamno-di-lipidic, RL2 - mono-rhamno-mono-lipidic, RL3 - di-rhamno-di-lipidic and RL4 - di-rhamno-mono-lipidic structures. RL1 and RL3 are usually classified as principal (common), while RL2 and RL4 are classed as atypical (uncommon) RLs (Tahzibi et al., 2004). The development of sensitive, high throughput analytical techniques, such as soft ionization mass spectrometry has led to the further discovery of a wide diversity of RL congeners and homologues (about 60) produced in different concentrations by various *Pseudomonas* species and other bacteria (Abdel-Mawgoud et al., 2010), (Table 1).

Variations in the nature and distribution of the different RL congeners identified in various RL mixtures might be attributed to diverse cultivation conditions as well as to strain-related differences (Abdel-Mawgoud et al., 2010; Déziel et al., 1999, 2000; Dubeau et al., 2009).

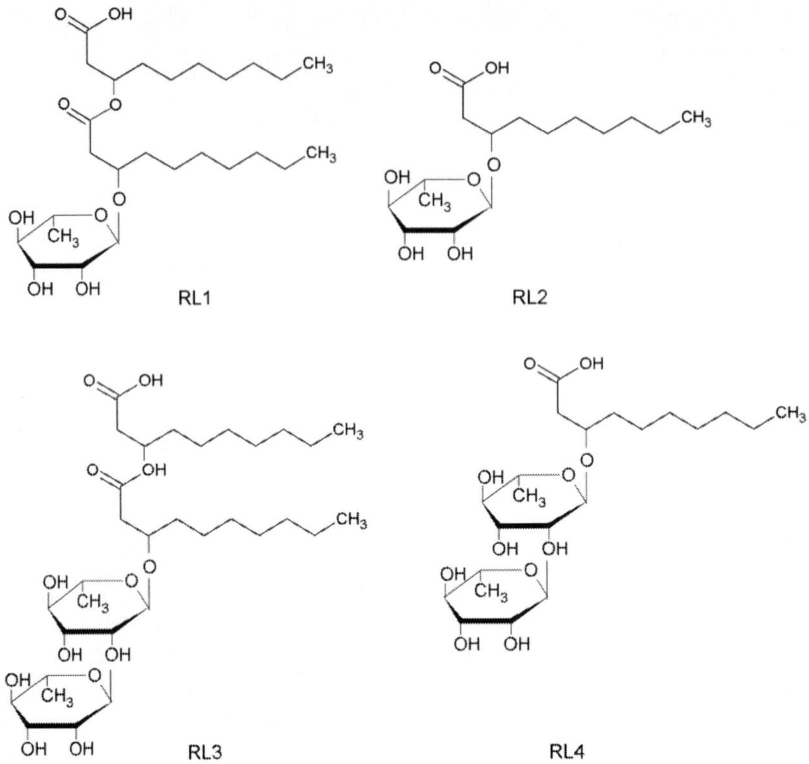

Figure 1. Structure of rhamnolipid congeners: RL1 (mono-rhamno-di-lipidic), RL2 (mono-rhamno-mono-lipidic), RL3 (di-rhamno-di-lipidic), RL4 (di-rhamno-mono-lipidic).

Table 1. Chemical structure of different identified rhamnolipid congeners and homologues (Abdel-Mawgoud et al., **2010)**

Number	Symbol	Molecular formula	Molecular weight
Mono-rhamno-mono-lipidic congener			
1	Rha-$C_{8:2}$	$C_{14}H_{22}O_7$	302.32
2	Rha-C_8	$C_{14}H_{26}O_7$	306.35
3	Rha-C_{10}	$C_{16}H30O_7$	334.41
4	Rha-$C_{12:2}$	$C_{14}H_{30}O_7$	358.43
5	Rha-C_{12}	$C_{18}H_{34}O_7$	362.46
6	Rha-$C_{14:2}$	$C_{20}H_{34}O_7$	386.48
Mono-rhamno-di-lipidic congener			
7	Rha-C_8-C_8	$C_{22}H_{40}O_9$	448.55
8	Rha-C_8-$C_{10:1}$	$C_{24}H_{42}O_9$	474.58
9	Rha-$C_{10:1}$-C_8	$C_{24}H_{42}O_9$	474.58
10	Rha-C_8-C_{10}	$C_{24}H_{44}O_9$	476.60
11	Rha-C_{10}-C_8	$C_{24}H_{44}O_9$	476.60
12	Rha-C_{10}-$C_{10:1}$	$C_{26}H_{46}O_9$	520.64
13	Rha-C_{10} -C_{10}	$C_{26}H_{48}O_9$	504.65
14	Rha-C_8-C_{12}	$C_{26}H_{48}O_9$	504.65
15	Rha-C_{12}-C_8	$C_{26}H_{48}O_9$	504.65

Number	Symbol	Molecular formula	Molecular weight
16	Rha-C_{10}-$C_{12:1}$	$C_{28}H_{50}O_9$	530.69
17	Rha-$C_{12:1}$-C_{10}	$C_{28}H_{50}O_9$	530.69
18	Rha-C_{10}-C_{12}	$C_{28}H_{52}O_9$	532.71
19	Rha-C_{12}-C_{10}	$C_{28}H_{52}O_9$	532.71
20	Rha-C_{10}-$C_{14:1}$	$C_{30}H_{54}O_9$	558.74
21	Rha-C_{12}-$C_{12:1}$	$C_{30}H_{54}O_9$	558.74
22	Rha-C_{10}-C_{14}	$C_{30}H_{56}O_9$	560.76
23	Rha-C_{12}-C_{12}	$C_{30}H_{56}O_9$	560.76
24	Rha-C_{12}-C_{14}	$C_{32}H_{60}O_9$	588.81
25	Rha-C_{14}-C_{14}	$C_{34}H_{64}O_9$	616.87
26	Rha-C_{14}-C_{16}	$C_{36}H_{68}O_9$	644.92
27	Rha-C_{16}-C_{16}	$C_{38}H_{72}O_9$	672.97
28	Rha-C_{10}-C_{10}-CH_3	$C_{27}H_{50}O_9$	518.68
29	decenoil-Rha-C_{10}-C_{10}	$C_{36}H_{64}O_{10}$	656.89
Di-rhamno-mono-lipidic congener			
30	Rha-Rha-C_8	$C_{20}H_{36}O_{11}$	452.49
31	Rha-Rha-C_{10}	$C_{22}H_{40}O_{11}$	480.55
32	Rha-Rha-$C_{12:1}$	$C_{24}H_{42}O_{11}$	506.68
33	Rha-Rha-C_{12}	$C_{24}H_{44}O_{11}$	508.60
34	Rha-Rha-C_{14}	$C_{26}H_{48}O_{11}$	536.65
Di-rhamno-di-lipidic congener			
35	Rha-Rha-C_8-C_8	$C_{28}H_{50}O_{13}$	594.69
36	Rha-Rha-C_8-C_{10}	$C_{30}H_{54}O_{13}$	622.74
37	Rha-Rha-C_{10}-C_8	$C_{30}H_{54}O_{13}$	622.74
38	Rha-Rha-C_{10}-$C_{10:1}$	$C_{32}H_{56}O_{13}$	648.78
39	Rha-Rha-C_{10}-C_{10}	$C_{32}H_{58}O_{13}$	650.79
40	Rha-Rha-C_8-$C_{12:1}$	$C_{32}H_{56}O_{13}$	648.78
41	Rha-Rha- $C_{12:1}$-C_{18}	$C_{32}H_{56}O_{13}$	648.78
42	Rha-Rha-C_{10}-$C_{12:1}$	$C_{34}H_{60}O_{13}$	676.83
43	Rha-Rha-$C_{12:1}$-C_{10}	$C_{34}H_{60}O_{13}$	676.83
44	Rha-Rha-C_{10}-C_{12}	$C_{34}H_{62}O_{13}$	678.84
45	Rha-Rha-C_{12}-C_{10}	$C_{34}H_{62}O_{13}$	678.84
46	Rha-Rha-C_{10}-$C_{14:1}$	$C_{36}H_{64}O_{13}$	704.89
47	Rha-Rha-C_{12}-$C_{12:1}$	$C_{36}H_{64}O_{13}$	704.89
48	Rha-Rha-$C_{12:1}$-C_{12}	$C_{36}H_{64}O_{13}$	704.89
49	Rha-Rha-C_{12}-C_{12}	$C_{36}H_{66}O_{13}$	706.90
50	Rha-Rha-C_{12}-C_{14}	$C_{38}H_{70}O_{13}$	734.95
51	Rha-Rha-C_{14}-C_{12}	$C_{38}H_{70}O_{13}$	734.95
52	Rha-Rha-C_{14}-C_{14}	$C_{40}H_{74}O_{13}$	763.00
53	Rha-Rha-C_{14}-C_{16}	$C_{42}H_{78}O_{13}$	791.06
54	Rha-Rha-C_{16}-C_{14}	$C_{42}H_{78}O_{13}$	791.06
55	Rha-Rha-C_{16}-C_{16}	$C_{44}H_{82}O_{13}$	819.11
56	Rha-Rha-C_{14}-C_{14}-C_{14}	$C_{54}H_{100}O_{15}$	989.36
57	Rha-Rha-C_{10}-C_{10}-CH_3	$C_{33}H_{60}O_{13}$	664.82
58	decenoil-Rha-C_{10}-C_{10}	$C_{42}H_{74}O_{14}$	803.03

Composition of RL Mixtures

RL biosurfactants are always produced as mixtures of different RL congeners, as observed with the various strains of *P. aeruginosa* (Abalos et al., 2001; Abdel-Mawgoud et al., 2009; Benincasa et al., 2004; Haba et al., 2003a; Mata-Sandoval et al., 1999; Pornsunthorntawee et al., 2008). The complexity of the composition of RL mixtures was found to depend on various factors such as: bacterial strain origin, type of carbon substrate (Benincasa et al., 2002; Benincasa and Accorsini 2008; Rahman et al., 2002), culture conditions (Costa et al., 2006), age of the culture (Haba et al., 2003a), and of course, the *P. aeruginosa* strain itself (Haba et al., 2003a, Rikalovic et al., 2013). Some RL congeners are predominant in all producing *P. aeruginosa* strains and are classified as the major RL structures, while others, produced only sometimes or with low abundance, are the minor RL structures. Both the major and the minor RL congeners contribute to the complete profile of RLs. It was previously reported that multiple factors, like individual strain, carbon sources, culture conditions and isolation procedures, were all possibly responsible for the diversity of the components in RL production (Déziel et al., 1999; Nitschke et al., 2010).

Detected RL congeners for several environmental isolates of *P. aeruginosa:* 2 KM, 3 KM and 10 TK (Górna et al., 2011), D1, D2 and D3 (Rikalovic et al., 2013), AT10 and 47T2 (Haba et al., 2003a) are listed in Table 2. It is noticeable that some of the RL structures are common to all, or almost all *P. aeruginosa* strains (Rha-C_8-C_{10}, Rha-C_{10}-C_{10}, Rha-C_{10}-$C_{12:1}$, Rha-C_{10}-C_{12}, Rha-Rha-C_8-C_{10}, Rha-Rha-C_{10}-C_{10} and Rha-Rha-C_{10}-C_{12}), while some are strain or group specific.

In the study of Price et al. (2009), RL mixtures produced by soil isolates of *Pseudomonas* spp. were analysed by MALDI-TOF MS (Matrix-Assisted Laser Desorption/Ionization – Time of Flight Mass Spectometry). Strains were grown on medium with glycerol as a carbon source. Only eight RL structures were detected (Rha-C_{10}-C_8, Rha-C_{10}-C_{10}, Rha-C_{10}-C_{12}, Rha-C_{10}-$C_{12:1}$, Rha-Rha-C_{10}-C_8, Rha-Rha-C_{10}-C_{10}, Rha-Rha-C_{10}-C_{12}, Rha-C_{10}-$C_{12:1}$). As these congeners were commonly detected in RL mixtures of *Pseudomonas* spp. (Aparna et al., 2012; Haba et al., 2003a; Nie et al., 2010; Price et al., 2009; Sarachat et al., 2010), these structures could be considered as *Pseudomonas*-specific RL.

Nie et al. (2010) investigated a novel RL biosurfactant-producing and polycyclic aromatic hydrocarbon (PAH) - degrading bacterium, *P. aeruginosa* strain NY3, isolated from petroleum-contaminated soil samples. Strain NY3 has been found capable of producing a structurally diverse RL mixture, with 25 RL structures in total detected by MALDI-TOF MS. In addition to the common RL, the authors detected ten new RL structures (Rha–C_8–$C_{8:1}$, Rha–C_{16}, Rha–$C_{16:1}$, Rha–$C_{17:1}$ and Rha–$C_{24:1}$, Rha–Rha–C_6–$C_{6:1}$, Rha–Rha–$C_{9:1}$, Rha–Rha–$C_{10:1}$–$C_{10:1}$, Rha–Rha–C_{24}, and Rha–Rha–$C_{24:1}$). Strain NY3 produced the mono-rhamnolipids Rha–C_{10}–C_{10} and di-rhamnolipids Rha–Rha–C_{10}–C_{10} as predominant components. This finding was consistent with previous reports for *P. aeruginosa* SP4 (Sarachat et al., 2010), *P. aeruginosa* LBI (Nitschke et al., 2010) and *P. aeruginosa* NRRL B-59182 (Rooney et al., 2009), *P. aeruginosa* 2 KM, 3KM, 10 TK (Górna et al., 2011) and *P. aeruginosa* D1, D2, D3, 67, NCAIM (P)B 001380 (Rikalovic et al., 2013).

In all studies of RL mixtures produced by *P. aeruginosa* strains, it was shown that, aside from their varying composition, mono-rhamnolipid Rha–C_{10}–C_{10} and di-rhamnolipid Rha–Rha–C_{10}–C_{10} were the predominant congeners.

Table 2. Frequently detected RL congeners produced by environmental *P. aeruginosa* isolates analyzed by HPLC MS analysis

	P. aeruginosa strain							
RL congener	2 KM	3 KM	10 TK	D1	D2	D3	AT10	47T2
Rha-$C_{8:1}$	-	-	-	-	-	-	+	-
Rha-C_{10}	-	-	-	+	+	+	-	-
Rha-$C_{12:2}$	-	-	-	+	+	+	+	-
Rha-C_8-C_8	+	+	+	-	-	-	-	-
Rha-C_8-C_{10}	+	+	+	+	+	+	-	+
Rha-$C_{8:1}$-C_{10}	-	-	+	-	-	-	-	-
Rha-C_{10}-C_{10}	+	+	+	+	+	+	+	+
Rha-C_{10}-$C_{12:1}$	-	+	+	+	+	+	+	+
Rha-C_{10}-C_{12}	-	+	+	+	+	+	+	+
Rha-Rha-C_8-C_8	+	+	-	-	-	-	-	-
Rha-Rha-C_8-C_{10}	+	+	+	+	+	+	-	+
Rha-Rha-$C_{8:1}$-C_{10}	+	+	+	-	-	-	-	-
Rha-Rha-$C_{8:1}$-C_{12}	-	-	-	-	-	-	-	+
Rha-Rha-C_{10}-C_{10}	+	+	+	+	+	+	+	+
Rha-Rha-C_{10}-C_{12}	+	+	+	+	+	+	+	+
Rha-Rha-C_{10}-$C_{12:1}$	-	+	+	-	-	-	+	+
Rha-Rha-C_{12}-C_{12}	-	-	-	+	+	+	-	-
Rha-Rha-$C_{12:1}$-C_{12}	-	-	-	-	-	-	-	+
Rha-Rha-C_{10}-$C_{14:1}$	-	-	-	-	-	-	-	+

Major Congeners and Ratio of Mono- and Di-RL

Only a few RL congeners appear as the major structures (Rha-C_8-C_{10}, Rha-C_{10}-C_{10}, Rha-C_{10}-$C_{12:1}$, Rha-C_{10}-C_{12}, Rha-Rha-C_8-C_{10}, Rha-Rha-C_{10}-C_{10} and Rha-Rha-C_{10}-C_{12}), while the remainder are considered as the minor RL structures. In the study of Górna et al. (2011), the composition of RL mixtures produced by three *P. aeruginosa* isolates from petroleum contaminated soil was analyzed and compared by HPLC MS. Two main RLs were found for each strain: Rha-C_{10}-C_{10} and Rha-Rha-C_{10}-C_{10}, ranging from 81-87%. However, some strain-specific differences were noted in the Rha-C_{10}-C_{10} to Rha-Rha-C_{10}-C_{10} ratios, which were 25.4 : 56.3 for Pa 2 KM, 16.1 : 64.9 for Pa 3 KM and 39.1 : 47.9 for Pa 10 TK. The lowest amount of mono-RLs was produced by strain Pa 3 KM and the highest by Pa 10 TK. On the other hand, the lowest amount of di-RLs was produced by strain Pa 10 TK and the highest by Pa 3 KM. Ratios of total di-/mono-RL fractions were 1.8, 3.6 and 1.0 for strain Pa 2 KM, Pa 3 KM and Pa 10 TK, respectively (Table 3).

Rikalovic et al. (2013) studied three strains of *P. aeruginosa* (D1, D2 and D3), which were isolated from mazut-contaminated soil (Beskoski et al., 2011). RL mixtures produced by these environmental isolates were analyzed using HPLC MS. The major mono- and di-RL congeners were Rha-C_{10}-C_{10} and Rha-Rha-C_{10}-C_{10} with total abundance for strain D1, D2 and D3 ranging from 83-88%. The ratio of total di-/mono-RL fractions were 4.2, 2.6 and 1.4 for strains D1, D2, D3, respectively (Table 3), which was previously observed as a typical value for *P. aeruginosa* (Dubeau et al., 2009).

Table 3. Comparison of *P. aeruginosa* strains isolated from contaminated soils: The major mono- and di-RL % abundances, ratio Rha-C_{10}-C_{10}/ Rha-Rha-C_{10}-C_{10}, ratio total di-/mono-RL

P. aeruginosa strain	The major mono-RL	The major di-RL	The major mono- and di-RL %	Ratio Rha-C_{10}-C_{10}/ Rha-Rha-C_{10}-C_{10}	Ratio total di-/mono-RL	Reference
2 KM	Rha-C_{10}-C_{10}	Rha-Rha-C_{10}-C_{10}	81.7	25.4 : 56.3	1.8	Górna et al., 2011
3 KM	Rha-C_{10}-C_{10}	Rha-Rha-C_{10}-C_{10}	81.0	16.1 : 64.9	3.6	Górna et al., 2011
10 TK	Rha-C_{10}-C_{10}	Rha-Rha-C_{10}-C_{10}	87.0	39.1 : 47.9	1.0	Górna et al., 2011
D1	Rha-C_{10}-C_{10}	Rha-Rha-C_{10}-C_{10}	86.7	17.4 : 69.3	4.2	Rikalovic et al., 2013
D2	Rha-C_{10}-C_{10}	Rha-Rha-C_{10}-C_{10}	87.1	25.4 : 61.7	2.6	Rikalovic et al., 2013
D3	Rha-C_{10}-C_{10}	Rha-Rha-C_{10}-C_{10}	83.6	33.3 : 50.3	1.4	Rikalovic et al., 2013

Temporally-Dependent RL Profiles

Haba et al. (2003a) monitored production of different RL structures during 96h of fermentation . The most abundant pair of homologues in the mixture was Rha-C_{10}-C_{10}/Rha-Rha-C_{10}-C_{10}, which accounted for up to 50.5%, the mono-RL homologue appeared first (23.8%), at 24 h of growth, whereas the di-RL homologue was detected at 48 h of incubation, and reached maximum levels between 72 and 96 h of incubation. Similarly, production of this RL was reported to be the most abundant when the carbon sources were corn oil by *P. aeruginosa* UG2 (Mata-Sandoval et al., 2001), or mannitol by *P. aeruginosa* 57RP (Déziel et al., 1999), sunflower oil and its waste by *P. aeruginosa* D1-D3, 67, NCAIM (P) B 001380 (Rikalovic et al., 2013), glucose by *P. aeruginosa* Pa 2 KM, 3 KM, TK 10 (Górna et al., 2011), or glycerol by *P. aeruginosa* NY3 (Nie et al., 2010).

This indicates that RL production varies temporally and that only the predominant homologues are produced constantly and in high concentrations, while production of other RL congeners varies during fermentation.

Physico-chemical Properties of RL Mixtures

Determination of surface activity of fermentation broth is a common method for identification and screening of potential producers of surface active compounds (Górna et al., 2011). It has been reported that the ratio and composition of the homologues, the presence of unsaturated bonds, the branching and length of the alkylic chain, or the size of the hydrophilic head group of the surfactant can all affect the CMC (critical micelle concentration) values (Dubeau et al., 2009; Haba et al., 2003b). Also, Rosen (1974) established the close relationship between the surface-active properties of a surfactant and its chemical structure . RL mixtures are considered as a single product and each mixture is expected to have different properties because of differences in their molecular composition. However, it is difficult to understand the amphipathic behaviour of RL due to the presence of one or two hydrophilic moieties and one or two hydrophobic groups with different alkyl chain lengths (Haba et al., 2003a). As shown in Table 4, the minimum value of coefficient of surface activity (γst) produced by the surface-active compound in water was higher for RLAT10 (26.8 mN/m) than for RL47T2; however, the CMC value of RL47T2 was lower than that of RLAT10, being 108.8 and 150 mg/l, respectively. The effect may be due to the higher hydrophobicity of the components, where in the case of RLAT10, the proportion of di-RL was lower (37.8%) than that found in RL47T2 (32.8 mN/m). A similar effect was reported by Mata-Sandoval et al. (1999). Alternatively, the effect may be due to unsaturation in the hydrophobic group; RLAT10 contained 35.7% unsaturated fatty acids whereas RL47T2 contained 17.9%.

Table 4. Physico-chemical properties of RLs from different environmental isolates of *P. aeruginosa* after 96 h of cultivation

P. aeruginosa strain	CMC mg/l	Surface activity, mN/m	Di-RL fraction, %	Mono-RL fraction, %	Unsaturated fatty acids, %
D1	162.0	39.6	80.7	19.3	1.3
D2	143.0	39.8	72.6	27.4	1.7
D3	135.0	38.8	59.1	41.9	6.1
67	131.0	40.5	56.2	43.8	2.9
NCAIM (P) B 001380	131.0	40.1	57.7	42.3	2.4
AT10	150.0	26.8	37.8	62.2	35.7
47T2	108.8	32.8	› 37.8	› 62.2	17.9

Statistically significant correlations were found for strain D1-D2, 67 and NCAIM, between CMC values and the ratio of total mono-/di-RL (P= 0.033) and also between CMC values and the ratio of total Rha-C_{10}-C_{10}/Rha-Rha-C_{10}-C_{10} (P= 0.015), (Rikalovic et al., 2013). Lower CMC arises from a higher mono-/di-RL (lower di-/mono-RL) ratio, indicating that mono-RL start to form micelles at lower concentrations than di-RL.

Physiological Functions of Rhamnolipids

Although numerous studies have been performed in the field of RLs, the exact physiological functions these molecules perform in or for the producing bacteria are still not precisely defined. However, some of their proposed functions, which are related to physico-chemical properties of RL (surface activity, wetting ability, detergency, and other amphipathic related properties) are: promotion of the uptake and biodegradation of poorly soluble substrates, immune modulators and virulence factors (i.e. antimicrobial agents), (Abdel-Mawgoud et al., 2010). These properties have encouraged the use of RL compounds in environmental bioremediation of contaminated soils and polluted waters. Aside from these mentioned roles of RLs, it has been found that these molecules play roles in the swarming process, acting both as surface wetting agents and as chemotaxis stimuli (regulating the cell population density-dependent control of genes expression in quorum sensing), in the architecture of biofilms (the formation of water channels in mature biofilms) produced by *P. aeruginosa*, and in cell dispersal from biofilms (Dusane et al., 2010).

Rhamnolipids and Quorum Sensing

Quorum sensing (QS) is the mechanism by which bacteria engage in cell-to-cell communication using diffusible molecules based on a critical cell density (Williams et al., 2007). QS signalling molecules control diverse physiological processes some of which are inter-related and under the control of QS systems. In *P. aeruginosa*, RL and exopolysaccharide production, antibiotic resistance and biofilm synthesis are controlled by QS (Dusane et al., 2010).

P. aeruginosa possesses two interrelated QS systems, namely the *las* and *rhl*, which regulate different processes including RL expression, enzyme production, pyocyanin pigment production and maintenance of biofilm architecture. Production of RL is governed by three QS molecules: *Pseudomonas* autoinducer 1 (PAI-1, N-(3-oxododecanoyl) homoserine lactone also known as 3-oxo-C12-HSL), *Pseudomonas* autoinducer 2 (PAI-2, N-butyryl homoserine lactone also known as C4-HSL), and *Pseudomonas* quinolone signal (PQS, 2-heptyl-3-hydroxy-4-quinolone), (Dusane et al., 2010). In *P. aeruginosa*, the *las* operon consists of two transcriptional activator proteins, the LasR and LasI, which direct the synthesis of PAI-1 autoinducer. Induction of the *lasB* gene that encodes the elastase enzyme and other virulence genes requires the expression of LasR and PAI-1 autoinducer. The production of RL is regulated by the *rhl* system. The synthesis of RLs takes place under the coordinated guidance of *rhlAB* genes that encode a group of enzymes termed the rhamnosyltransferases. RL is synthesized by two enzymes - rhamnosyltransferase 1 and rhamnosyltransferase 2. The *rhl* system consists of transcriptional activator proteins RhlR and RhlI, which regulates the synthesis of a QS molecule PAI-2 autoinducer. The transcriptional activator RhlR activates the transcription of *rhlAB* operon genes, coding for rhamnosyltransferase 1, and another gene, *rhlC* encoding for the rhamnosyltransferase 2. With increasing bacterial cell density, the induction of the *las* QS system takes place, resulting in an increase in the concentration of PAI-1 autoinducer molecule. This QS molecule (PAI-1) then binds to the transcriptional activator site LasR and forms the LasR–PAI-1 complex. The LasR–PAI-1 complexes induce genes controlled by the *las* QS system, including a negative regulator gene *rasL*, *rhlR* and *pqsH*, required for PQS production. The activity of these signals is dependent upon their ability to dissolve in and freely diffuse through the aqueous solution (Dusane et al., 2010).

P. aeruginosa produces RL biosurfactant, which enhances the solubility of PQS in aqueous solutions. PQS acts as a link between the *las* and *rhl* QS systems. PQS, either directly or indirectly, induces the *rhlI* gene which directs the production of PAI-2 autoinducer molecule that binds to and activates RhlR. The operon *rhlAB* that encodes these enzymes responsible for RL production is controlled at the transcriptional and translational levels by RhlR and PAI-2. The RhlR–PAI-2 complex induces genes controlled by the *rhl* QS system for the production of RL. The *las* system controls the expression of transcriptional activator RhlR.

Along with this, an important gene, *rhlG*, is involved in the synthesis of the β-hydroxyacid moiety of RLs. A QS hierarchy therefore exists in *P. aeruginosa las* and *rhl* systems for the synthesis of RL (Dusane et al., 2010).

The expression of the *rhlAB* operon and the production of RLs are regulated, in addition to QS signals, by environmental/nutritional factors (Abdel-Mawgoud et al., 2011), which implies significant complexity in RL biosynthesis. Restriction in the availability of a number of nutrients, except the carbon source, is known to promote the production of RLs. For instance, the transcription of *rhlAB* and the production of RLs are inversely proportional to the concentration of iron available to the bacterial cells. An explanation is provided by the well-established link between iron availability and QS in *P. aeruginosa*, which is that the expression of *lasIR* and *rhlIR* is enhanced by Fe limitation and/or repressed by Fe supplementation.

Production of RLs is inhibited by the presence of NH_4^+, glutamine, asparagine, and arginine as nitrogen source and promoted by NO_3^-, glutamate, and aspartate. Elevated C/N and C/P ratios are conductive for RL production, while divalent cations inhibit biosurfactant production (Chayabutra et al., 2001; Soberón-Chávez et al., 2005).

Improving Rhamnolipid Production by Environmental Isolates of *P. aeruginosa*

The great potential for the use of biosurfactants as environmentally compatible soil-washing agents has resulted in demand for the improved production of biosurfactants. Three main postulates are: high yield, low cost and ecologic production of RLs, as major biosurfactants produced by *P. aeruginosa.*

The quest for the cost-efficient production of biosurfactants and alternative producing strains is associated with efficient biosurfactant analysis (Hayd et al., 2008). Applied methods range from indirect analysis based on physical properties of RLs (determination of surface tension and hemolytic test), (Abdel-Mawgoud et al., 2011; Siegmund and Wagner 1991), colorimetric measurement (cetyltrimethylammonium bromide agar test, anthrone method and orcinol test), (Chandrasekaran and BeMiller 1980; Hodge and Hofreiter 1962; Siegmund and Wagner 1991) to a sophisticated large scale analysis of sample composition by mass spectrometry, (Déziel et al., 1999; 2000; Haba et al., 2003a; Nie et al., 2010; Price et al., 2009). HPLC coupled with MS presents, nowadays, the most precise method for RL identification and quantification (Hayd et al., 2008), albeit MALDI-TOF MS approaches developed for high-throughput screening of naturally occurring mixtures of RLs from *Pseudomonas* spp. can be effective too (Price et al., 2009).

RL biosurfactant produced by *P. aeruginosa* isolates from hydrocarbon rich environments (e.g. petroleum- or mazut-contaminated soil, or various industrial wastes including mineral metal cutting oil) were studied in terms of productivity, composition, physico-chemical properties and potential application in bioremediation. The following section summarizes reported data giving an insight into improved production of RL from environmental isolates of *P. aeruginosa*.

Production of Rhamnolipids by Environmental Strains - Carbon and Nitrogen Source

RLs are secondary metabolites and their production coincides with the onset of the stationary phase. Therefore, all cultivation strategies for the microbial production of RLs aim at inducing RL biosynthesis by limiting at least one medium component, for example, the nitrogen or the phosphorous source (Soberón-Chávez et al., 2005).

Table 5. Production of rhamnolipids by environmental isolates of *P. aeruginosa* growing on different carbon and nitrogen sources

Production strain	Carbon and nitrogen source	RL yield, g/l	Strain Origin	Reference
Pseudomonas sp. JAMM	OOME, $NaNO_3$	1.4	Contaminated soil and water	Mercade et al., 1993
P. aeruginosa GS3	Molasses and corn steep liquor	0.25	Soil	Patel and Desai 1997
Pseudomonas 47T2 NCBI 40044	Sunflower frying oil, $NaNO_3$	2.70	Contaminated soil and water	Haba et al., 2000
P. aeruginosa LBI	Soap stock, $NaNO_3$	12.0	Hydrocarbon-contaminated soil	Benincasa et al., 2002
P. aeruginosa DS10–129	Sunflower and soybean oil, mineral salt	4.31	Gasoline and diesel spilled gas station soil	Rahman et al., 2002
P. aeruginosa PA1	Glycerol, $NaNO_3$	3.16	Wastewater from oil wells	Santa Anna et al., 2002
P. aeruginosa AT10	Soybean residual fatty acids	16.5	Contaminated soil and water	Haba et al., 2003a
P. aeruginosa BS2	Distillery and crud whey waste	0.92	Oily sludge	Dubey and Juwarker 2001
P. aeruginosa LBI	Brazilian nut oil,$NaNO_3$, yeast extract	9.9	Hydrocarbon-contaminated soil	Costa et al., 2006
P. aeruginosa HR	Glycerol, $NaNO_3$	4.20	Oil wells	Rashedi et al., 2006
P. aeruginosa NY3	Glucose, peptone , NH_4Cl	0.27	Petroleum-contaminated soil	Nie et al., 2010
Pseudomonas sp. 2B	Molasses, Peptone, NH_4Cl	4.97	Petroleum-contaminated soil	Aparna et al., 2012
P. aeruginosa D1	Sunflower oil, Peptone, NH_4Cl	1.73	Mazut-contaminated soil	Rikalovic et al., 2013
P. aeruginosa D2	Sunflower oil, Peptone, NH_4Cl	1.71	Mazut-contaminated soil	Rikalovic et al., 2013
P. aeruginosa 67	Sunflower oil, peptone, NH_4Cl	3.34	Petroleum-contaminated soil	Rikalovic et al., 2013
P. aeruginosa NCAIM (P) B 001380	Sunflower oil, Peptone, NH_4Cl	0.43	Mineral cutting oil	Rikalovic et al., 2013
P. aeruginosa NCAIM (P) B 001380	Sunflower frying oil, Peptone, NH_4Cl	1.30	Mineral cutting oil	Rikalovic et al., 2013

Cultivation strategies applied to RL production involve batch, fed-batch continuous, and integrated microbial/enzymatic processes. The reported biotechnological cultivation strategies applied to the production of RLs are: (fed-) batch cultivation under growth-limiting conditions, batch cultivation under resting cell conditions, semi-continuous production with immobilized cells (excluding any nitrogen source), continuous cultivation and production with free cells and solid state fermentation (Abdel-Mawgoud et al., 2011).

Table 5 shows production of RLs by environmental *P. aeruginosa* strains on different carbon and nitrogen sources. Reported data indicate that production of RLs is possible from simple carbon sources like glycerol and glucose or complex carbon sources (olive oil, sunflower oil) or wastes (crud whey, distillery waste, molasses, corn step liquor, SOME, OOME, frying oil, soap stock) and mineral ($NaNO_3$, NH_4Cl) or combined nitrogen source ($NaNO_3$ and yeast extract or NH_4Cl and peptone). This suggests that biosurfactant production can be enhanced to economically viable values using renewable feedstock as the carbon source, reiterating the importance of the utilization of industrial by-products and agricultural wastes as cost-effective alternative substrates for microbial growth and biosurfactant production (Dubeau and Juwaker 2001; Maneerat 2005; Moldes et al., 2007; Patel and Desai 1997; Rahman et al., 2002; Raza et al., 2007a; Rikalovic et al., 2013).

RHAMNOLIPIDS IN BIOREMEDIATION

Bioremediation typically involves augmentation of soil or other media, contaminated with pollutants with nutrients and sometimes microorganisms, to improve processes for biodegradation of the contaminants (Singh et al., 2007). Bioremediation options range from simply monitoring a contaminated site to aggressive treatment of the site (Maier and Soberon-Cháves 2000). Biosurfactants, especially RLs, have been studied and show potential in bioremediation of organics, as organic flushing agents, as metal flushing agents, and in biodegradation of organics in metal-organic co-contaminated systems (Maier and Soberon-Cháves 2000). *P. aeruginosa*, as the major producers of RLs, play a role in bioremediation processes including oil recovery, de-emulsification of oil emulsions and biosorption of heavy metals (Singh et al., 2007, 2010).

Biodegradation of Hydrocarbons: Challenges and Limits

Release of hydrocarbons into the environment, whether accidentally or due to human activities, is a main cause of water and soil pollution (Das and Chandran 2011). The threat of petroleum contamination has inevitably increased over the last few years as a result of rapid industrial development (Górna et al., 2011). Among many potential means of hydrocarbon removal, bioremediation processes have been recognized as efficient and relatively inexpensive methods with a high application potential (Singh et al., 2007; Ward et al., 2003). Many recent studies in this field seems to support this statement (Beskoski et al., 2011; Bordoloi and Konwar 2009; Jovancicevic et al., 2008). Biodegradation by natural, indigenous populations of microorganisms represents one of the primary mechanisms by which petroleum and other hydrocarbon pollutants can be removed from the environment (Urlici

2000) and it is believed to be non-invasive and relatively more cost-effective than other remediation technologies (Leahy and Colwell 1990).

One of the important factors that limits biodegradation of oil pollutants in the environment is their limited availability to microorganisms. Petroleum hydrocarbon compounds bind to soil components, and they are difficult to remove or degrade (Barathi and Vasudevan 2001).

Rhamnolipids in Biodegradation of Hydrocarbons

The biodegradation rate of a contaminant in soil depends on its bioavailability to the metabolizing organisms, which is influenced by factors such as desorption, diffusion and dissolution (Singh et al., 2007). Many of the most persistent contaminants exhibit low water solubility and hence, bioavailability of contaminants can often be improved by addition of emulsifiers (Banat et al., 2000).

This impedes their sufficient uptake by bacteria, and thus, it is not obvious how these hydrophobic molecules enter the cells to sustain bacterial growth. Proposed methods of entry for medium- and long-chain-length *n*-alkane hydrophobic molecules into bacterial cells include access either by a surfactant-facilitated process (emulsification and solubilization) or by direct bacterial adhesion to hydrocarbon droplets (surfactant mediated increase of cell surface hydrophobicity), (Abdel-Mawgoud et al., 2010). RLs have been mostly studied for their potential to mediate the assimilation of hydrophobic substrates like hydrocarbons in liquid cultures (Beal and Betts 2000; Koch et al., 1991; Noordman and Janssen 2002) and enhance the biodegradation of poorly soluble molecules by causing alterations on the cell surface (Al-Tahhan et al., 2000; Shreve et al., 1995; Zhang and Miller 1994).

Numerous studies have confirmed that biosurfactants, especially RLs, can affect the biodegradation of hydrocarbons, both aliphatic and aromatic (Miller 1995a). Furthermore, it was shown that addition of RLs to pure cultures enhances biodegradation of hexadecane, octadecane, *n*-parafines and phenatrene (Shreve *at al* 1995; Zhang and Miller 1994, 1995), as well as degradation in soil systems in the presence of hexadecane, tetradecane, pristane and creosote or hydrocarbon mixture (Maier and Soberon-Cháves 2000). Also, there are reports that added RLs can inhibit degradation processes (Maier and Soberon-Cháves 2000). Regardless of whether the biodegradation process is enhanced or inhibited, effects are strain-specific in the sense of strain characteristics and response to environmental conditions (Zhang and Miller 1994). A potential reason for inhibition of degradation is the possibility that RL is actually favoured as a carbon source for bacterial metabolism (Maslin and Maier 2000).

Two mechanisms of hydrocarbon biodegradation facilitated by RL have been proposed. The first model assumes that RLs, due to their physico-chemical properties, increase hydrocarbon solubility and consequently their bioavailability (Sherve et al., 1995). The second mechanism hypothesizes an interaction with the bacterial cell, which makes the cell surface more hydrophobic and easily accessible to the hydrophobic substrate (Sherve et al., 1995, Zhang and Miller 1994). Analysis of interaction of RLs and cell surface showed that RLs cause the loss of lipopolysaccharides, an important hydrophilic part of the cell surface, from the cell wall, which results in the cell surface becoming more hydrophobic (Al-Tahhan et al., 2000). Further studies showed that the second mechanism is more important for processes of *in situ* bioremediation. The reason for this is that a smaller amount of RL is needed to alter the cell surface compared to the amounts of biosurfactant needed to increase hydrocarbon solubility (Maslin and Maier 2000). Franzetti et al. (2010) discuss three mechanisms of interaction between microorganisms and hydrocarbons: access to water-solubilized hydrocarbons, direct contact of cells with large oil drops and contact with pseudosolubilized or emulsified oil. The recent report by Cameotra and Singh (2009) throws more light on the uptake mechanism of *n*-alkane by *P. aeruginosa* and the role of RLs in the process of internalization of hydrocarbon inside the cell for subsequent degradation. Although much work has been done by many groups to explain the role of biosurfactants in the degradation of water-immiscible substrates, most processes still remain unclear (Das and Chandran 2011).

Uptake of Hydrocarbons by Environmental Isolates of *P. aeruginosa* - Producers of Rhamnolipids

The capability of biosurfactants and biosurfactant-producing bacterial strains to enhance organic contaminant availability and biodegradation rates was reported by several authors (Déziel et al., 1996; Inakollu et al., 2004; Rahman et al., 2003).

Bioremediation of oil sludge using autogenous biosurfactants produced by a microbial consortium of *Pseudomonas* sp, isolated from oily sludge, has been recently reported (Cameotra and Singh, 2008). The biosurfactant produced by a consortium member was identified as being a mixture of 11 RL congeners. The consortium degraded 91% of the hydrocarbon content of soil contaminated with 1% (v/v) crude oil sludge in 5 weeks. The data substantiated the use of a crude biosurfactant for hydrocarbon remediation.

In another study, Noordman and Jansen (2002) studied the biodegradation of hexadecane by five biosurfactant-producing bacterial strains (*P. aeruginosa* UG2, *Acinetobacter calcoaceticus* RAG1, *Rhodococcus. erythropolis* DSM 43066, *R. erythropolis* ATCC 19558, and strain BCG112) in the presence and absence of externally added biosurfactants. The degradation of hexadecane by *P. aeruginosa* was stimulated only by the RL biosurfactant produced by the same organism, suggesting that *P. aeruginosa* has a mode of hexadecane uptake different from those of the other organisms.

Beal and Betts (2000) investigated the mechanisms for biosurfactant-enhanced hexadecane uptake into *P. aeruginosa*. Two strains of *P. aeruginosa* were studied, one producing RLs (PG201), the other being RL deficient (UO299). RLs produced by PG201 acted to increase the solubility of *n*-hexadecane in the culture medium, while during growth on *n*-hexadecane, rates of *n*-hexadecane uptake and mineralization were higher in PG201 than in UO299. Also, the cell surface hydrophobicity of both PG201 and UO299 increased 50.5 and 33.7% respectively, compared with that observed in water-soluble growth substrates (7-8%). These studies support the hypothesis of alkane transport into microbial cells by direct contact with larger alkane droplets and by pseudosolubilization. Also, it appears that both mechanisms occur simultaneously.

The usefulness of biosurfactant producing strains in bioremediation of sites highly contaminated with crude petroleum-oil hydrocarbons was confirmed by Das and Mukherjee (2007), by using three biosurfactant producing strains: *Bacillus subtilis* DM-04, *P. aeruginosa* M and *P. aeruginosa* NM (all strains originated from petroleum contaminated soil) to remediate petroleum crude-oil contaminated soil samples.

In the study of Nie et al. (2010), a novel RL biosurfactant-producing and PAH-degrading bacterium, *P. aeruginosa* strain NY3, isolated from petroleum-contaminated soil samples was investigated. Strain NY3 was characterized by its extraordinary capacity to produce structurally diverse RLs and among these compounds were ten new RLs.

Rhamnolipids As Flushing Agents for Organic Pollution

Some biodegradation processes are too slow or unfeasible, so it can be necessary to remove organics from soil using *ex situ* soil washing or *in situ* soil flushing, also called pump and treat. This can be the situation with nonaqueous phase liquids (NAPL) and soil phase organics like polycyclic aromatic hydrocarbons (PAH). In these cases, the low aqueous solubility of organics represents an obstacle for their removal by water alone and addition of surfactants can enhance the process. Although various synthetic surfactants have been studied in this regard, biosurfactants are of interest due to their environmentally-friendly nature (biodegradabile and non-toxic), (Maier and Soberon-Cháves, 2000).

For NAPL, addition of surfactants to a flushing solution can enhance flushing efficiency, either by mobilization of NAPL or by an increase in solubilization of the NAPL. Thus, to be effective, a surfactant must have good solubilization capacity and/or be able to reduce interfacial tension. RL has been shown to have a solubilization capacity (expressed as a molar solubility ratio, MSR) for the model NAPL, hexadecane, which is 20 times greater than the MSR for hexadecane alkyl benzyl sulfonate (Thangmani and Sherve, 1994). In studies examining the use of RLs for removal of residual hexadecane from soil, it was shown that RL (20%) was more effective than either SDS (negligible removal) or Tween 80 (6% removal), (Bai et al., 1997). It was further shown that NAPL removal could be optimized by altering pH and ionic strength to achieve a maximum reduction of surface tension, and the final result was 60% removal of hexadecane residues by RL (Bai et al 1998).

Similar results were obtained for RL solubilization of solid phase materials. For example, rhamnolipid-octadecane MSR was ten times higher than MSR for Triton-X-114-octadecane (Miller, 1995a), five times higher than MSR for Corexit 0600-octadecane (Thai and Maier, 1992) and from 1.7 to 2.8 times higher than for 13 different synthetic surfactants which were tested with phenanthrene (Edwards et al., 1991; Theim 1994; Zhang et al., 1997). Furthermore, in a comparison of the removal of a hydrocarbon mixture (undecane, pentadecane, hexadecane, octadecane, pristane, naphtalene, phenetrene and pyrene) from soil, RLs were more effective than Triton X-100 or Tween 60 for all hydrocarbon components (Scheibenbogen et al., 1994). Finally, RL-enhanced removal of phenanthrene, pyrene and polychlorinated biphenyls and a variety of PAH from soil have been reported (Maier and Soberon-Cháves, 2000).

Biosurfactants in Bioremediation of Heavy Metals

Due to their extremely toxic nature even at trace concentrations and their non-biodegradability (unlike organic pollutants), heavy metals are a persistent threat to the environment. Bioremediation of heavy metals is currently attracting considerable attention, not only because of its innovation, but also because of potential application in industry (Singh et al., 2010). Viable or dead biomass (bacteria, algae, fungi and plants) is able to modify or remove heavy metals present in their environment, which offers economical alternatives to the sorption techniques (Gupta et al., 2000). Simple microbial metabolic activity alone has proven to be inadequate, especially at sites with metal and organic compound co-contamination (Singh and Cameotra 2004). Industrial uses of biosurfactants, including their role in heavy metal pollution removal, have been reviewed earlier (Banat 1995, 2000; Eliora and Rosenberg 2001, 2002; Makar and Cameotra 2002; Siegmund 2002).

According to Miller (1995b), biosurfactants enhance desorption of heavy metals from soils in two ways: 1 - complexation of the free form of the metal residing in solution. This decreases the solution phase activity of the metal and, therefore, promotes desorption according to Le Chatelier's principle and 2 - direct contact of biosurfactant to sorbed metal at the solid solution interface under conditions of reduced interfacial tension, which allows biosurfactants to accumulate at the solid solution interface. The effect of applied biosurfactant depends on many factors: soil composition, pH, cation exchange capacity, particle size, time and type of contamination, geological layout, etc. (Mulligan et al., 2001a).

Juwarkar et al. (2007) conducted column experiments to evaluate the potential of environmentally compatible RL biosurfactants produced by *P. aeruginosa* strain BS2, which was isolated from oil sludge, to remove Cd and Pb from artificially contaminated soil. The strain used in this study is considered as a di-RL producing strain. Results have shown that di-RL removed not only the leachable or available fraction of Cd and Pb but also the bound metals, compared to tap water, which removed the mobile fraction only. Washing of contaminated soil with tap water revealed that ≈2.7% of Cd and 9.8% of Pb in contaminated soil was in freely available or weakly bound forms, whereas washing with RL removed 92% of Cd and 88% of Pb after 36 h of leaching. This indicated that di-RL selectively favours mobilization of metals in the order of Cd > Pb. The microbial population of the contaminated soil was increased after removal of metals by biosurfactant indicating the decrease of toxicity of metals to soil microbiota. This study shows that biosurfactant technology can be an effective and non-destructive method for bioremediation of cadmium- and lead-contaminated soil.

Wang and Mulligan (2004) evaluated the feasibility of using RL foam to remove Cd and Ni from a sandy soil. They reported that the use of foam had a significant effect on the mobility of biosurfactant flowing in a porous medium and made a more uniform and efficient contact of biosurfactant with the metals. Application of RL foam increased efficiency and allowed removal of 73.2% and 68.1% of Cd and Ni, respectively, whereas the RL solution flushed only 61.7% and 51% of Cd and Ni, respectively.

In batch washing experiments designed to evaluate the feasibility of using biosurfactants to remove heavy metals from sediments, surfactant from *Bacillus subtilis*, RLs from *P. aeruginosa*, and sophorolipid from *Torulopsis bombicola* were evaluated on sediment containing 110 mg/kg copper and 3300 mg/kg zinc (Mulligan et al., 2001b). A single washing with 0.5% RL removed 65% of the copper and 18% of the zinc, whereas 4% sophorolipid removed 25% of the copper and 60% of the zinc. Surfactin was less effective, removing 15% of the copper and 6% of the zinc. It was postulated that metal removal by the biosurfactants occurred through sorption of the surfactant onto the soil surface and complexation with the metal, detachment of the metal from the soil into the soil solution, and association with surfactant micelles. The RL and surfactin removed organically-bound copper, and the sophorolipid removed carbonate and oxide-bound zinc. These initial results support the use of biosurfactants to remove heavy metals from sediments (Mulligan et al., 2001b).

Avramovic et al. (2012) studied *P. aeruginosa* NCAIM (P) B001380 chromium (VI) tolerance. The tested strain was isolated from mineral cutting oil and identified as a RL producer (Rikalovic et al., 2012). That study showed that *P. aeruginosa* NCAIM (P) B001380 is chromium tolerant and has potential for application in heavy metal bioremediation. Maximum RL production was found in the late-stationary phase coinciding with the maximum Cr (VI) removal capacity. The effect of Cr (VI) on the distribution of mono-RLs and di-RLs and their ratio was studied by electrospray ionization mass spectrometry, and an increase was observed in a di-RL/mono-RL ratio for RL isolated from medium with 100 mg/l of Cr (IV) compared to control, when the RLs were produced in the absence of Cr (IV).

Rhamnolipids in Bioremediation of Co-contaminated Sites

It is estimated that 37% of sites contaminated with organics are also contaminated with metals, which can cause inhibition of biodegradation of organic compounds. RLs have been demonstrated to enhance hydrocarbon degradation in systems co-contaminated with organics and toxic metals (Maier and Soberon-Cháves 2000). One advantage in the case of co-contaminated soil is that biosurfactants can potentially be produced *in situ* using the organic contaminants as substrates for their production, which subsequently would lead to remediation of both the contaminants along with great reduction of the remediation cost.

Sandrin et al. (2000) studied the effectiveness of RL biosurfactants in the remediation of a Cd and naphthalene co-contaminated site. They observed that reduced cadmium toxicity by addition of *P. aeruginosa* RL led to enhanced naphthalene biodegradation by *Burkholderia* sp. NCBI U37342. These authors suggested that reduction of metal toxicity by RL might involve a combination of RL complexation of Cd and RL interaction with the cell surface to alter Cd uptake resulting in enhanced rates of bioremediation. In another co-contaminant study, it was observed that the inhibition of phenanthrene mineralization in the presence of Cd was reduced by the pulsed addition of RL (Maslin and Maier 2000).

Dahrazma and Mulligan (2007) reported the higher rate of removal of Cu and Ni from sediments by adding 1% NaOH to the solution of RL. Many metals mostly exist in the environment organic fraction, and adding OH^- to the sediment solubilizes this fraction; thus, more metals are available for removal by a RL biosurfactant.

Song et al. (2009) studied simultaneous Cr (VI) reduction and phenol degradation in a reactor containing *P. aeruginosa* CCTCC AB91095. *P. aeruginosa* utilized phenol and its metabolites as its carbon and energy source for Cr (VI) reduction. Cr (VI) inhibited both Cr (VI) reduction and phenol degradation when the Cr (VI) concentration exceeded the optimum value (20 mg/l), whereas phenol enhanced both Cr (VI) reduction and phenol degradation below the optimum initial concentration of 100 mg/l. Cr (III) was the predominant product of Cr (VI) reduction in cultures after incubation for 24 h. Both Cr (VI) reduction and phenol degradation were influenced by the amount of inocula. The study showed that the use of *P. aeruginosa* was promising for the reduction of toxic Cr (VI) and degradation of organic pollutants simultaneously in the mineral liquid medium.

To remove the toxic metals from a co-contaminated site, complexion agents can be used. Beside the strength of the metal complexation, complexion agents must be environmentally compatible. However, some organic ligands which are proposed for metal complexation (NTA, EDTA and DTPA) due to their high stability constants, have demonstrated toxic effects such as carcinogenic properties and low biodegradability (Maier and Soberon-Cháves 2000). On the other hand, RLs, which are readily degraded, have shown very low toxicity together with high affinity for a variety of metals including cadmium, copper, lanthanum, lead and zinc (Maier and Soberon-Cháves 2000). Efficient removal of Zn and Cu from co-contaminated soil with a 12.6% oil content using RLs was also demonstrated (Mulligan et al., 1999).

Extracellular Polymeric Substance and Sequestration of Heavy Metals

Many bacteria in the environment exist in surface-attached communities called biofilms. Biofilm bacteria are usually embedded in an extracellular polymeric substance (EPS) matrix composed of polysaccharides, proteins and nucleic acids (Flemming and Wingender, 2001). A proposed mechanism that contributes to increased resistance is binding and sequestration of heavy metals by EPS components, such as negatively charged phosphate, sulfate, and carboxylic acid groups. It has been reported that biofilms are capable of removing heavy metal ions from bulk liquid (Huang et al., 2000), and the use of biofilms to remove heavy metals from wastewater has been investigated (von Canstein et al., 1999).

Among EPS, bacterial exopolysaccharides, in particular, have been reported to help in biosorption of heavy metals (Avramovic et al., 2012; Kiliç and Dönmenz, 2008). Kiliç and Dönmenz (2008) investigated three different chromium-resistant microorganisms (*P. aeruginosa*, *Micrococcus* sp. and *Ochrobactrum* sp.), following exopolysaccharide production at different pH levels, temperatures, Cr (VI) concentrations, and incubation periods. Results of this study indicated an effective removal of Cr (VI) that coincided with an increase of exopolysaccharide production, suggesting binding of the heavy metal to exopolysaccharides. *P. aeruginosa* demonstrated a high capacity to remove Cr (VI), and accordingly, had the greatest EPS formation compared to *Micrococcus* sp. or *Ochrobactrum* sp..

Kazy et al. (2002) studied extracellular polysaccharides of a copper-sensitive and a copper-resistant *P. aeruginosa* strain. The extent of exopolysaccharide synthesis by the resistant strain (4.78 mg/mg of cell dry weight) was considerably higher than its sensitive counterpart (2.78 mg/dry of cell dry weight). The presence of Cu (II) in the growth medium caused a dramatic stimulation (approximately 4-fold) in exopolysaccharide synthesis by the Cu-resistant strain, while in a similar condition, the Cu-sensitive strain failed to exhibit such a response. The polymer of the resistant strain showed elevated Cu (II) binding (320 mg/g of EPS) compared to that of the sensitive type (270 mg/g of EPS). The overall observations show the potential of the EPS from the Cu-resistant strain for its deployment in metal bioremediation.

Pseudomonas strains have the potential to produce many different types of EPS constituents, including, for example, alginate, levan, marginalan, cellulose, in addition to various heteropolysaccharides and protein polysaccharide complexes. Extracellular polysaccharides (alginate and exopolysaccharides Pel and Psl) with biosurfactant properties (Franklin et al., 2011) are produced as a response and adaptation of bacteria to changed environmental conditions. While the greatest numbers of environmental and clinical *P.*

aeruginosa isolates produce Psl or Pel, alginate is synthesized only by some *P. aeruginosa* strains. Whereas alginate and Pel structure are well characterized, the chemical composition of Pel is thought to be a polysaccharide rich in glucose, but different to cellulose (Figure 2), (Franklin et al., 2011). Mucoid strains secreted alginate into the growth medium, and this polysaccharide is not covalently bonded with the bacterial cell surface. Psl forms a matrix, which is connected with biofilm and additionally has a role in cell-to-cell and cell-to-surface interactions in the process of biofilm formation (Franklin et al., 2011).

Alginate, the most studied *Pseudomonas* polysaccharide, consists of a linear polymer of 1,4-linked mannuronic acid and its epimer, guluronic acid. *P. aeruginosa* alginates are highly O-acetylated, influencing the water and metal-binding capacities of the polymer. The mannuronate and guluronate confer anionic characteristics to the exopolysaccharide (Sabra and Zeng, 2009). Generally, alginate production *in vitro* is increased under nutrient limiting conditions or exposure to increased concentrations of sodium chloride, ethanol and oxygen, or to desiccation.

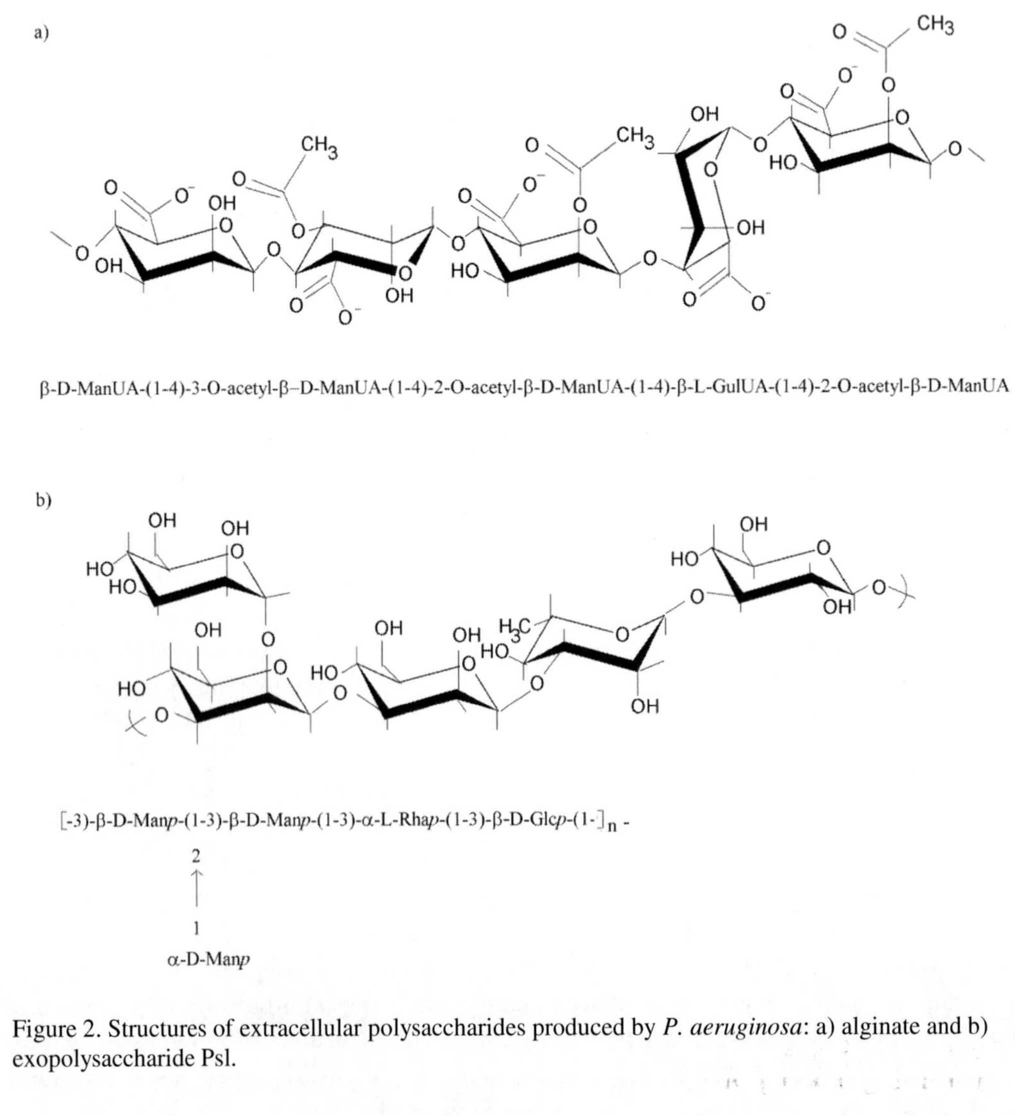

Figure 2. Structures of extracellular polysaccharides produced by *P. aeruginosa*: a) alginate and b) exopolysaccharide Psl.

Interaction of EPS and metals may depend on the abundance and chemistry of EPS, which in turns varies with bacterial strain and growth conditions. Consequently, the influence of EPS on metal adsorption was studied using various microbial strains as single culture or biofilms through EPS characterization and quantification, identification of metal binding sites or modelling adsorption behaviours. Changes in EPS composition could be observed during exposure to toxic organic solvents or heavy metals. The chemical composition of EPS from *P. putida* G7 was studied in order to explain the previously reported high cadmium and lead binding capacity of this strain. Crude exopolymer is composed of protein, LPS (lipopolysaccharide) and polysaccharide, which was found to be composed of glucose, rhamnose, ribose, N-acetylgalactosamine and glucuronic acids were found. It is assumed that the negative charge of glucuronic acid would play a major role in the metal binding through electrostatic attraction (Kachany et al., 2001). Apart from an increase in total EPS content, the change in carbohydrate composition of *P. putida* biofilms exposed to Cr was observed. The greatest increases were for glucose, N-acetylgalactosamine, mannose and rhamnose (Priester et al., 2006). The functional groups on bacterial surfaces have been suggested to be directly responsible for the reactivity of bacterial cells. Binding of contaminants may lead to stabilization (short term effect) or immobilization (long term effect). Reduction of chromate and immobilization of Cr (III) by formation of Cr (III) phosphate using mixed microbial cultures from granular biofilms was demonstrated (Nancharaiah et al., 2010). Zn sorption by *P putida* biofilms could be attributed mainly to Zn-phosphoryl complexes formed, while Cd adsorption occurs mainly by Cd carboxyl complexes.

In the study of Teitzel and Parsek (2003), the effects of the heavy metals copper, lead, and zinc on biofilm and planktonic *P. aeruginosa* were examined. It was determined that biofilms were from 2 to 600 times more resistant to heavy metal stress than free-swimming cells. When planktonic cells at different stages of growth were examined, it was found that logarithmically growing cells were more resistant to Cu and Pb stress than stationary-phase cells. However, biofilms were observed to be more resistant to heavy metals than either stationary-phase or logarithmically growing planktonic cells. The exterior of the biofilm was preferentially killed after exposure to elevated concentrations of Cu, and the majority of living cells were near the substratum. A potential explanation for this is that the extracellular polymeric substances that encase a biofilm may be responsible for protecting cells from heavy metal stress by binding the heavy metals and retarding their diffusion within the biofilm.

Compative Proteomics as an Approach to Monitor Bioremediation Processes

It has become clear that the response to a variety of environmental stresses, including heavy metal exposure, is characterized by the induction of a large set of proteins. Analysis of proteomes, the entire set of proteins expressed by an organism, cell or organelle, show that protein production is highly complex, temporal and under constant flux, in response to various stimuli. On the other hand, an understanding of the growth conditions governing the expression of the proteome in specific conditions is essential for developing rational strategies for successful bioremediation (Zhao et al., 2008). Regarding this, analysis of differential protein expression as a response to heavy metal or organic pollutants exposure in microorganism would shed light on understanding of the organism's response to stress. With this knowledge, it would be possible to control the production of desirable metabolites useful for environmental protection. Proteomics, together with bioinformatics, has a crucial role in studying structures and functions of proteins. Proteomic platforms that enable profiling of a high number of proteins across large sets of complex samples hold a great potential to study mechanisms of response to exposure to different pollutants, giving at the same time an opportunity to monitor differential expression of proteins/enzymes that could serve as unique environmental biomarkers. Differential proteomics would not only help understanding of the dynamic of the synthesis of certain proteins, but would also facilitate engineering of new proteins which could meet even more complex demands of biotechnological application in future time. Generally, the application of proteomics in environmental bioremediation research provides a global view of the protein compositions of the microbial cells and offers a promising approach to address the molecular mechanisms of bioremediation or how to dissect the cellular responses to environmental stimuli, such as stress response, induction and expression of regulatory proteins/enzymes in response to organic pollutants and heavy metals.

The molecular basis of heavy metal resistance of *Pseudomonas* using a large scale proteomics approach, despite having good potential for application in environmental protection, has not yet been enough either understood or focused. Recently published comparative proteomics analysis of *P. putida* indicated that the core molecular response to chromate was comprised of seven up-regulated proteins belonging to six different functional categories including transcription, inorganic ion transport/metabolism, and amino acid transport/metabolism (Thompson et al., 2010), whereas differential cellular responses to Cd and Cu were found in *P. putida KT244,* supporting restricted internalization of the metals by using different sets of binding proteins and efflux pumps, in addition to mechanisms to protect against oxidative stress (Miller et al., 2009). A proteomics study of chromium resistant *P. aeruginosa* suggested an upregulation of biosynthesis of enzymes responsible for exopolysaccharide production (Kiliç et al., 2010), whereas differentially expressed proteins in cadmium stressed psychrotolerant *P. putida* 710A and alkaliphilic *P. monteilli* 97AN, identified using only the *in-silico* approach, were found to be enzymes involved in energy mediating metabolic pathways; this revealed their significance to overcome the energy requirement of cells under cadmium-induced stress (Jain et al., 2010).

Conclusion

The roles of rhamnolipids in the processes of hydrophylization of organic compounds and as excellent heavy metals bioligands indicate the broad potential for their commercial application. The main problems related to the cost of rhamnolipid production on a large scale could be partially solved by using renewable substrates such as agroindustrial waste. Additionally, the development of methods for the selection and rapid characterization of environmental strains together with robust analytical platforms for the analysis of rhamnolipid mixtures and studies of complex regulation of rhamnolipid biosynthesis, will enable the successful application of these compounds in bioremediation. Furthermore, environmental strains of *P. aeruginosa*, as rhamnolipid producers, which are naturally present in contaminated environments and resistant to heavy metals and toxic organic compounds, are a logical choice in the strategy of *in vivo* bioremediation of environments contaminated with hydrocarbons and/or heavy metals. Most importantly, these microbial products, as natural, nontoxic, biodegradable compounds, with possible production on renewable substrates, are part of the concept of green chemistry and important factors in the process of bioremediation.

ACKNOWLEDGMENT

This work was supported by the project III43004 of the Ministry of Science and Technological Development of Serbia.

REFERENCES

Abalos, A., Pinazo, A., Infante, M.R., Casals, M., Garcia, F., Manresa, A., 2001. Physicochemical and antimicrobial properties of new rhamnolipids produced by *Pseudomonas aeruginosa* AT10 from soybean oil refinery wastes. *Langmuir* 17, 1367–1371.

Abdel-Mawgoud, A.M., Aboulwafa, M.M., Hassouna, N.A.H., 2009. Characterizations of rhamnolipid produced by *Pseudomonas aeruginosa* isolate Bs 20. *Appl. Biochem. Biotechnol.* 157, 329–345.

Abdel-Mawgoud, A.M., Lépine, F., Déziel, E., 2010. Rhamnolipids: diversity of structures, microbial origins and roles. *Appl. Microbiol. Biotechnol.* 86, 1323–1336.

Abdel-Mawgoud, A.M., Hausmann, R., Lépine, F., Műller, M.M., Déziel, E., 2011. Rhamnolipids: Detection, analysis, biosynthesis, genetic regulation and bioengineering of production. In: Biosurfactants, Microbiology Monographs 20, G. Soberón-Chávez (ed.), Springer-Verlag Berlin Heidelberg doi: 10.1007/978-3-642-14490-5_2

Al-Tahhan, R.A., Sandrin, T.R., Bodour, A.A., Maier, R.M., 2000. Rhamnolipid induced removal of lipopolysaccharide from *Pseudomonas aeruginosa*: effect on cell surface properties and interaction with hydrophobic substrates. *Appl. Environ. Microbiol.* 66, 3262–3268.

Aparna, A., Srinkethan, G., Smitha, H., 2012. Production and characterization of biosurfactant produced by a novel *Pseudomonas* sp. 2B. *Colloids Surf. B* 95, 23– 29.

Avramovic, N.S., Nikloic-Mandic, S.D., Karadzic, I.M., Influence of rhamnolipids from *Pseudomonas aeruginosa* NCAIM (P) B 001380 on Cr (IV) removal capacity in liquid medium, *J. Serb. Chem. Soc., in press*

Bai, E., Brusseau, M.L., Miller, R.M., 1997. Biosurfactant-enhanced removal of hydrocarbon from soil. *J. Contam. Hydrol.* 25, 157-170.

Bai, G., Brisseau, M.L., Miller, R.M., 1998. Influence of cation type, ionic strenght and pH on solubilisation and mineralization of residual hydrocarbon by a biosurfactant. *J. Contam. Hydrol.* 30, 265-279.

Banat, I.M., 1995. Characterization of biosurfactants and their use in pollution removal—state of the art, *Acta. Biotechnol.* 3, 251–267.

Banat, I.M., Makkar, R.S., Cameotra, S.S., 2000. Potential commercial applications of microbial surfactants. *Appl. Microbiol. Biotechnol.* 53, 495–508.

Barathi, S., Vasudevan, N., 2001. Utilization of petroleum hydrocarbons by *Pseudomonas fluorescens* isolated from a petroleum-contaminated soil, *Environ. Intern.* 26, 413–416.

Beal, R., Betts, W.B., 2000. Role of rhamnolipid biosurfactants in the uptake and mineralization of hexadecane in *Pseudomonas aeruginosa. J. Appl. Microbiol.* 89, 158–168.

Benincasa, M., Contiero, J., Manresa, M.A., Moraes, I.O., 2002. Rhamnolipid production by *Pseudomonas aeruginosa* LBI growing on soapstock as the sole carbon source. *J. Food Eng.* 54, 283-288.

Benincasa, M., Abalos, A., Oliveira, I., Manresa, A., 2004. Chemical structure, surface properties and biological activities of the biosurfactant produced by *Pseudomonas aeruginosa* LBI from soapstock. *Antonie Van Leeuwenhoek* 85, 1–8.

Benincasa, M., Accorsini, F.R., 2008. *Pseudomonas aeruginosa* LBI production as an integrated process using the wastes from sunflower-oil refining as substrate. *Biores. Technol.* 99, 3843-3849.

Beskoski, V.P., Gojgic-Cvijovic, G., Milic, J., Ilic, M., Miletic, S., Solevic, T., Vrvic, M.M., 2011. Ex situ bioremediation of a soil contaminated by mazut (heavy residual fuel oil) - A field experiment. *Chemosphere* 83, 34-40.

Bodilis, J., Nsigue-Meilo, S., Besaury, L., Quillet, L., 2012. Variable Copy Number, Intra-Genomic Heterogeneities and Lateral Transfers of the 16S rRNA Gene in *Pseudomonas.* PLoS ONE 7(4): e35647. doi:10.1371/journal.pone.0035647

Bordoloi, N.K., Konwar, B.K., 2009. Bacterial biosurfactant in enhancing solubility and metabolism of petroleum hydrocarbons. *J. Hazard Mat.* 170, 495–505.

Cameotra, S.S., Singh, P., 2008. Bioremediation of oil sludge using crude biosurfactants. *Inter. Biodeter. Biodegrad.* 62: 274–280

Cameotra, S.S., Singh, P., 2009. Synthesis of rhamnolipid biosurfactant and mode of hexadecane uptake by *Pseudomonas* species. *Microbial Cell Fact* 8, 1–7.

Chandrasekaran, E.V., BeMiller, J.N., 1980. In: Methods in Carbohydrate Chemistry, Whistler RL (ed.). Academic Press, New York, p 89.

Chayabutra, C., Wu, J.,Kwang, J.K., 2001. Rhamnolipid production by *Pseudomonas aeruginosa* under denitrification: Effects of limiting nutrients and carbon sources. *Biotechnol. Bioeng.* 72**,** 25-33.

Colwell, R.R. Walker, J.D., Cooney, J.J., 1977. Ecological aspects of microbial degradation of petroleum in the marine environment. *Crit. Rev. Microbiol.* 5, 423–445.

Costa, S.G.V.A.O., Nitschke, M., Hadaad, R., Eberlin, M.N., Contiero, J., 2006. Production of *Pseudomonas aeruginosa* LBI rhamnolipids following growth on Brazilian native oils. *Process Biochem.* 41, 483-488.

Dahrazma, B., Mulligan, C.N., 2007. Investigation of the removal of heavy metals from sediments using rhamnolipid in a continuous flow configuration. *Chemosphere* 69, 705-711.

Das, K., Mukherjee, A.K., 2007. Crude petroleum-oil biodegradation efficiency of *Bacillus subtilis* and *Pseudomonas aeruginosa* strains isolated from a petroleum-oil contaminated soil from North-East India. *Bioresource Technol.* 98, 1339–1345.

Das, N., Chandran, P., 2011. Microbial Degradation of Petroleum Hydrocarbon Contaminants: An Overview. *Biotechnol. Research. Int.*Article ID 941810, doi:10.4061/2011/941810

Déziel, E., Paquette, G., Villemur, R., Lépine, F., Bisaillon, J.G., 1996. Biosurfactant production by a soil *Pseudomonas* strain growing on polycyclic aromatic hydrocarbons. *Appl. Environ. Microbiol.* 62, 1908–1912.

Déziel, E., Lépine, F., Dennie, D., Boismenu, D., Mamer, O.A., Villemur, R., 1999. Liquid chromatography/mass spectrometry analysis of mixtures of rhamnolipids produced by *Pseudomonas aeruginosa* strain 57RP grown on mannitol or naphthalene. *Biochim. Biophys. Acta* 1440, 244–252.

Déziel, E., Lépine, F., Milot, S., Villemur, R., 2000. Mass spectrometry monitoring of rhamnolipids from a growing culture of *Pseudomonas aeruginosa* strain 57RP. *Biochim. Biophys. Acta* 1485, 145–152.

Dimitrijevic, A., Velickovic, Rikalovic, M., Avramovic, N., Milosavic, N., Jankov, R., Karadzic, I., 2011. Simultaneous production of exopolysaccharide and lipase from extremophylic *Pseudomonas aeruginosa* san-ai strain: a novel approach for lipase immobilization and purification. *Carbohyd. Polym.* 83, 1397–1401.

Dubeau, D., Déziel, E., Woods, D., Lépine, F., 2009. *Burkholderia thailandensis* harbors two identical rhl gene clusters responsible for the biosynthesis of rhamnolipids. *BMC Microbiology* 9: 263, doi:10.1186/1471-2180-9-263

Dubey, K., Juwarkar, A., 2001. Distillery and crud whey wastes as viable alternative sources for biosurfactant production. *World J. Microbiol. Biotechnol.* 17, 61–69.

Dusane, D.H., Zinjarde, S.S., Smita S., Venugopalan, V.P., Mclean, R.J.C.; Weber, M.M., Rahman, P.K.S.M., 2010. Quorum sensing: implications on rhamnolipid biosurfactant production. *Biotechnol. Genet. Eng. Rev.* 27, 159-184.

Edwards, D.A., Luthy, R.G., Liu, Z., 1991. Solubilization of polycyclic aromatic hydrocarbons in micellar nonionic surfactant solutions. *Environ. Sci. Technol.* 25, 127-133.

Eliora, Z.R., Rosenberg, E., 2001. Natural role of biosurfactants. *Environ. Microbiol.* 3, 229–236.

Eliora, Z.R. Rosenberg, E., 2002. Biosurfactant and oil bioremediation. *Curr. Opin. Biotechnol.* 13, 249–252.

Flemming, H.C., Wingender, J., 2001. Relevance of microbial extracellular polymeric substances (EPSs)-Part II: Technical aspects. *Water Sci. Teschnol.* 43, 9-16.

Foght, J.M., Westlake, D.W., Johnson, W.M., Ridgway, H.F., 1996. Environmental gasoline-utilizing isolates and clinical isolates of *Pseudomonas aeruginosa* are taxonomically indistinguishable by chernotaxonornic and molecular techniques. *Microbiology* 142, 2333-2340.

Franklin, M.T., Nivens, D.E., Weadge, J.T., Lynne-Howell, L., 2011. Biosynthesis of *P. aeruginosa* extracelullar polisacharides alginate, Pel and Psl. *Front. Microbiol.* 2, 1-16 doi: 10.3389/fmicb.2011.00167

Franzetti, A., Gandolfi, I., Bestetti, G., Smyth, T.J., Banat, I.M., 2010. Production and applications of trehalose lipid biosurfactants. *Eur. J. Lipid Sci. Tech.* 112, 617–627.

Freire, D.M.G., 2002. Production of biosurfactants from *Pseudomonas aeruginosa* PA1 isolated in oil environments. *Braz. J. Chem. Eng.* 19, 159-166.

Górna, H., Lawnicazak, Ł., Zgoła-Greśkowiak, A., Kaczorek, E., 2011. Differences and dynamic changes in cell surface properties of three *Pseudomonas aeruginosa* strain isolated from petroleum-pollutes soil as a response to various carbon sources and the external addition of rhamnolipids. *Biores. Technol.* 102, 3028-3033.

Gupta, R., *Ahuja*, P., *Khan*, S., *Saxena, R.K.*, *Mohapatra*, H., 2000. *Microbial biosorbents: meeting challenges of heavy metal pollution in aqueous solutions. Curr. Sci. 78, 967–973.*

Haba, E., Espuny, M.J., Busquets, M., Manresa, A., 2000. Screening and production of rhamnolipids by *Pseudomonas aeruginosa* 47T2 NCIB 40044 from waste frying oils. *J. Appl. Microbiol.* 88, 379-387.

Haba, E., Abalos, A., Jauregui, O., Espuny, M.J., Manresa, A., 2003a. Use of liquid chromatography-mass spectroscopy for studying the composition and properties of rhamnolipids produced by different strains of *Pseudomonas aeruginosa. J. Surfactants. Deterg*. 6, 155–161.

Haba, E., Pinazo, A., Jauregui, O., Espuny, M.J., Infante, M.R., Manresa, A., 2003b. Physicochemical characterization and antimicrobial properties of rhamnolipids produced by *Pseudomonas aeruginosa* 47T2 NCBIM 40044. *Biotechnol. Bioeng*. 81, 316–322.

Hasanuzzaman, M., Ueno, A., Ito, H., Ito, Y., Yamamoto, Y., Yumoto, I., Okuyama, H., 2007. Degradation of long-chain *n*-alkanes (C36 and C40) by *Pseudomonas aeruginosa* strain WatG. *Inter. Biodet. Biodeg*. 59, 40–43.

Heyd, M., Kohnert, A., Tan, T.-H., Nusser, M., Kirschhöfer, F., Brenner-Weiss, G., Franzreb, M., Berensmeier, S., 2008. Development and trends of biosurfactant analysis and purification using rhamnolipids as an example. *Anal. Bioanal. Chem.* 391, 1579–1590.

Hodge, J.E., Hofreiter, B.T., 1962. In: Whistler RL, Wolfrom ML (eds) Methods in carbohydrate chemistry, vol 1. Academic., New York, p 380.

Huang, Y.-B., Wang, W.-H., Peng, A., 2000. Accumulation of Cu(II) and Pb(II) by biofilms grown on particulate in aquatic systems. *J. Environ. Sci. Health Part A Environ.* Sci. En. 35, 575-592.

Inakollu, S., Hung, H., Shreve, G.S., 2004. Biosurfactant enhancement of microbial degradation of various strructural classes of hydrocarbon in mixed waste systems. *Environ. Eng. Sci.* 21, 463–469.

Jain, S., Rani, A., Marla, S., Goel, R., 2010. Differential proteomic analysis of psychrotolerant *Pseudomonas putida* 710a and alkalophilic *Pseudomonas monteilli* 97 a for cadmium stress. *Intel. J. Bio. Med. Res*. 1, 234-241.

Jovancicevic, B., Antic, M., Pavlovic, I., Vrvic, M., Beskoski, V., Kronimus, A., Schwarzbauer, J., 2008. Transformation of petroleum saturated hydrocarbons during soil bioremediation experiments. *Water Air Soil Pollut.* 190, 299–307.

Juwarkar, A.A., Dubey, K.V., Nair, A., Singh, S.K., 2008. Specific amphiphilic nature of biosurfactants is responsible for their mechanisms of action in bioremediation. *Indian J. Microbiol.* 48, 142-146.

Juwarkar, A.A., Nair, A., Dubey, K.V. Singh, S.K., Devotta, S., 2007. Biosurfactant technology for remediation of cadmium and lead contaminated soils. *Chemosphere.* 68, 1996–2002.

Kachany, S.C., Levery, S.B., Kim, J.S., Reuhs, B.L., Lion, L.W., Ghiorse, W.C., 2001. Structure and carbohydrate analysis of the exopolysaccahride capsule of *Pseudomonas putida* G7. *Environ. Microbiol.* 3, 774-784.

Karadzic, I., Masui, A., Fujiwara, N., 2004. Purification and characterization of a protease from *Pseudomonas aeruginosa* growth in cutting oil. *J. Biosci. Bioeng.* 98, 145-152.

Karadzic, I., Masui, A., Izrael-Zivkovic, L., Fujiwara, N., 2006. Purification and characterization of an alkaline lipase from *Pseudomonas aeruginosa* isolated from putrid mineral cutting oil as metal working fluid. *J. Biosc. Bioeng.* 102, 82-89.

Kazy, S.K., Sar, P., Singh, S.P., Sen, A.K., D'Souza, S.F., 2002. Extracellular polysaccharides of a copper-sensitive and a copper-resistant *Pseudomonas aeruginosa* strain: synthesis, chemical nature and copper binding. *World J. Microbiol. Biotechnol.* 18, 583–588.

Kiewitz, C., Tummler, B., 2000. Sequence diversity of *Pseudomonas aeruginosa*: impact on population structure and genome evolution. *J. Bacteriol.* 182, 3125-3135.

Kiliç, N.K., Dönmenz, G., 2008. Environmental conditions affecting exopolysaccharide production by *Pseudomonas aeruginosa*, *Micrococcus* sp., and *Ochrobactrum* sp.. *J. Hazard. Mat.* 154, 1019-1024.

Kiliç, N.K, Stensballe, A., Otzen, D.E., Dönmenz, G., 2010. Proteomic changes in response to chromium (VI) toxicity in *Pseudomonas aeruginosa*. *Biores. Technol.* 101, 2134-2140.

Ko, S.H., Lebeault, J.M., 1999. Effect of a mixed culture on co-oxidation during the degradation of saturated hydrocarbon mixture. *J. Appl. Microbiol.* 87, 72–79.

Koch, A.K., Käppeli, O., Fiechte,r A., Reiser, J., 1991. Hydrocarbon assimilation and biosurfactant production in *Pseudomonas aeruginosa* mutants. *J. Bacteriol.* 173, 4212–4219.

Leahy, J.G., Colwell. R.R., 1990. Microbial degradation of hydrocarbons in the environment. *Microbiol. Rev.* 54, 305–315.

Maier, R.M., Soberon-Chávez, G., 2000. *Pseudomonas aeruginosa* rhamnolipids: biosinthesis and potential application. *Appl. Microbiol. Biotechnol.* 54, 625-633.

Makkar, R.S., Cameotra, S.S., 2002. An update on the use of unconventional substrates for biosurfactants production and their new applications. *Appl. Microbiol. Biotecnol.* 58, 428–434.

Maneerat, S., 2005. Production of biosurfactants using substrates from renewable-resources. *Songklanakarin J. Sci. Technol.* 27, 675-683.

Maslin, P., Maier, R.M., 2000. Rhamnolipid-enhanced mineralization of phenanthrene in organic-metal co-contaminated soils. *Bioremediat. J.* 4, 295–308.

Maier, R.M., 2003. Biosurfactants: Evolution and Diversity in Bacteria. *Adv. Appl. Microbiol.* 52, 101-121

Mata-Sandoval, J., Karns, J., Torrents, A., 1999. High-performance liquid chromatography method for the characterization of rhamnolipid mixtures produced by *Pseudomonas aeruginosa* UG2 on corn oil. *J. Chromatogr.* 864, 211-220.

Mata-Sandoval, J., Karns, J., Torrens, A., 2001. Effect of nutritional and environmental conditions on the production and composition of rhamnolipids by *P. aeruginosa* UG2. *Microbiol. Res.* 155, 249-256.

Mercade, M.E., Manresa, M.A., Robert, M., Espuny, M.J., De Andres, C., Guinea, J., 1993. Olive oil mill effluent (OOME): New substrate for biosurfactant production. *J. Bioresour. Technol.* 43, 1-6.

Miller, R.M., 1995a. Surfactant-enhanced bioavailability of slightly soluble organic compounds. Eds. H.D. Skipper, R.F. Turco RF in Bioremediation: science and applications. Soil Science Society of America Madison (USA), pp 322–354.

Miller, R.M., 1995b. Biosurfactant facilitated remediation of contaminated soil. *Environ. Health Perspect.* 103, 59–62.

Miller C.D., Pettee B., Zhang, C., Pabst, M., McLean, J.E., Anderson, A.J., 2009. Copper and cadmium: responses in *Pseudomonas putida* KT2440. *Lett. Appl. Microbiol.* 49, 775-783.

Moldes, A.B., Torrado, A.M., Barral, M.T., Domianguez, J.M., 2007. Evaluation of biosurfactant production from various agricultural residues by *Lactobacillus pentosus. J. Agric. Food Chem.* 55: 4481–4486

Moore, E.R.B., Tindal, B.J., Martins Dos Santos, V.A.P., Pieper, D.H., Ramos, J.L., Palleroni, N.J., 2006. Nonmedical *Pseudomonas.* In Prokaryotes, Eds. M. Dworkin, S. Falkow, E. Rosenberg, K.H. Schleifer, E. Stackebrandt, vol.6. Springer, New York (USA), pp 646-703.

Mulligan, C.N. Yong, C.N. Gibbs, B.F., 1999. On the use of biosurfactants for the removal of heavy metals from oil contaminated soil. *Environ. Prog.* 18, 31–35.

Mulligan, C.N., Yong, R.N., Gibbs, B.F., 2001a. Enhanced remediation of contaminated soil: a review. *Eng. Geol.* 60, 371-338.

Mulligan, C.N., Yong, R.N., Gibbs, B.F. 2001b. Heavy metal removal from sediments by biosurfactants. *J. Hazard Mat.* 85, 111-125.

Muthusamy, K., Gopalakrishnan, S., Kochupappy Ravi, T., Sivachidambaram, P., 2008. Biosurfactants: properties, commercial production and application. *Curr. Sci.* 94, 736-744.

Nancharaiah, Y.V., Dodge, C., Venugopalan, V.P., Narasimhan, S.V., Francis, A.J., 2010. Immobilization of Cr(VI) and its reduction to Cr(III) phosphate by granular biofilms comprising a mixture of microbes. *Appl. Environ. Microbiol.* 76, 2333-2338.

Nie, M., Yin, X., Ren, C., Wang, Y., Xu, F., Shen, Q., 2010. Novel rhamnolipid biosurfactants produced by a PAH-degrading bacterium *P. aeruginosa* strain NY3. *Biotechnol. Adv.* 28, 635-643.

Nitschke, M., Costa, S.G., Contiero. J., 2010. Structure and applications of a rhamnolipid surfactant produced in soybean oil waste. *Appl. Biochem. Biotechnol.* 160, 2066–2074.

Noordman, W.H., Janssen, D.B., 2002. Rhamnolipid stimulates uptake of hydrophobic compounds by *Pseudomonas aeruginosa. Appl. Environ. Microbiol.* 68, 4502–4508.

Pacwa-Płociniczak, M., Płaza, G.A., Piotrowska-Seget, Z., Cameotra, S.S., 2011. Environmental applications of biosurfactants: Recent Advances. *Int. J. Mol. Sci.* 12, 633-654.

Patel, R.M., Desai, A.J., 1997. Biosurfactant production by *Pseudomonas aeruginosa* GS3 from molasses. *Lett. Appl. Microbiol.* 25, 91–94.

Pornsunthorntawee, O., Wongpanit, P., Chavadej, S., Abe, M., Rujiravanit, R., 2008. Structural and physiochemical characterization of crude biosurfactant produced by *Pseudomonas aeruginoa* SP4 isolated from petroleum-contaminated soil. *Biores. Technol.* 99, 1589 -1595.

Price, N.P.J., Ray, K.J., Vermillion, K., Kuo, T.M., 2009. MALDI-TOF mass spectrometry of naturally occurring mixtures of monorhamnolipids and dirhamnolipids. *Carbohydr. Res.* 344, 204–209.

Priester, J.H., Olson, S.G., Webb, S.M., Neu, M.P., Hersman, L.E., Holden, P.A., 2006. Enhanced exopolymer production and chromium stabilization in *Pseudomonas putida* unsaturated biofilms. *Appl. Environ. Microbiol.* 72, 1988-1996.

Rahman, K.S.M., Rahman, T.J., McClean, S., Marchant, R., Banat, I.M., 2002. Rhamnolipid biosurfactants production by strains of *Pseudomonas aeruginosa* using low cost raw materials. *Biotechol. Prog.* 18, 1277-1281.

Rahman, K.S.M., Rahman, T.J., Lakshmanaperumalsamy, P., Marchant, R., Banat, I.M., 2003. The potential of bacterial isolates for emulsification with range of hydrocarbons. *Acta Biotechnol.* 4, 335–345.

Rashedi, H., Jamshidi, E., Mazaheri, A.M., Bonakdarpour, B., 2006. Isolation and production of biosurfactant from *Pseudomonas aeruginosa* isolated from Iranian southern wells oil. *J. Iran. Chem. Eng.* 25, 25-30.

Raza, Z.A., Rehman, A., Khan, M.S., Khalid, Z.M., 2007. Improved production of biosurfactant by a *Pseudomonas aeruginosa* mutant using vegetable oil refinery wastes. *Biodegradation* 18, 115-121.

Rikalovic, M., Gojgic-Cvijovic, G., Vrvic, M.M., Karadzic, I., 2012. Production and characterization of rahamnolipid from strain *Pseudomonas aeruginosa* san ai. *J. Serb. Chem. Soc.* 77, 27–42.

Rikalovic, M.G., Abdel-Mawgoud, A.M., Déziel, E., G., Gojgic-Cvijovic, G.Dj., Nestorović, Z., Vrvic, M.M. Karadzic, I.M., 2013. Comparative analysis of rhamnolipids from novel environmental isolates of *Pseudomonas aeruginosa*. *J. Surfactants Deterg.* 16, 673-682.

Rooney, A.P., Price, N.P., Ray, K.J., Kuo, T.M., 2009. Isolation and characterization of rhamnolipid-producing bacterial strains from a biodiesel facility. *FEMS Microbiol. Lett.* 295, 82–87.

Rosen, M., 1974. Relationship of structure properties in surfactants: II. Efficiency in surfaces or interfacial tension reduction. *J. Am. Oil Chem. Soc.* 51, 461-481.

Rosenberg, E., Ron, E.Z., 1999. High and low-molecular-mass microbial surfactants. *Appl Microbiol. Biotechnol.* 52, 154–162.

Sabra, B., Zeng, A.P., 2009. Microbial production of alginates: physiology and process aspects In: Alginates: Biology and Applications, Rehm, B.H.A. (ed), Springer Verlag Berlin Heidelberg (Germany), pp 153-174.

Sandrin, R.T., Andrea, M.C., Maier, R.M., 2000. A rhamnolipid biosurfactant reduces cadmium toxicity during naphthalene biodegradation. *Appl. Environ. Microbiol.* 66, 4585–4588.

Santa Anna, L.M., Sebastian, G.V., Menezes, E.P., Alves, T.L.M., Santos, A.S., Pereira, N. Jr., Freire, D.M.G., 2002. Production of biosurfactants from *Pseudomonas aeruginosa* PA1 isolated in oil environments. *Braz. J. Chem. Eng.* 19, 159-166.

Sarachat, T., Pornsunthorntawee, O., Chavadej, S., Rujiravanit, R., 2010. Purification and concentration of a rhamnolipid biosurfactant produced by *Pseudomonas aeruginosa* SP4 using foam fractionation. *Bioresour. Technol.* 101, 324–30.

Satyanarayana, T., 2012. Microorganisms in Environmental Management: Microbes and Environment, in: Bhavdish, N., Prakash, A. (eds), Springer-Verlag Berlin Heidelberg (Geramany)

Scheibenbogen, K., Zytner, R.G., Lee, H., Trevors, J.T., 1994. Enhenced removal of selected hydrocarbons from soil by Pseudomonas aeruginosa UG2 biosurfactant and some chemical surfactants. *J. Chem. Tech. Biotechnol.* 59, 53-59.

Shreve, G.S., Inguva, S, Gunnam, S., 1995. Rhamnolipid biosurfactant Enhancement of hexadecane biodegradation by *Pseudomonas aeruginosa*. *Mol. Mar. Biol. Biotechnol.* 4, 331–337.

Singh, P., Cameotra, S.S., 2004. Potential applications of microbial surfactants in biomedical sciences. *Trends Biotechnol.* 22, 142-146.

Singh, A., Van Hamme, J.D., Ward, O.P., 2007. Surfactants in microbiology and biotechnology: Part 2. Application aspects. *Biotechol. Adv.* 25. 99-121.

Singh, V., Chauhan, P.K., Kanta, R., Dhewa, T., Kumar, V., 2010. Isolation and characterization of Pseudomonas resistant to heavy metals contaminants, *Int. J. Pharm. Sci. Rev. Res.* 3, 164-167.

Siegmund, I., Wagner, F., 1991. New method for detecting rhamnolipids excreted by *Pseudomonas* species during growth on mineral agar. *Biotechnol. Tech.* 5, 265-268.

Siegmund, L., 2002. Biological amphiphiles (microbial biosurfactants). *Curr. Opp. Coll. Inter. Sci.* 7, 12–20.

Soberon-Chavez, G., Lepine, F., Deziel, E., 2005. Production of rhamnolipids by *Pseudomonas aeruginosa. Appl. Microbiol. Biotechnol.* 68, 718–725.

Song, H., Liu, Y., Xu, W., Zeng, G., Aibibu, N., Xu, L., Chen, B., 2009. Simultaneous Cr(VI) reduction and phenol degradation in pure cultures of *Pseudomonas aeruginosa* CCTCC AB91095. *Biores. Technol.* 100, 5079–5084.

Spangenberg, C., Montie, T.C., Tummler, B., 1998. Structural and functional implications of sequence diversity of *Pseudomonas aeruginosa* genes *oriC*, *ampC* and *fliC*. *Electrophoresis* 19, 545-550.

Spiers, A.J., Buckling, A., Rainey P.B., 2000. The causes of *Pseudomonas* diversity. *Microbiology* 146, 2345–2350.

Stover, C.K., Pham, X.Q., Erwin, A.L., Mizoguchi, S.D., Warrener, P., Hickey, M.J., Brinkman FS, Hufnagle, W.O., Kowalik D.J., Lagrou, M., Garber, R.L., Goltry, L., Tolentino E., Westbrock-Wadman, S., Yuan, Y., Brody, L.L., Coulter, S.N., Folger, K.R., Kas, A., Larbig, K., Lim, R., Smith, K., Spencer, D., Wong, G.K., Wu, Z., Paulsen, I.T., Reizer, J., Saier, M.H., Hancock, R.E., Lory, S., Olson, M.V., 2000. Complete genome sequence of *Pseudomonas aeruginosa* PAO1, an opportunistic pathogen. Nature 406, 959–964.

Thangamani, S., Shreve, G.S., 1994. Effect of anionic biosurfactan on hexadecane partitioning in multiphase systems. *Environ. Sci. Technol.* 28, 1993-2000.

Tahzibi, A., Kamal, F., Assadi, M.M., 2004. Improved production of rhamnolipids by a *Pseudomonas aeruginosa* mutant. *Iran. Biomed. J.* 8, 25-31.

Thai, L.T., Maier, W.J., 1992. Solubilisation adn biodegradation of octadecane in the presence of two commercial surfactants. Proccesdings of the 47th Annual Purdue University Industrial Waste Conference Ann Arbor Press, Chelsea, Mich

Teitzel, G.M., Parsek, M.R., 2002. Heavy Metal Resistance of Biofilm and Planktonic *Pseudomonas aeruginosa. Appl. Environ. Microbiol.* 69, 2313–2320.

Thompson, D.K., Chourey, K., Wickham, G.S., Thieman, S.B., VerBerkmoes, N.C., Zhang, B., McCarthy, A.T., Rudisill, M.A., Shah, M., Hettich. R.L., 2010. Proteomics reveals a core molecular response of *Pseudomonas putida* F1 to acute chromate challenge. *BCM Genomics* 11:311, doi:10.1186/1471-2164-11-311

Tiehm, A., 1994. Degradation of polycyclic aromatic hydrocarbons in the presence of synthetis surfactants. *Appl. Environ. Microbiol.* 60, 258-263.

Ulrici, W., 2000. Contaminant soil areas, different countries and contaminant monitoring of contaminants, In: Environmental Process II. Soil Decontamination Biotechnology, H. J. Rehm and G. Reed (eds) vol. 11, pp. 5–42.

Van Hamme, J.D., Singh, A., Ward, O.P., 2006. Surfactants in microbiology and biotechnology: Part 1. Physiological aspects. *Biotechnol. Adv*. 24, 604-620.

von Canstein, H.Y.L, Timmis, K.N., Deckwer, W.-D., Wagner-Döbler, I., 1999. Removal of mercury from chloralkali electrolysis wastewater by a mercury-resistant *Pseudomonas* strain. Appl. Environ. Microbiol. 65**,** 5279-5284.

Wang, S., Mulligan, C.N., 2004. Rhamnolipid foam enhanced remediation of cadmium and nickel contaminated soil. Water Air Soil Pollut. 157, 315–330.

Ward, O.P., Singh, A., Van Hamme, J., 2003. Acclerated biodegradation of petroleum hydrocarbon waste. *J. Ind. Microbiol. Biotechnol*. 30, 260–270.

Williams, P., Winzer, K., Chan, W.C., Camara, M., 2007. Look who's talking: communication and quorum sensing in the bacterial world. *Philos. T. Roy. Soc. B: Biol. Sci.* 362, 1119-1134.

Winsor, G.L., Lam, D.K., Fleming, L., Lo, R., Whiteside, M.D., Yu, N.Y., Hancock, R.E., Brinkman, F.S., 2011. *Pseudomonas* Genome Database: improved comparative analysis and population genomics capability for *Pseudomonas* genomes. Nucleic Acids Res. 39, D596–D600.

Wongsa, P., Tanaka, M., Ueno, A., Hasanuzzaman, M., Yumoto, I., Okuyama, H., 2004. Isolation and characterization of novel strains of *Pseudomonas aeruginosa* and *Serratia marcescens* possessing high efficiency to degrade gasoline, kerosene, diesel oil and lubricating oil. *Curr. Microbiol.* 49, 415-422.

Zhang, Y.M., Miller, R.M., 1994. Effect of a *Pseudomonas* rhamnolipid biosurfactanton cell hydrophobicity and biodegradation of octadecane. *Appl. Environ. Microbiol.* 60, 2101–2106.

Zhang, Y.M., Miller, R.M., 1995. Effect of rhamnolipid (biosurfactant) structure on solubilization and biodegradation of n-alkanes. *Appl. Environ. Microbiol.* 61, 2247–2251.

Zhang, Y., Maier, W.J., Miler, R.M., 1997. Effect of rhamnolipids on the dissotiation, bioavilability and biodegredation of phenanthrene. *Environ. Sci. Technol.* 31, 2211-2217.

Zhao, B., Poh, C.L., 2008. Insights into environmental bioremediation by microorganisms through functional genomics and proteomics. *Proteomics* 8, 874–81.

Yahr, T.L., Parsek, M.R., 2006. *Pseudomonas aeruginosa*. In: *The Prokaryotes*. Dworkin, M., Falkow, S., Rosenberg, E., and Schleifer, K.-H., and Stackebrandt, E. (eds). New York, Springer Science+Business Media, LLC, pp 704–713.

In: Bioremediation: Processes, Challenges and Future Prospects ISBN: 978-1-62948-513-3
Editors: Jesús Bernardino Velázquez-Fernández et al. © 2014 Nova Science Publishers, Inc.

Chapter 16

SOME ENZYMES USEFUL FOR BIOREMEDIATION TECHNOLOGIES ON ORGANIC POLLUTANTS

Geovanny Naim Melchor Partida[1], Abril Bernardette Martinez Rizo[1], Saé Muñiz-Hernández[2], Adela L. Rendón-Ramírez[3] and Jesús Bernardino Velázquez-Fernández[4,*]

[1] Posgrado en Ciencias Biológico Agropecuarias, Autonomous University of Nayarit, Mexico
[2] National Institute of Cancerology, Mexico
[3] Biophysics Unit, (CSIC, UPV/EHU), Spain
[4] Autonomous University of Nayarit, Mexico

ABSTRACT

Biological resilience has been used to decrease negative effects of pollutants on the environment. Bioremediators use several strategies to be tolerant and biodegrade pollutants. Within the several processes involved in bioremediation, enzymes are the main agents mediating biodegradation of organic pollutants. Organisms have recruited different kinds of enzymes to survive in polluted environments. The knowledge on these enzymes' capability would allow improvements on choice and use of bioremediators strategies. Among the enzymes recruited for biodegradation, we can find peroxidases, glutathione S-transferases and dioxygenases; all of them have shown to be useful for bioremediation. Among other functions, peroxidases depict scarce selectivity and oxidize several organic compounds. Glutathione S-transferases catalyze the conjugation of glutathione to electrophilic atoms, which could be helpful for organochlorine compounds biodegradation. Dioxygenases catalyze the addition of two oxygen atoms, avoiding mono-oxygenated metabolites that could be more toxic than the parental compound. In this chapter, we review findings and features of these enzymes that make them valuable for bioremediation purposes.

---

* Corresponding Author address: CEMIC03, Posgrado en Ciencias Biológico Agropecuarias y Pesqueras, Cd. De la Cultura Amado Nervo, Tepic, Nayarit, Mexico. Email: jesusbvf@gmail.com.

Keywords: Enzymes; bioremediation; glutathione S-transferases; dioxygenases

INTRODUCTION

During the last decades, the planet earth has suffered the adverse effects of pollution. Each year, thousands of new organic chemicals are synthesized, reaching the environment and becoming a hazard to human and environmental health. Their release could be unintentional or caused by their use. Whichever the reason, the environment has had to cope with several organic pollutants. Some of them are persistent, i.e., they are not easily degraded because of their physical or chemical properties or the absence of organisms able to biotransform them. The Stockholm Convention enlists several of the worst persistant pollutants (Stockholm Convention, 2012). Typically, these pollutants are toxic, chemically stable, accumulable, and move throughout ecosystems. International regulations aim to restrict or prohibit the environmental release of these pollutants. Nevertheless, some of the chemicals are already present in the environment, hereby placing a permanent hazard. Therefore, these persistant organic pollutants (POPs) are an unavoidable problem that could become worse if no measure is taken to reduce them or their adverse environmental effects, i.e., to remediate the pollution.

Among the strategies used to remediate environment pollution are the use of organisms or their derivatives to alleviate, confine, immobilize or mitigate the adverse effects. Such strategies are grouped into the so-called "bioremediation" technology. This technology uses the natural resilience mechanism of organisms. Among the resilience mechanisms, several may be found such as biodegradation, biotransformation, confinement, immobilization, translocation, among others. Biodegradation occurs thanks to biotransformation or physico-chemical properties of biomolecules. The enzymatic action could lead to either biodegradation or toxic activation of pollutants. Nevertheless, there is still little information of adequate enzymes for bioremediation. This chapter reviews information on some enzymes that can be good candidates for bioremediation purposes.

IMPORTANCE OF ENZYMES IN NATURE

Most reactions in life are catalyzed by enzymes. Because of them, life reactions occur in short times and not in years. Several functions in organisms are due to enzyme action (Brett, 2003). Tolerance and resilience to xenobiotics (exogenous compounds) are one of these functions and frequently are mediated by several enzymes, each one catalyzing different reactions. In other words, because of these enzymes, tolerance or degradation of xenobiotics could be achieved. The tolerance mechanisms have evolved along centuries and involve enzymes that have been recruited due to the capability and resistance they confer to organisms. Therefore, thanks to mutations throughout generations, organisms have evolved to cope with xenobiotics via biotransformation. These mutations could lead to a broader promiscuity of enzymes or to any other change that would confer an evolutive advantage to the organism. Hence, recruited enzymes often come from promiscuous enzymes (Copley, 2009). For instance, dioxygenases and monooxygenases with low specificity are implicated in the two-phase (anaerobic reduction and aerobic oxidation) biotransformation of organohalide pollutants (Wacket, 1995).

The mutations conferring evolutive advantages would be observed after several generations. Since generation time of most bacteria and some fungi is below a month, most resilience mutations can be observed in human-life span. This is specially true for organic synthetic pollutants. Because they are new for the earth ecosystems, i.e., they have appeared on the earth since less than a century, organisms with a generation time of more than a decade (some animals, for instance) have not developed or evolved into resilience mutations. It could be supposed that those organisms that have developed mechanisms of resilience, have also recruited enzymes able to biotransform or biodegrade them. For these reasons, it has been proposed that enzymes for bioremediation should be searched for in those resilient/tolerant organisms (Velázquez-Fernández et al., 2012)

According to the reactions they catalyze, enzymes would be classified and named. Lyases and hydrolases catalyze breakage of compounds. Hydrolases use water for their activity. Isomerases and transferases catalyze isomerization and transfer reactions, respectively. Oxydoreductases catalyze reactions involving electron transfer. Synthases catalyze formation or synthesis reactions. Interestingly, enzymes implicated in organic pollutants biotransformation fall into the group of hydrolases, oxydoreductases, and transferases. Those biotransformation enzymes could be useful for bioremediation since they have naturally evolved to defend organisms against xenobiotics. Nevertheless, the natural features of enzymes must be improved or enhanced in order to take advantage of them in bioremediation technologies.

Advantanges and Disadvantages of Enzymes Utilization

Bacteria display more diversity in enzymatic action. This, together with their low duplication time, makes bacteria the most adequate and used organisms for bioremediation (Wackett, 2004). Bacterial mutation could occur in time periods as short as weeks. Thus, enzyme recruitment in bacteria could occur in time intervals which are detectable during the human-life span. All the above contribute to make bacteria able to cope with any class of pollutants and to biotransform several kinds of them. Because of this, bacteria are a proper source to search for biotransformation enzymes.

Bioremediation could be achieved with either enzymes or with organisms able to bioremediate or biodegrade pollutants, i.e., the bioremediator or biodegrader. Enzymes could be used in bioreactors to allow them to perform biotransformation of pollutants. For this purpose, they are chosen based on some features such as stability and promiscuity, among others. Thus, some advantages of using enzymes arise from their proper selection, whereas the use of organisms could be improved by genetic engineering. In other words, transgenic organisms could also result in adequate bioremediators. This is specially true when several steps (and so, several enzymes) are needed for a proper biodegradation of pollutants. For instance, to improve biodegradability, the use of transgenic plants for PCB's (Van Aken et al., 2010) or even for a broader range of pollutants has been suggested (Abhilash et al., 2009)

Most POPs are aromatic or halogenated. Their chemical stability and the reactivity of the byproducts generated by their degradation should be considered when choosing a bioremediator. To be considered as good candidates for biorremediation of POPs, enzymes should be able to catalyze degradation of these structures without being inhibited or affecting the biodegrader organism (Sutherland, 2004). Unfortunately, few enzymes meet these requirements, and none of them alone could accomplish proper full degradation of pollutants. In addition, enzymes should have low specificity, i.e., they should be promiscous; otherwise, they could not degrade the large diversity of pollutant structures, but only substrates or structures for which they were recruited. In fact, enzymes promiscuity allows de-halogenation or a breakdown of different pollutant structures (Wackett, 2004).

The use of organisms for bioremediation (bioremediators) has several advantages as compared with other physicochemical methods applied for environmental remediation. Some advantages are: 1) the cost is low, 2) their use is environmentally friendly, 3) aesthetic, 4) easy-acepted by people, 5) can be applied *in situ* or *ex situ*, 6) lower times than natural attenuation, 7) low energy/input requirements, 8) organisms could replace the enzymes if destroyed. Nevertheless, for the use of bioremediators the following is necessary: 1) to know what kind of pollutant is to be treated, 2) to know a specific organism or consortium that is efficient against that pollutant, 3) the organisms must be able to survive or live in such ecosystem to bioremediate, 4) to meet all the growth or maintenance requirements for organisms, 5) to test if such organisms or consortia can actually bioremediate in the specific conditions. Regarding the latter, it is worth to note that results of biodegradation in the lab could not agree with those when testing the bioremediator in the environment. This could be caused by the difficulties in pollutant-enzymes interaction. In other words, either the pollutant is not bioavailable to the organism or the adequate enzyme is intracellular, and the pollutant may not enter the cell, compromising severely the biodegradation.

When it is said that the bioremediation strategy should be "tailored" to the polluted ecosystem, it is implied that the above requirements should be met. Among the disadvantages of using bioremediators, there could be: 1) longer times than physicochemical remediation methods; 2) they must be tailored; 3) should a non-indigenous organism be added (bioaugmentation), the ecosystem's balance could be altered; 4) specific physical or chemical requirements (pH, oxygenation, temperature) could be difficult to maintain, or be non-compatible with the ecosystem to be remediated or more resources were needed, which increase the economic or energetic cost; 5) should the enzyme be induced, the knowledge and needs to induce it are a must; 6) if organisms were able to grow, then the resultant biomass could be an issue of disposal, worse if it contains toxic compounds.

The use of enzymes isolated from organisms could avoid some of the disadvantages. For instance, enzymes are chosen to be stable and promiscuous, and enzyme concentration could be manipulated in order to enhance biodegradation rates. Thus, bioremediation time could be improved by using enzymes. In addition, no bioaugmentation would happen with enzymes, no biomass would be generated, and specific requirements of induction or growth should be met. Nevertheless, if the enzyme requires a pH buffering or temperature regulation, probably, it would be more difficult to apply than using bioremediators. Similarly to the use of bioremediators, when using enzymes, tests must be performed to evaluate biodegradation efficiency. The employed enzyme should be isolated and immobilized, which avoids problems of bioavailability and could enhance interactions between the enzyme and the pollutant. Still, efficiency should be evaluated since enzyme immobilization alters stability (usually improves it) and decreases catalytic efficiency. Another disadvantage of using enzymes is the cost of production and purification, which could be improved by using genetic engineering.

As the enzyme bioreactor should be designed according to the pollutant, it is imperative to have enough knowledge on adequate enzymes that could be used for this derivative bioremediation. Whether bioremediator or enzyme is used, information on enzyme characteristics would help to improve bioremediation efficiency.

When using enzymes in bioreactors, they seem to be more stable and have a larger half-life than inside the cell, but the immobilization step is still needed to avoid losing enzyme. Besides, thanks to biotechnology, nowadays, it is possible to produce large amounts of enzymes for bioreactores. This is a key advantage of enzymes over other kinds of bioremediation strategies. Enzymatic bioremediation could be accomplished either *ex situ* or *in situ*; thus, transportation of polluted environmental matrix is a choice and not a *sine qua non* condition (McGuinees, 2007). Furthermore, enzymatic bioremediation could be easily improved thanks to genetic engineering that could enhance catalytic features like promiscuity, velocity, affinity or efficiency (Sutherland, 2004). Still, the cost of enzyme production is higher than that of microorganism cultures, but it avoids biomass production that could happen when using microorganism cultures. Another disadvantage of enzymatic bioremediation is the lack of knowledge on systematic control and use of enzymes due to their large diversity. Each and every enzyme has its own special features on stability and factors affecting enzymatic velocity that could limit their suitability for bioremediation. Stability of enzymes could be affected by pH, temperature, ionic force, or kaotropes (Mathews, 2004). The enzymatic velocity depends on internal or external factors. Internal factors are substrate selectivity and affinity, allosteric or cooperative regulation and the kinetics parameters. Among the external factors, we can find the co-substrate presence, inhibitor presence or other environmental factors. All these factors should be studied for each enzyme before choosing it for bioremediation. Moreover, enzymes shall be engineered to improve its features and be useful for these biotechnological purposes.

Recalcitrance of pollutants is mainly caused by their chemical stability due, in turn, to features like aromaticity and halogenation. Aromatic rings are often cleaved by adding one or two atoms of oxygen by oxygenases or peroxidases. Dehalogenation could occur by reduction, oxidation, group transference, hydrolysis or elimination (Wacket et al., 2004). It has been described that the position and number of chlorine atoms affect greatly the biodegradation of compounds (Furukawa et al., 1978; Alexander, 1999). In addition, ring cleavage is easier, the less chlorine atoms are substituted in the aromatic group.

Among enzymes, oxydoreductases, hydrolases, and transferases seem to be the most adequate for POPs bioremediation. By using oxydoreductases, bacteria and plants could take pollutants as final electron acceptors, thus biodegrading them. Among them, monooxygenases and dioxygenases could catalyze the insertion of one or two oxygen atoms into aromatic rings, respectively. Both reactions could lead to hydroxylation of aromatic rings (Yuan, 1995). Oxygenases are ubiquitous in nature and play an important role in the metabolism of a broad range of compounds. These enzymes generally use NADH or NADPH cofactors to provide reducing potential for the supply of electrons to the substrate and can be metal-, heme- or flavin-dependent. In dioxygenase reactions both atoms from a single molecule of O_2 are incorporated into the organic substrates (Ullrich and Hofrichter 2007).

Monooxygenases, like those from the Cytochrome P450 (CYP) superfamily, have been shown to be promiscuous and able to degrade several kinds of structures. It has been proposed to use this enzyme in transgenic plants to enhace phytoremediation (Abhilash et al., 2009). Nevertheless, some of the reactions catalyzed by CYP lead to more toxic compounds and eventually to cell damage. For this reason, CYP does not seem to be a proper candidate for bioremediation. However, in plants, CYP could display peroxidase activity allowing organisms to cope with some pollutants (Kvesitadze et al., 2001). Contrastingly, peroxidases and dioxygenases rarely activate toxic chemicals (produce more toxic compounds than the parental) and may be good candidates. In the next parragraphs we describe three families of enzymes that seem to be the most adequate candidates for bioremediation purposes.

Dioxygenases

Dioxygenases are enzymes that introduce two oxygen atoms in substrates. This seems to be one of the first steps in aromatic ring cleavage. Hydroxyquinone dioxygenase, hidroxyquinol dioxigenase, catechol dioxygenases are good examples of these enzymes that appear to be proper candidates for bioremediation. The present book includes a chapter reviewing these enzymes. Additionally to hydroxyaromatic compounds, it has been suggested that dioxygenases could be involved in DDT degradation (Nadeau et al., 1994). Many reviews have documented the microbial metabolism of natural and synthetic hydrocarbons (Husain, 2008). Bacterial aromatic hydrocarbon dioxygenases are multicomponent enzyme systems that add dioxygen to the aromatic nucleus to form arene-cis-diols, for the oxidation of benzene to cis-1,2-dihydroxycyclohexa-3,5- diene (benzene cis-diol) by toluene dioxygenase (Gibson, 1970). Aromatic hydrocarbon dioxygenases belong to a large family known as aromatic-ring-hydroxylating dioxygenases.

Dioxygenases are non-heme iron-containing enzymes important in the biosynthesis of plant signaling compounds such as abscisic acid, gibberellins, and ethylene, as well as of secondary metabolites, notably flavonoids and alkaloids. Plant dioxygenases fall into two classes: lipoxygenases and 2-oxoacid-dependent dioxygenases. The latter catalyze hydroxylation, epoxidation, and desaturation reactions; some enzymes catalyze more than one type of reaction in successive steps in a biosynthetic pathway (Prescott and John, 1996).

There are two principal reasons for the current interest in the structure and function of aromatic hydrocarbon dioxygenases (biocatalysis and biodegradation). First, aromatic hydrocarbons are common contaminants of soils and groundwaters. The removal of these compounds from polluted environments by microorganisms represents a potential solution to the environmental problems posed by these pollutants. And the second reason for interest in aromatic hydrocarbon dioxygenases is related to industry's search for environmentally benign procedures for the synthesis of useful chemicals. The dioxygenases fulfill this 'green chemistry' requirement as they are a source of new enantiopure arene-cis-diols that are not attainable by conventional chemical synthesis.

Results from a large number of recent studies, using in vivo and in vitro approaches for the generation of hybrid and mutant dioxygenases, have confirmed the conclusion that a very small number of amino acid differences in biphenyl dioxygenases are responsible for large differences in polychlorinated biphenyl (PCBs) congener specificity (Brühlmann, 1999). These aspects of biodegradation may be important to consider when constructing strains for bioremediation purposes.

The similarity of the reactions catalyzed by naphthalene dioxygenase and cytochrome P450 are notable, as revealed by a biomimetic model for the oxidation of olefins to cis-diols. The remarkable number of substrates oxidized to single enantiomers by toluene and naphthalene dioxygenases has led to their use in asymmetric syntheses of many biologically active products. Only a few of these have been targeted for commercial use.

Finally, the recent emergence of sphingomonads as a source of new Rieske non-heme iron oxygenases with novel organizational and regulatory properties complements the list of organisms that have potential use in the development of bioremediation technology.

GLUTATHIONE S-TRANSFERASES

Glutathione S-transferase (GST) is a family of enzymes catalyzing the transfer from a glutathionyl group to electrophilic substituents of biomolecules or xenobiotics. They use glutathione as a co-substrate, so it should be added for enzymatic activity to be performed. GST could catalyze de-chlorination of dichloromethane (Willeumier, 1996) and of some halogenated compounds (Bezalel, 1997; McGuiness, 2007). They seem to be quite promiscuous and work in a broad range of pH. Another advantage of GSTs is that their activity does not generate toxic products like epoxides by CYP acitivity (McGuiness, 2007). GSTs can be produced by animals, plants, fungi, and bacteria. It has been described that glutathione S-transferases could exhibit glutathione peroxidase activity. All these factors make GST an adequate candidate for bioremediation.

GST catalyzes the conjugation of glutathione to a variety of endogenous electrophilic compounds as part of the defense mechanism against toxic organisms (Cota, 2010). GST is therefore a good candidate for bioremediation of contaminated environments due to its promiscuity, stability, and its ability to degrade organic compounds such as pesticides and environmental polyphenyls. *Sphingobium chlorophenolicum* has demonstrated the ability to biodegrade PCP through various enzymes, being GST one of them (Huang et al., 2008). GST has been shown to participate in the biodegradation of organophosphorus pesticides (Abel et al., 2004), triazine herbicides (Hatton et al., 1999), and it has been suggested for xenobiotics (herbicides, explosives, PCBs) in general in plants (Schröder et al., 2007, Abhilash et al., 2009).

GST is a promiscuous enzyme that provides an excellent starting point for the development of new enzymes when a new activity becomes important for fitness or survival of certain organisms in contaminated environments (Copley, 2009). It has been pointed out the importance of certain enzymes with biodegradability for their application in bioremediation of organic pesticides (Velazquez et al., 2012).

For bioremediation purposes, GST has been produced from bacteria such as *Burkholderia xenovorans* (McGuiness et al., 2007), *Sphingomonas chorophenolica* (Whenva, 1997). In addition, GST has been shown to be involved in the resistance to some aromatic pollutants in *Saccharomyces cerivisiae* (Choi, 1998) or to organophosphorus pesticides (Abel, 2004). GST has been shown to be influenced by the presence of pollutants, either increasing or decreasing its activity. Most types of the GST superfamily are expressed in the cytosol (Dixon, 2010), thus, they are hydrosoluble.

GSTs are a ubiquitous group of multifunctional dimeric proteins involved in the detoxification of a large variety of both endobiotic and xenobiotic compounds. GSTs catalyze the conjugation of the sulfur atom of glutathione with a variety of electrophilic substrates and have been purified from a wide range of organisms (McGuinness, 2007)

The basic structure of GST has two domains, N and C-terminal. The N-terminal domain constitutes about one third of the protein and consists of a β-α-β-α-β-α-β structure. The N-terminal domain provides a binding site for GSH, which is located at the end of a beta sheet, the GSH molecule interacting with the protein through electrostatic bonds and hydrogen bonds (Barreto, 2006).

GST is a homodimer with an approximate native and a subunit molecular mass of 53 kDa and 23.4 kDa, respectively. The enzyme has a pI value of 4.8, a pH optimum of 8.0, and apparent activation energy (Ea) of 1.42 kcal mol^{-1}. The purified GST acts readily on 1-chloro-2,4-dinitrobenzene (CDNB) with almost negligible peroxidase activity and the activity is inhibited by Cibacron Blue (IC50 0.252 μM) and hematin (IC50 3.55 μM). *M. mucedo* GST displays a non-Michaelian behavior. At low (0.1-0.3 mM) and high (0.3-2 mM) substrate concentration, Km (GSH) was calculated to be 0.179 and 0.65 mM, whereas Km (CDNB) was 0.531 and 11 mM and k(cat) was 39.8 and 552 s^{-1} (Hamed et al., 2005).

GST has been studied in various detoxification processes of endogenous and xenobiotic compounds, they have since been found to have additional important roles as transport proteins and in protection against oxidative stress (Ding et al., 2009). Two *Saccharomyces cerevisiae* genes, GTT1 and GTT2 (glutathione transferase 1 and 2) encode for functional GST enzymes. Gtt1p and Gtt2p exhibit GST activity with CDNB as substrate (Choi 1998). Glutathione S-transferases, even though they catalyze transference of glutathione to electrophilic pesticides, can also show hydrolytic and peroxidase activities. Interestingly, GST can also catalyze dehalogenation of rings (Velazquez et al., 2012).

Our group has investigated the GST role in a phytoremdiation process. Firstly, we observed that endosulfan concentration in soil is decreased when *Ocimum basilicum* is grown in it. Endosulfan belongs to the group of organochlorine pesticides, it is widely distributed in Mexico and was recently included in the Stockholm convention. We observed that GST ativity in *Ocimum basilicum* plant increases in an organ-dependent way, i.e., it increases in roots and leaves, when exposed to endosulfan in the soil. So far, we believe that GST of this plant has an important role in the bioremediation process of organochlorine. At this moment, we are attempting to purify the enzyme to measure its actual effect on the organlchlorine pollutant; in other words, to evaluate its participation in a bioremediation process of this compound and in a solid medium.

PEROXIDASES

Peroxidases are enzymes that catalyze oxidation-reduction of hydrogen peroxide and a variety of hydrogen donors. Peroxidases are relatively stable at high temperatures and their activity can be easily measured using simple chromogenic reactions (Perez et al., 2008). Most peroxidases are promiscuous but their activity requires co-substrates. Peroxidases usually are metallo-enzymes, hence, they can be inhibited by azide, cyanide, and chelants such as EDTA or EGTA. Peroxide is needed as a co-substrate for peroxidase activity, this is a great limitation in using peroxidases. In addition, peroxide can react with most functional groups, thus, its application in bioremediation could lead to noncontrolled reactions and unexpected products.

The studied model for peroxidases is the horseradish peroxidase that has been extensively used in research. Peroxidases comprise a polypeptide of approximately 308 amino acids Peroxidases can have a molecular mass of 38 to 60 kDa and an isoelectric point (pI) of 3.8, as in the case of those involved in decolorization (Kim, 1999). The enzyme is optimally active at pH 6.0 and between 30 °C and 40 °C (Shin et al., 2005). It contains two metal iron centers (III), protoporphyrin IX, and two Ca atoms, both essential for enzyme structure and function. The heme enzyme is integrated into the coordinate bonding between the side chain of histidine and the heme Fe atom; the second axial site is unoccupied but available for enzymatic action (Veitch, 2004)

Peroxidase is an enzyme that might degrade toxic persistent organic compounds. Peroxidases have been shown to polymerize phenolic compounds, thereby removing them from solution by precipitation. Others have studied the role of root surface-associated peroxidases as a defense against fungal root pathogens; however, their use in detoxification of organic pollutants in vivo at the root surface has not been studied (Adler, 1994).

Peroxidase activity has been used to degrade dyes (Raghukumar, 2000), polycyclic aromatic hydrocarbons (Baborová et al., 2006), organochlorine pesticides (Davila-Vazquez et al., 2004); and it has been suggested that peroxidases might mineralize explosives and DDT (Aust, 1995). The production of ligninolytic enzymes and their use in bioremediation has attracted great attention in the past. The main factors affecting the production of ligninolytic enzymes are the availability and chemical form of nutrients, temperature, the presence of inhibitors or inducers, and the effect of interspecific interactions. Ligninolytic enzymes are promiscuous, which makes them a proper candidate for bioremediation. Nevertheless, peroxidases need to be studied further to shed light on all their requirements.

Peroxidases can be detected in plant roots such as *Eichhornia crassipes* (C. Mart) Solms-Laub (water hyacinth), *Lycopersicon esculentum* L. (tomato) (Adler et al., 1994), and in fungi. In these, main peroxidase activity is aimed at degrading lignin. Ligninolytic activity by lignin peroxidases (LiP) relies not only on laccase (Lacc) but on several other enzymes. LiP and manganese peroxidase (MnP) has been found in *Phanerochaete chrysosporium* (Boominathan & Reddy, 1992; Tanga et al., 2006), sphingomonas (Harayama, 1997), white rot fungus and grape vine pathogen *Fornitiporia mediterranea* (Morgenstern et al., 2010). LiP has been detected in *Phanerochaete chrysosporium*. The edible fungus *Pleurotus ostreatus* also produces several kinds of peroxidases. Copper and cadmium seem to induce Lacc production by *P. ostreatus*. Copper and mercury show contrary effects, the former increases Lacc activity, while the latter decreases Lacc activity and stability (Baldrian and Gabriel, 2002). Fungus *Agaricus bisporus* also produces Lacc, which is inducible by benzyl and veratryl alcohol (Kumar et al., 2011)

Ligninolytic activity could be used to bioremediate municipal solid waste by composting (Tanga et al., 2006) and to isolate phenolic compounds by polimerization such as those from plants (Adler et al., 1994). Moreover, LiP from sphingomonas and white rot fungi have been suggested to be capable of degrading PAHs (Harayama, 1997).

It has been shown that ligninolytic enzyme activities could also be used to degrade polycyclic aromatic hydrocarbons (PAHs) by the fungus *Irpex lacteus*. During growth in sterile soil, the fungus synthesized LiP and laccase but not MnP. *I. lacteus* efficiently removed three- and four-ringed PAHs from liquid media and artificially-spiked soil. The variety of ligninolytic enzymes, robust growth, capability of soil colonization, and resistance to inhibitory action of soil bacteria make *I. lacteus* a suitable fungal organism to be used in bioremediation (Novotný 2000). The peroxidase in the white-rot fungus *I. lacteus* has been reported to be an efficient degrader of PAHs, polychlorinated biphenyls (PCBs), and pentachlorophenol. Baborováa (2006) showed that this fungus can produce ligninolytic enzymes laccase, lignin peroxidase, and manganese peroxidase (MnP), the latter being the main one produced. Degradation of four representatives of PAHs (phenanthrene, anthracene, fluoranthene, and pyrene) was tested and the enzyme showed the ability to degrade them in vitro. Major degradation products of anthracene were identified. (Baborová et al., 2006).

White-rot fungi produce an unusual enzyme system, characterized by a specialized group of peroxidases that catalyzes the degradation of the complex plant polymer lignin. This ligninolytic system shows a high degree of nonspecificity and oxidizes a very large variety of compounds in addition to lignin. Among these compounds are numerous environmental pollutants. One white-rot fungus, *Phanerochaete chrysosporium*, has been studied in great detail with regard to ligninolytic enzymes and the degradation of anthropogenic chemicals. It has been widely promoted as a bioremediation agent (Paszczynski, 1995). Although ligninolytic enzymes are regulated in vitro by nitrogen content, *P. chrysosporium* produces Mn-peroxidase and lignin peroxidase under N-limitation. Other fungal species respond in the opposite way. Mn-peroxidase is generally induced by the presence of manganese, hydrogen peroxide, and lignin; laccase is induced by a range of simple phenolic compounds (Leonowicz et al., 2001) and certain metal ions. Regulation in soil is more complex, but the basic concepts are valid: the decomposition of plant litter increases with Mn content of soils (Baldrian, 2008). It has been shown that *Aspergillus niger* isolated from soil of a leather tanning effluent polluted with chromium showed higher peroxidase activity to remove chromium than the other fungal isolates (Srivastava, 2006).

Versatile peroxidase (VP) from the white rot fungus *Bjerkandera adusta* UAMH 8258 was used to study the transformation of several pesticides, including some highly halogenated as the wood preservative pentachlorophenol (PCP). Several polymers were obtained from the peroxidase oxidation of bromoxynil. In all cases, dehalogenation reactions were found to be mediated by the versatile peroxidase (Davila-Vazquez et al., 2004).

Conclusion

In the past years, factors affecting the biodegradation of pollutants have been elucidated. Although further research is needed, the current knowledge indicates that promiscous enzymes such as peroxidases, dioxygenases and glutathione S-transferases could be very useful for bioremediation of a wide range of pollutants.

References

Abel E.L., Bammler T.K., Eaton D.L., 2004. Biotransformation of Methyl Parathion by Glutathione S-Transferases. *Toxicol. Sci.* 79:224-232.

Abhilash P.C., Jamil S., Singh N., 2009. Transgenic plants for enhanced biodegradation and phytoremediation of organic xenobiotics. *Biotechnol. Advances* 27, 474-488.

Adler P.R., Arora R., El Ghaouth A., Glenn D.M., Solar J.M., 1994. Bioremediation of phenolic compounds from water with plant root surface peroxidases. *J. Environ. Qual.* 23(5):1113-1117

Alexander M., 1999. Biodegradation and Bioremediation 2nd. Edition. Edit. Academic Press. USA.

Aust S.D., 1995. Mechanisms of Degradation by White Rot Fungi. *Environ. Health Perspect.* 103(S5) :59-61

Baborová P., Möder M., Baldrian P., Cajthamlová K., Cajthaml T., 2006. Purification of a new manganese peroxidase of the white-rot fungus *Irpex lacteus*, and degradation of polycyclic aromatic hydrocarbons by the enzyme. *Res. Microbiol.* 157(3):248-53.

Baldrian P. Gabriel J., 2002. Copper and cadmium increase laccase activity in *Pleurotus ostreatus. FEMS Microbiol. Lett.* 206:69-74

Baldrian P., 2008. Wood-inhabiting ligninolytic basidiomycetes in soils: Ecology and constraints for applicability in bioremediation. *Fungal Ecol.* 1(1):4–12

Barreto P.P.L. Glutatión transferasas de clase omega en *Saccharomyces cerevisiae*: estudio bioquímico y funcional. Universitat de Lleida. Lleida. 2006.

Boominathan K., Reddy C.A., 1994. cAMP-mediated differential regulation of lignin peroxidase and manganese-dependent peroxidase production in the white-rot basidiomycete *Phanerochaete chrysosporium*. *Proc. Nat. Acad. Sci. USA*. 89:5586-5590.

Brühlmann F., Chen W., 1999. Tuning biphenyl dioxygenase for extended substrate specificity. *Biotechnol Bioeng* 63, 544-551.

Choi J.H., Lou W., Vancura A., 1998. A Novel Membrane-bound Glutathione S-Transferase Functions in the Stationary Phase of the Yeast *Saccharomyces cerevisiae*. *J. Biol. Chem.* (273)45:29915–29922.

Copley S., 2009. Evolution of Efficient Pathways for Degradation of Anthropogenic Chemicals. *Nat. Chem Biol.* 5(8):559-566

Cota-Ramírez E.L., 2010. Determination of Glutation S-Transferase in *Ocimum basilicum*. Autonomous University of Nayarit. Mexico [In spanish]

Davila-Vazquez G., Tinoco R., Pickard M.A., Vazquez-Duhalt R., 2005. Transformation of halogenated pesticides by versatile peroxidase from *Bjerkandera adusta*. *Enzym. & Microb. Technol.* 36(2–3) 223–231.

Ding Y., Ortelli F., Rossiter L., Hemingway J., Ranson H., 2009. The Anopheles gambiae glutathione transferase supergene family: annotation, phylogeny and expression profiles. BMC Genomics, *Biomed central*. 4:1-16.

Furukawa K., Tonomura K., Kamibayashi A., 1978. Effect of Chlorine Substitution on the Biodegradability of Polychlorinated Bipenyls. *Appl. Environm Microbiol.* 35(2) :223-227

Gibson D.T., Cardini G.E., Maseles F.C., Kallio R.E., 1970. Incorporation of oxygen-18 into benzene by *Pseudomonas putida*. *Biochemistry* 9, 1631-1635.

Hamed R.R., Abu-Shady M.R., El-Beih F.M., Abdalla A.M., Afifi O.M., 2005. Purification and characterization of a glutathione S-transferase from *Mucor mucedo*. *Pol. J. Microbiol.* 54(2):153-60.

Harayama S., 1997. Polycyclic aromatic hydrocarbon bioremediation design. *Curr. Opinion Biotechnol.* 8(3) :268–273.

Hatton P.J., Cummins I., Cole D.J., Edwards R., 1999. Glutathione transferases involved in herbicide detoxification in the leaves of *Setaria faberi* (giant foxtail). *Physiologia Plantarum* 15 :9-16

Huang Y., Xun R., Chen G., Xun L., 2008. Maintenance Role of a Glutathionyl-Hydroquinone Lyase (PcpF) in Pentachlorophenol Degradation by *Sphingobium chlorophenolicum*. *J. Bacteriol.* 190(23):7595–7600.

Husain S., 2008. Literature overview: microbial metabolism of high molecular weight polycyclic aromatic hydrocarbons. *Remed J* 18, 131–161

Kim S.J., Shoda M., 1999. Purification and characterization of a novel peroxidase from *Geotrichum candidum dec* 1 involved in decolorization of dyes. *Appl. Environ. Microbiol.* 65(3):1029-35.

Kumar V.V., Kirupha S.D., Periyaraman P., Sivanesan S., 2011. Screening and induction of laccasse activity in fungal species and its application in dye decolorization. *African J. Microbiol. Res.* 5(11) :1261-1267

Kvesitadze G., Gordeziani M., Khatisashvili G., Sadunishvili T. Ramsden J.J. 2001. *J. Biol. Physics & Chem.* 1 :49-57

McGuinness M.C., Mazurkiewicz V., Brennan E., Dowling D.N., 2007. Dechlorination of Pesticides by a Specific Bacterial Glutathione-S-transferase, BphKLB400: Potential for Bioremediation. *Eng. Life Sci.* 7(6):611–615.

Morgenstern I., Robertson D.L., Hibbett D.S., 2010. Characterization of three mnp genes of *Fomitiporia mediterranea*. *Appl. Environ. Microbiol.* 76(19) :6431-6440.

Nadeau L.J., Menn FM., Breen A., Sayler G.S., 1994. Aerobic degradation of 1,1,1,-Trichloro-2,2-Bis(4-Chlorophenyl)Ethane (DDT) by *Alcaligenes eutrophus* A5. *Appl. Environ. Microbiol.* 60(1):51-55

Novotný C., Erbanová P., Cajthaml T., Rothschild N., Dosoretz C., Sasek V., 2000. *Irpex lacteus*, a white rot fungus applicable to water and soil bioremediation. *Appl. Microbiol. Biotechnol.* 54(6):850-3.

Paszczynski A., Crawford R.L., 1995. Potential for Bioremediation of Xenobiotic Compounds by the White-Rot Fungus *Phanerochaete chrysosporium*. Biotechnol. Progress 11(4):368–379

Pérez A.D.B., Regalado G.C., Rodríguez C.N.A., 2008. Purification of a recombinant peroxidase in radish by affinity chromatography. Autonomous University of Querétaro. [in Spanish].

Prescott A.,G., John P., 1996. DIOXYGENASES: Molecular Structure and Role in Plant Metabolism. *Annu Rev Plant Physiol Plant Mol Biol.* 47, 245-271.

Raghukumar C., 2000. Fungi from marine habitats: an application in bioremediation. *Mycol. Res.* 104(10):1222–1226.

Schröder P., Scheer C.E., Diekmann F., Stampfl A., 2007. How Plants Cope with Foreign Compounds. Translocation of xenobiotic glutathione conjugates in roots of barley (*Hordeum vulgare*). *Env. Sci. Pollut. Res.* 14(2) :114-122

Shin K., Kim Y. H., Lim J., 2005. Purification and Characterization of Manganese Peroxidase. *J. Microbiol.* 43(6):503-509

Srivastavaa S., Thakur I.S., 2006. Evaluation of bioremediation and detoxification potentiality of *Aspergillus niger* for removal of hexavalent chromium in soil microcosm. *Soil Biol. Biochem.* 38(7):1904–1911.

Tanga L., Guang-Ming Z., Guo-Li S., Yi Z., Guo-He H., Jian-Bing L., 2006. Simultaneous amperometric determination of lignin peroxidase and manganese peroxidase activities in compost bioremediation using artificial neural networks. *Anal. Chim. Acta*. 579(1):109-16.

Ullrich R., Hofrichter M., 2007. Enzymatic hydroxylation of aromatic compounds. *Cell Mol Life Sci* 64, 271–293.

Van Aken B., Correa P.A., Schnoor J. L., 2010. Phytoremediation of Polychlorinated Biphenyls : New Trends and Promises. *Environ. Sci. Technol.* 44(8):2767-2776.

Veitch N. C. 2004. Horseradish peroxidase: a modern view of a classic enzyme. *Phytochemistry* 65:249–259.

Velázquez-Fernández J.B., Ramírez-Sandoval M., Domínguez-Ojeda D., Martínez-Rizo A.B. Chapter 12: Biodegradation and bioremediation of organic pesticides. In : Soundararajan RP. Pesticides: Recent Trends in Pesticide Residue Assay. Edit. Intech. 2012. Croatia.

Wacket L.P., 1995. Recruitment of Co-metabolic Enzymes for Environmental Detoxification of Organohalides. *Environm. Health Perspect.* 103 (Suppl. 5) :45-48

Index

B

C

D

E

F

G

H

I

K

L

M

N

O

P

Q

R

S

T

U

V

W

Y

Z